LAWS OF THE SEA

Interdisciplinary Currents

Laws of the Sea assembles scholars from law, geography, anthropology, and environmental humanities to consider the possibilities of a critical ocean approach in legal studies.

Unlike the United Nations' monumental Convention on the Law of the Sea, which imagines one comprehensive constitutional framework for governing the ocean, *Laws of the Sea* approaches oceanic law in plural and dynamic ways. Critically engaging contemporary concerns about the fate of the ocean, the collection's twelve chapters range from hydrothermal vents through the continental shelf and marine genetic resources to coastal communities in France, Sweden, Florida, and Indonesia. Documenting the longstanding binary of land and sea, the chapters pose a fundamental challenge to European law's "terracentrism" and its pervasive influence on juridical modes of knowing and making the world. Together, the chapters ask: is contemporary Eurocentric law—and international law in particular—capable of moving away from its capitalist and colonial legacies, established through myriad oceanic abstractions and classifications, toward more amphibious legalities?

Laws of the Sea will appeal to legal scholars, geographers, anthropologists, cultural and political theorists, as well as scholars in the environmental humanities, political ecology, ocean studies, and animal studies.

Irus Braverman is Professor of Law and Adjunct Professor of Geography at the University at Buffalo, the State University of New York. Her books include *Planted Flags: Trees, Land, and Law in Israel/Palestine* (2009), *Zooland: The Institution of Nature* (2012), and *Coral Whisperers: Scientists on the Brink* (2018) as well as the coedited volume *Blue Legalities: The Laws and Life of the Sea* (2020). Braverman's monograph, *Settling Nature: The Conservation Regime in Palestine-Israel*, is forthcoming.

The human impact on the atmosphere is a matter of intense common concern. The human impact on the ocean, humanity's other ultimate common good, must also be studied at the empirical and legal and conceptual levels, enabling a more sophisticated legal response. This ground-breaking book is a major contribution to that study.

—Philip Allott, *Professor Emeritus of International Public Law,*
Cambridge University

These thought-provoking, imaginative essays push beyond conventional representations of the oceans as distinctive legal spaces. The authors connect deeply researched case studies to new analytical approaches to maritime legal geographies.

—Lauren Benton, *Barton M. Biggs Professor of*
History and Professor of Law, Yale University

We are at a critical juncture in ocean governance. This collection raises important questions that highlight both the explicit and the less explicit choices in future ocean governance, including whether the existing legal architecture should be fixed or remade. The unique timing of this collection makes the questions tackled in this book not only academically interesting for the multiple disciplines represented, but of immense practical importance for our shared future.

—Lisa Campbell, *Rachel Carson Distinguished Professor of*
Marine Affairs and Policy, Duke University

The sea is a space of law—and more, exactly, laws, plural. As the contributors to this book teach us across a range of powerful near-shore, open-ocean, deep marine, and aquabiotic cases, legal abstractions now saturate, slice up, and, sometimes, sicken the sea itself. Tuning to how law in fact operates as amphibious—mixing land and sea—this book is a brief for re-mapping sea law in ways at once more empirical and more just.

—Stefan Helmreich, *Elting E. Morison Professor of*
Anthropology, Massachusetts Institute of Technology

This is a wonderful series of critiques of traditional law of the sea. The authors show the utter inadequacy of formal distinctions such as those between territorial sea and the high seas to the interdependent reality of the ocean space. From a number of interdisciplinary perspectives, they develop a pertinent critique of the extractive bias in the UN Convention on the Law of the Sea. They also demonstrate the destructive consequences of deep-sea mining on vital life forms on the abyssal ocean floor and implore the law's ignorance of the importance of ice for the living conditions of Arctic peoples. Although much information on marine ecosystems has been collected since the conclusion of the Law of the Sea Treaty in 1982, very little is still known of how present and planned activities affect biological diversity and seabed features such as hydrothermal vents. The essays in this collection present a kaleidoscopic image of the innumerable ways in which human and marine lives are dependent on each other. The result is a pertinent warning against allowing superficial economic interests to determine the ways in which human activities relate to the ocean world.

—Martti Koskenniemi, *Emeritus Professor of International Law, University of Helsinki*

The wateriness of law is longstanding. Colonial domination, slavery, and indentured labor were enabled by an amphibious assault of power/knowledge. But in most accounts, the watery spaces and beings of this planet remain over-determined by land-based notions of sovereignty, and laws "grounded" on islands, shorelines, and continental contiguities. This collection of timely and evocative essays shatters the land/sea binary, and breaks down the borders of life/non-life. Moving across the seabed to river deltas, marine genetic resources, and fisheries, the book is a rich collection of the new forms of knowledge and epistemic practices needed in order to appreciate amphibious legal geographies. This book, then, is a raft that may help life and non-life survive the toxic legacies of western legal abstraction.

—Stewart Motha, *Professor of Law, Birkbeck, University of London*

LAWS OF THE SEA

Interdisciplinary Currents

Edited by Irus Braverman

Routledge
Taylor & Francis Group
a GlassHouse Book

Cover image: Getty/Kevin Schafer

First published 2023
by Routledge
4 Park Square, Milton Park, Abingdon, Oxon OX14 4RN

and by Routledge
605 Third Avenue, New York, NY 10158

A GlassHouse Book

Routledge is an imprint of the Taylor & Francis Group, an informa business

British Library Cataloguing-in-Publication Data
A catalogue record for this book is available from the British Library

Library of Congress Cataloging-in-Publication Data
Names: Braverman, Irus, 1970- editor.
Title: Laws of the sea: interdisciplinary currents/edited by
Irus Braverman.
Description: Abingdon, Oxon; New York, NY: Routledge, 2022. |
"This collection assumed oceanic qualities from its very beginning.
Born on the cusp of the Covid pandemic, the original idea of
organizing a four-day in-person workshop in Puerto Rico to convene
scholars from three continents so as to contemplate the interfaces of
sea and law from critical interdisciplinary perspectives and utilizing
myriad methodological tools soon became inconceivable. Rather than
postpone our workshop for an indefinite period of time, contributors
offered to ride the momentum and meet over Zoom. We were then
novice Zoom users, but have become adapt in the intricacies of this
mode of communication as time lingered on (and on, and on). Our
first daylong meeting was in September 2020, and then we met again
in January 2021. Our final workshop took place over Zoom in May
2021 and lasted three days"–ECIP acknowledgement. | Includes
bibliographical references and index.
Identifiers: LCCN 2021062881 (print) | LCCN 2021062882
(ebook) | ISBN 9781032070575 (hardback) | ISBN 9781032070629
(paperback) | ISBN 9781003205173 (ebook)
Subjects: LCSH: Law of the sea–Congresses. | Law of the
sea–Economic aspects–Congresses. | Law of the sea–Political
aspects–Congresses.
Classification: LCC KZA1141.L3925 2022 (print) | LCC KZA1141
(ebook) | DDC 341.4/5–dc23/eng/20220429
LC record available at https://lccn.loc.gov/2021062881
LC ebook record available at https://lccn.loc.gov/2021062882

ISBN: 978-1-032-07057-5 (hbk)
ISBN: 978-1-032-07062-9 (pbk)
ISBN: 978-1-003-20517-3 (ebk)

DOI: 10.4324/9781003205173

Typeset in Bembo
by Deanta Global Publishing Services, Chennai, India

To Keren Shavit-Buckley
with a sea of gratitude

CONTENTS

ACKNOWLEDGMENTS

This collection assumed oceanic qualities from its very start. Born on the cusp of the Covid pandemic, the original idea of organizing a four-day in-person workshop in Puerto Rico to assemble scholars from three continents quickly became inconceivable. Rather than postpone this meeting indefinitely, the scholars offered to ride the wave and meet over Zoom. We were then novice Zoom users, but have become adept in the intricacies of this mode of communication as time lingered on. Our first daylong workshop took place in September 2020 and then we met again in January 2021. Our final workshop over Zoom was in May 2021 and lasted three days, the third of which was dedicated to discussing themes that weave across the papers as well as to a "wavewriting" performance by one of the participants, which appears here in written form as the collection's Afterword. In between the workshops, multiple discussions among the participants enriched the exchange on myriad levels and scales.

Like the ocean, then, this collection was conceived in a fluid process. The two years of coming together for this project were certainly not easy. But the difficult circumstances enhanced the intensity and productivity of our interactions, building what felt like a global intellectual community of support and care for oceans. As one of the participants phrased it: "This was a bit of an academic experiment in the time of Covid." It was clear to us that the work we were doing for this project was urgent and that our critical message must therefore be foregrounded and radicalized.

I would like to thank this collection's contributors for their personal commitment to this project through what were often rough waters (oceanic puns fully intended from now on). I am grateful also to Naor Ben-Yehoyada and Itamar Mann, with whom I first conceived the idea for this project during a walk on the beach in Tel Aviv; to Renisa Mawani, who helped identify potential contributors and was an engaged interlocutor throughout; to Bethany Berger, for her involvement and advice; and to Frode Flemsæter and Katrina Rønningen from Ruralis—the Institute for Rural and Regional Research in Norway—who reached out to solicit my participation in their grant proposal and who were supportive throughout the entire project. Special thanks to the National Humanities Center, where I visited as a Hurford Family fellow during the fall of 2021, for providing me with the intellectual, social, and financial support to push this collection to the finish line, and to Matthew Booker and Jessica Hurley in particular. This collection would not have been possible without the generous funding by the Research Council of Norway as part of the research project *Funding Future Welfare: Bioeconomy as the "New Oil" and the Sharing of Benefits from Natural Resources* ("Bioshare," grant no. 294867), led by Ruralis.

Last but not least, I am grateful to Matthew Delaney, Timothy Conti, and Anita Gesel of the University at Buffalo's School of Law for their patient support in all matters technical and financial and to John Mondo for his wizard-like library skills. Emily Liao provided comprehensive administrative assistance, Daniel Piersa helped with research, and Margaret Drzewiecki was indispensable on all editorial matters—my gratitude extends to each of them for the many hours they have put into this project. I would be amiss not to mention Colin Perrin from Routledge, a superb editor of the rare and critically endangered kind.

I dedicate this collection to my good friend Keren Shavit-Buckley, whose life circumstances, like mine, brought her from Jerusalem to Buffalo, New York. Our brief encounter while fussing over our daughters' hair buns before their ballet performance was the beginning of an unassuming relationship that has flourished over the years. Jerusalemites are often mocked by all other Israelis for "not knowing the sea." But that didn't seem to apply to Keren and me. During our high school years (which happened, alas, in different decades), we both used to escape school—and the sharp, hard, cold, white rocks of Jerusalem—to the soft, warm, and always-changing shores of the Mediterranean Sea. Later on, we would each develop our love of the sea in different directions. When we couldn't make it to the ocean during Covid, we met every Sunday afternoon at the marina for a stormy walk on Lake Erie, which was the closest we could get. To you, Keren, with a sea of gratitude for reminding me the value of friendship, so far away from home.

CONTRIBUTORS

Elizabeth DeLoughrey is Professor in the Department of English and the Institute of the Environment at UCLA, US. She is the author of *Routes and Roots: Navigating Caribbean and Pacific Literatures* (2007), and *Allegories of the Anthropocene* (2019), which examines climate change and empire in the literary and visual arts. She is co-editor of the volumes *Caribbean Literature and the Environment: Between Nature and Culture* (2005), *Postcolonial Ecologies: Literatures of the Environment* (2011), and *Global Ecologies and the Environmental Humanities: Postcolonial Approaches* (2015) and of numerous journal issues on critical ocean and island studies.

Vito De Lucia is Professor of International Law at the faculty of Law, UiT The Arctic University of Norway and full member of the Norwegian Centre for the Law of the Sea (NCLOS). His research interests lie at the intersection of critical theory, international law, and ecology. His current research agenda focuses on ocean commons, on the ongoing negotiations toward a global treaty on marine biodiversity in areas beyond national jurisdiction, and on Arctic governance. He is author of *The Ecosystem Approach in International Environmental Law* (2019).

Florian Grisel is Associate Professor of Socio-Legal Studies at the University of Oxford, UK. His research focuses on the emergence and evolution of private governance. His latest book, *The Limits of Private Governance: Norms and Rules in a Mediterranean Fishery*, was published in 2021.

Henry Jones is Associate Professor of Law at Durham University, UK. He writes on the history and spatiality of international law, with a particular focus on early modern European colonialism. He is currently conducting research toward a book on international law and geography.

Jeffrey S. Kahn is Associate Professor of Anthropology at the University of California, Davis, US. He is the author of *Islands of Sovereignty: Haitian Migration and the Borders of Empire* (2019).

Annet Pauwelussen is Assistant Professor in Marine Governance with the Environmental Policy Group of Wageningen University, the Netherlands. She is also PI and research fellow with the Ocean Nexus program at the University of Washington, working on gender and shifting notions of nature in reef restoration. Her anthropological research explores epistemological and ontological complexity in human–sea relations and marine conservation. It builds on over a decade of ethnographic engagement with coastal and maritime communities, sea people and marine scientists in Southeast Asia.

Andreas Philippopoulos-Mihalopoulos is Professor of Law & Theory at the University of Westminster, UK, and Director of *The Westminster Law & Theory Lab*. His books include *Absent Environments* (2007), *Niklas Luhmann: Law, Justice, Society* (2009), and *Spatial Justice: Body Lawscape Atmosphere* (2014). His fiction book, *The Book of Water*, was published in Greek and English in 2021. His art practice includes performance, photography, and text, as well as sculpture and painting and his work has been presented at the Venice Art Biennale, the Venice Architecture Biennale, and the Tate Modern.

Elspeth Probyn is Professor of Gender & Cultural Studies at the University of Sydney, Australia. She is the author of several monographs, the most recent of which is *Eating the Ocean* (2016). She has also authored hundreds of chapters and articles and (co)edited several collections, including *Sustaining Seas: Oceanic Space & the Politics of Care* (2020).

Surabhi Ranganathan is Associate Professor of International Law at the University of Cambridge, UK. She is the author of *Strategically Created Treaty Conflicts and the Politics of International Law* (2015), and writes on the history and politics of the law of the sea, covering ocean depths and bottoms, global commons, marine infrastructures, and techno-utopian imaginaries.

Susan Reid is a writer, artist, creative researcher, lawyer, and PhD candidate in the Department of Gender and Cultural Studies at the University of Sydney, Australia. Her transdisciplinary research intervenes in dominant legal, extractivist, and oceanic imaginaries to explore possibilities for multibeing justice with the seas. Reid lives and works between unceded Gadigal Country and Yugambeh Country, by the ocean.

Philip Steinberg is UArctic Chair of Political Geography at Durham University, where he also directs IBRU: The Centre for Borders Research and the Durham Arctic Research Centre for Training and Interdisciplinary Collaboration

(DurhamARCTIC). From 2014 to 2019, he directed the Project on Indeterminate Change and the Environment: Law, the Anthropocene, and the World (the ICE LAW Project), whose leadership collectively authored the chapter in this volume. Other members of the ICE LAW Project authorship team are Claudio Aporta, Gavin Bridge, Aldo Chircop, Kate Coddington, Stuart Elden, Greta Ferloni, Stephanie C. Kane, Timo Koivurova, Jessica Shadian, and Anna Stammler-Gossmann.

Shannon Switzer Swanson is a marine social ecologist studying coastal communities in Southeast Asia and Oceania, primarily in the Philippines and Indonesia. She is a PhD Candidate in the Emmett Interdisciplinary Program in Environment and Resources at Stanford University, US.

Aron Westholm is a researcher at the Department of Law, University of Gothenburg, Sweden. His research primarily concerns ocean governance law and the relationship between administrative systems and the natural environment.

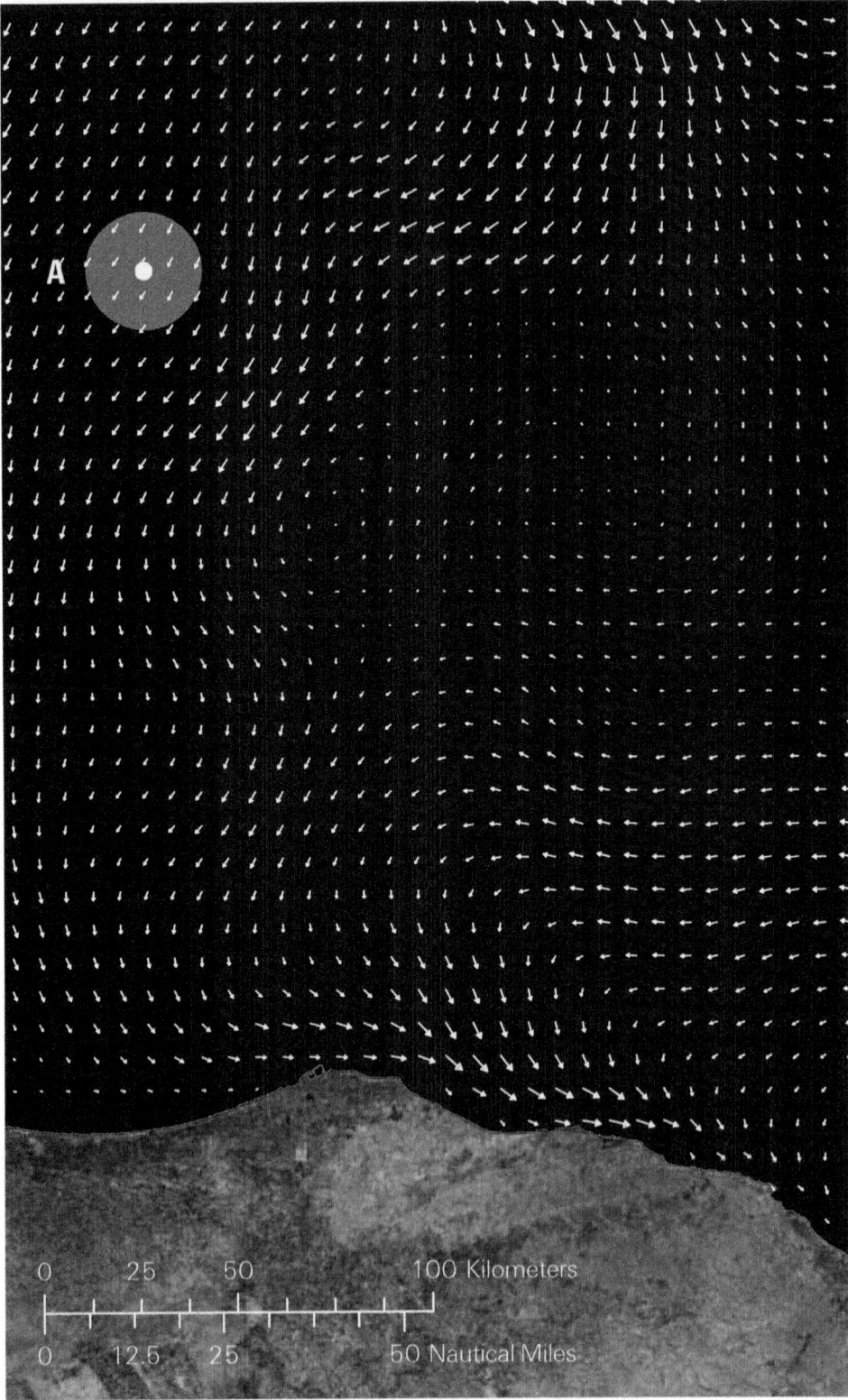

A

0	25	50		100 Kilometers
0	12.5	25		50 Nautical Miles

INTRODUCTION

Amphibious Legal Geographies: Toward Land–Sea Regimes

Irus Braverman

> We sweat and cry salt water, so we know that the ocean is really in our blood.
>
> —Teresia Teaiwa[1]

Most legal thinkers and practitioners view law as fundamentally terrestrial. Indeed, law—in its Eurocentric iteration at least—ultimately imagines itself as beginning and ending on *terra firma*. Land is perceived as a fully historicized, mapped, and regulated space that stands in stark opposition to the seemingly a-temporal, empty, and unruly sea.[2] This collection traces some of the juridical thinking that has enshrined the land/sea divide into contemporary governmental infrastructures, disciplinary traditions, and regulatory apparatuses, and charts the disastrous implications that such a legal fixation on the land/sea binary has wrought on human and other-than-human lifeworlds.

Ultimately, the question we ask here is whether contemporary Western[3] law is capable of pushing beyond its mythical imaginaries and its imperial and colonial legacies—which are arguably founded upon the land/sea binary, among other

FIGURE I.1   In spring 2011, 72 migrants attempted to flee Libya departing from Tripoli and destined for an Italian island in the Mediterranean Sea. They got halfway to their destination only to then run out of fuel and drift at sea for 14 days in what, at the time, was likely the most surveilled maritime zone in the world. To piece together the chain of events which ultimately resulted in 63 deaths, the Forensic Oceanography project produced a report on what has since become known as the "Left to Die Boat." The report was submitted as evidence in a lawsuit filed on behalf of the two survivors. The claim targets French and Spanish military ships for criminal neglect of the United Nations Convention on the Law of the Sea. By synthesizing several types of data, diagrams were created to elucidate the sequence of events beginning at the boat's departure, through the point at which the boat began to drift, and to the eventual landing south of Tripoli 14 days later. Reprinted with permission by SITU Research.

DOI: 10.4324/9781003205173-1

juxtaposed divides—to correspond with, rather than ignore or attempt to fix, the dynamic and relational nature of oceanic entities. This collection's offering of what I refer to as "amphibious legal geographies" is a call to recognize the shared material and symbolic qualities of both land and sea and our continued existence as an "edge species"[4] that thrives not only *by* the sea but also *with* the sea. Focusing on the ecotonal[5] qualities of law—namely, law's overlapping and fluid existences in multiple ecologies—*Laws of the Sea* envisions pluralistic, relational, dynamic, and just legal geographies that move beyond the land/sea binary toward land–sea regimes—that is, "maritime and terrestrial processes crossing the land/sea divide."[6]

Assembling scholars from law, geography, anthropology, and environmental humanities, this collection's 12 chapters explore the juxtaposition between land and sea, providing insights into its manifestations in juridical ways of knowing and making the world and being made by it. The chapters utilize a variety of intersecting methodologies to explore two central broad themes. The first theme studies one of the most powerful technologies of Western law—abstraction—that produces what Susan Reid calls in this collection an "architecture of exploitation."[7] The productive, and extractive, forces of law are studied in the collection's chapters through multiple oceanic geographies: hydrothermal vents (Ranganathan), the continental shelf (Jones), the seabed (Reid), areas beyond national jurisdiction (De Lucia), the genes in marine bodies (Braverman), and works of speculative fiction (DeLoughrey).

As the collection proceeds, a second broad theme emerges, building on the first: when one rethinks the abstraction of law as played out on the ground, the "ground" itself shifts and fundamental divisions between land and sea that serve as the foundations of Western law are undermined. The collection's call for a mode of legal reasoning that spans the edge between land and sea is supported through a range of conceptual and empirical studies. These include discussions of the seawater's edge (Arctic ice in Steinberg et al.'s contribution), the borders of mobility and access (Probyn on maritime chokepoints), and the legal geographies of coastal communities in Miami (Kahn), Marseille (Grisel), Gothenburg (Westholm), and Berau (Pauwelussen & Switzer Swanson).

Emerging from the collection's coastal chapters is the realization that humans live "around the sea like ants or frogs around the pond," as ancient Greek philosopher Plato put it.[8] When applied in the legal context, adopting amphibious, multispecies perspectives that see "the lives of other species and their habitats as worthy of our moral imagination and as inseparable constituents of our social worlds"[9] will move us closer toward what I have referred to elsewhere as "more-than-human legalities."[10] Attending to the amphibiousness of law—alluded to in this collection's discussions of heterolegalities (De Lucia) and hinterlands (Kahn), and captured elsewhere in notions such as wet ontologies,[11] shoals,[12] tidalectics,[13] littoral societies,[14] and land–sea regimes[15]—calls attention not only to the lively materialities underlying every legal construct,[16] but also, specifically, to the importance of breaching the land/sea divide in order to envision alternative modes of planetary governance that take their cues from local and Indigenous knowledges. This introduction first explores the power of abstraction that is so inherent to law's extractivist logics and then moves to discuss the amphibious legal geographies straddling the divisions between land and water, living and nonliving, and human and other-than-human.

The Legal Production of the Land/Water Binary

"A first step in this process," as John Gillis states in his archeological challenge to the Garden of Eden myth, "is to recognize that land and water are not opposites but inseparable parts of an ecological continuum."[17] Evolutionary biologists, too, are realizing that what they once saw as "fundamental differences" between marine and terrestrial taxa might in fact be the product of anthropocentric projections. Here, from a scientific review published in 2008: "It is conceivable that the starkness of the contrast between marine and terrestrial systems has remained a persistent motif in part by virtue of our being terrestrial organisms which renders [marine] pelagic systems inherently inaccessible to us."[18] Relatedly, the concept "hypersea"—defined by paleontologists Mark and Dianna McMenamin as the "body fluids of land eukaryotes, and the organisms through which they flow"—was coined to highlight the existence of the ocean inside us.[19] These recognitions of terraqueous continuities were illustrated in Rachel Carson's trailblazing 1961 text *The Sea Around Us*,[20] and were expressed in marine biologist Sylvia Earle's more recent statement that "the most valuable thing we extract from our oceans is our existence."[21]

The land/sea binary has also been criticized by legal historians. Pointing to its continued prominence in private and public international law,[22] Lauren Benton and Nathan Perl-Rosenthal challenge the juridical land/sea divide (represented by the slash) and call instead for land-sea regimes (represented by the hyphen) that emerge across the divide.[23] Similarly, this collection focuses on juridical processes that expose but then resist the reification of land/sea stratifications and the essentialization of land and oceans as fundamentally distinct materialities, setting the stage for the emergence of plural, yet integrative, legal and political administrations that do justice to vernacular and Indigenous knowledges.[24]

Ironically, it is precisely the ocean's seemingly natural state as free from law that enables its deep saturation by law. To understand this process, one must first unearth law's historic and contemporary fixation with land and its respective alienation of the sea—law's "terracentric" perspective, as historian Marcus Rediker calls it.[25] This fixation started with the ancient Hebrews, was passed on to the early Christians, then to Greek and Roman understandings of earth,[26] and finally was "strengthened by the rise of the modern nation-state in the late eighteenth century."[27] The representation of land and sea as juxtaposed juridical spaces was particularly central in Dutch jurist Hugo Grotius's 1609 tract *Mare Liberum*, "The Free Sea," where he constructs the high seas' existence as a space beyond national and imperial claims to sovereignty for imperial ends. The *mare liberum* doctrine eventually enabled Britain's ascendancy as a maritime empire, and was marked since then "by overlapping histories of colonial and racial dispossession."[28] As Maori lawyer Moana Jackson put it: "The freedom to use the seas has become a license to abuse."[29]

Of course, the high seas are anything but unregulated. "The law is all over,"[30] critical legal scholars like to remind those who still think that law can be contained in and constrained by its differentiation from social and political norms.[31] Precisely because it appears to be absent from this space, law is especially "all over" the oceans. Law's phantom existence at sea makes for particularly powerful legalities. Legal historian Fahad Ahmad Bishara commented about this in the context of the Indian Ocean:

> law was both everywhere and nowhere. It was hardly ever visible but it left its mark on every actor, artifact, and action … from the visible heights of formal political power to the unnoticed depths of everyday life. Law was a central part of how this Indian Ocean world took its shape.[32]

Utilizing a critical legal geography[33] approach, this collection sheds light on the coproductive relationship between law and space as it becomes manifest in the sea, pointing to the ways in which laws make the ocean and thus to how the ocean becomes subsumed, saturated, and even strangulated by law. As we witness the slow strangulation[34] of the sea from pollution, heat, acidification, and deoxygenation,[35] we must also consider how law is making the sea sick and, vice versa, what the sea's illness means for law. If the death of the sea as we know it could prompt the death of law as we know it (and the demise of its human progenitors), then this volume affirms the reverse: that the death of law as we know it might bring about hope for biodiverse life at sea. Indeed, our commitment to addressing the fraught state of the ocean entails an unmaking of existing modes of law, and a rethinking of the relationship between property, territory, sovereignty, and jurisdiction in particular. Jackson writes along these lines that such a reform "cannot be achieved through international legal legerdemain or by a mere refinement of existing conventions. It can be done only by addressing the fundamental reasons for the exploitation and the complementary rationale of the law that permitted it."[36]

Haunting the contemporary law of the sea is Grotius's legacy, which was elaborated and developed, with notable differences,[37] in Nazi jurist Carl Schmitt's 1942 thesis *Land and Sea*.[38] There, Schmitt brought the juxtaposition between land and sea to its ultimate expression as the basic law of the planet. The human, for Schmitt, is a land-being. In his words: "Among the four traditional elements … the earth is the element that is prescribed for humans and that most strongly defines the human."[39] Needless to say, this definition denies humanity from all those defined as non-land dwellers and, more broadly, from "cosmopolitan" people who were understood as lacking grounded roots (Jewish people in particular, given the context in which Schmitt was writing). Schmitt's main argument revolves around the fundamental tensions between such sea and land powers. He

writes: "World history is a history of the battle of sea powers against land powers and of land powers against sea powers."[40] The sea/land battle also manifests
in the battle of the great fish, the Leviathan, against the great land animal, the
Behemoth. In the political arena, England was perceived by Schmitt as holding
the power of the Leviathan over the world's oceans against the land-dwelling
people of Germany, who held the power of the Behemoth.[41]

Despite its outlandish elementality and explicit antisemitism, Schmitt's work is
frequently referenced by scholars across the political spectrum.[42] His reification of
the land/sea divide, so enshrined in Western theology and political science, also
extends to disciplines such as Black and Indigenous studies. Tiffany Lethabo King
argues against this tendency, claiming that the land/sea binary has historically set
apart the Black studies literature as focusing on diaspora and fluidity from that of
Indigenous studies scholarship that was seen as particularly rooted in land. She
offers, instead, a heightened focus on shoals as liminal spaces that "confound the
binary and dialectical thinking that would separate ocean from land and render
Black people and Indigenous people as an antagonism."[43] This realization is especially acute for geographically dispersed, seafaring people like the Bajau in the Asia
Pacific. Because of the Bajau's sea-based, mobile, and in some cases stateless way of
life, Indonesia has often denied their Indigenous claims, as this volume's chapter
by Pauwelussen and Switzer Swanson documents.[44] Following a similar logic, even
the most ambitious global instrument on the rights of Indigenous peoples—the
UN Declaration on the Rights of Indigenous Peoples (UNDRIP)—references
Indigenous marine entitlements only in passing.[45]

While we recognize the forces that connect such seemingly bifurcated land/
water existences, our collection's legal geography orientation means that we also
realize that this ideological bifurcation has become embodied in bureaucratic
structures. Such structures were diligently constructed on top of the imagined
land/sea binary over the last two centuries at least, reifying and replicating it to
the point of its seeming inevitability and naturalness (the international seabed
regime depicted by Susan Reid and Surabhi Ranganathan in this volume is a
case in point). To collapse these massive infrastructures, one must, as geographer
Philip Steinberg suggests elsewhere,

> develop an epistemology that views the ocean as continually being recon-
> stituted by a variety of elements: the non-human and the human, the bio-
> logical and the geophysical, the historic and the contemporary. Only then,
> can we think with the ocean in order to enhance our understanding of and
> visions for the world at large.[46]

"On the Waves, There Is Nothing but Waves":[47] Legal Abstractions at Sea

A focus on ocean geographies entails a systematic engagement with modern
international law, and our collection's focus on amphibious legal geographies

indeed attends to the minutiae of such law, what Jeffrey Kahn calls elsewhere a "legal anthropology of the technical."[48] It is arguably no coincidence that oceans have been central to the development of modern international law, and that the ocean jurist Grotius is (controversially) considered the "father" of such law.[49] But while Grotius believed in an innate sense of ethics amidst humans, which translated into his jurisprudential focus on justice (but which also aligned well with his pragmatic orientation toward imperial ends), the last several decades have exposed international law's role in the creation of a neoliberal global economy that "obstructs rather than promotes the aspirations for social justice."[50] As international law scholar Janne Nijman puts it: "The neglect for humanity's common good … has produced a crooked international legal order that betrays the human sense of justice."[51] Jackson laments, similarly, that Eurocentric international law is "merely the subordinate bastard child of a liaison of convenience between imperial states and the law that serves their interests."[52] She explains that for many Indigenous people, the international law's rejection of indigenously defined rights is "arrogant legalistic nonsense."[53] The volume's chapters add to these voices by casting further light on the codependent and even coproductive relationship between international ocean law and extractivist logics.

Abstraction is a key feature of law, and of international law in particular. Henry Jones highlights elsewhere that: "From Grotius to UNCLOS, international law assisted in the process of striating the seas, making them featureless and amenable for capital."[54] The sea, in fact, "is where international law pioneered its practices of territorialization."[55] As Jones claims in this volume, the sea is especially susceptible for the line-drawing exercises inherent to international law and necessary for its extractivist enterprise. But whereas the practice of drawing lines on "empty" spaces in maps may have started at sea, "it is [by now] the signature style of colonial cartography on land."[56] Along similar lines, international law scholar Phillip Allott writes that "law conceives social reality as abstraction, then reabstracts it in the form of generalized legal relations."[57] For Allott, then, "To do law is necessarily to do theory."[58]

Marxist legal theorists help us think further about the power of abstraction. According to Evgeny Pashukanis, law "is that which regulates disputes between formally equal, abstract individuals," over equal and abstract things. He maintains that law only becomes universal under capitalism, where generalizable commodities with an abstract value can be exchanged and law both creates and secures this abstract value. In the ocean context, law's power to abstract promotes an extractive imaginary of this space.[59] Vito De Lucia emphasizes the importance of territorialization as the abstracting process through which concrete places are transformed into space and then into territory. Drawing on the work of political geographer Stuart Elden, De Lucia suggests in this volume that "territorialization requires the combination of the technical (for measuring) and the legal (for controlling) or, in other words, [of] knowledge and power."[60] Along the same lines, Philip Steinberg and his colleagues from the ICE LAW project rely on their observations of navigation law in the Arctic to point out in this volume that modern notions of land and sea, although seemingly opposed, are in fact grounded in a common understanding of territory as "the reduction of space to fixed points … [in a] two-dimensional field."[61]

However, unlike land, the ocean (except for the seabed, which is perceived as submerged land) is understood as having no terrain—namely, as a "territory beyond terra,"[62] where the points exist solely as mathematical abstractions. Steinberg et al. emphasize here that the reification of a binary between land (even if submerged) as territory and the ocean as a formless liquid abstraction, or non-territory, constrains the Western legal tradition's imagination. Concerned with the regulation of icebreaking, where the idealized binary between territorial land and formless ocean directly challenges any effort to preserve the structural coherence of sea ice, Steinberg et al. ask in their chapter: "How can one preserve *form* in a space that is legally constructed as formless?"[63]

Abstraction also occurs at the microscale, transforming multispecies marine lives—spliced by international treaties such as the Convention on Biological Diversity and the Nagoya Protocol—into "marine genetic resources" and "digital sequence information."[64] As I describe in my contribution to this volume, in this new *mare geneticum* frontier, situated life is severed from its ecological context so that it may be efficiently utilized for scientific and commercial ends.[65] Philosopher Sabina Leonelli refers to this project of governing data travel as "data regimes."[66] The debate over the definition of marine genetic resources in areas beyond national jurisdiction, which is ongoing as I write this text, is essentially a debate over the level of abstraction and decontextualization of such "resources."

From Abstraction to Extraction

Underlying its myriad technologies for abstraction is law's desire for order, performed through compulsive classification. "Law is necessarily a taxonomy," Allott instructs. "That is to say, law must classify material reality for its own purposes. It must determine kinds of persons, kinds of events, kinds of behavior, and kinds of places. But in the law, classifying is also prescribing."[67] Historically, extractivism was prescribed through one of two legal classifications. Whereas the "old" law of the sea operated around the freedom of the seas concept (Grotius's *mare liberum*), allowing limitless fishing and resource extraction, the "new" law of the post-war period relied on a different extractivist logic: that of enclosed national and international regimes (Selden's system of *mare clausum*).[68] Since the United Nations Convention on the Law of the Sea (UNCLOS) is all about delineating enclosures yet leaves open wide swaths of non-enclosed sea, it can be depicted as a merge of Grotius's free sea and Selden's enclosed sea concepts. For these reasons, some have characterized the modern history of ocean governance as a "seesaw"—an oscillation between free and enclosed regimes.[69]

In the seabed, the move from old to new law took place over a relatively short time. According to Surabhi Ranganathan and Susan Reid (both in this volume), deep seabed mining is where extractivism perfected itself by applying the freedom of the sea through the "common heritage of mankind" principle, which was originally conceived to benefit humanity.[70] Reid asks in her chapter: "If it is accepted that humans are materially embodied, ecologically interconnected, and reliant on the ocean for wellbeing, how could the environmental violences that result from multi-decadal commercial seabed mining activity benefit humanity 'as a whole'?"[71] Following Jacques Derrida, Reid invokes the notion of *eating well* to point to the ethics that emerge at the intersection of needing to protect the deep ocean and needing to enact violences upon it in order to source minerals.[72] Her work also points to the more invisible materialities of extractivism: the "waste" produced in its predatory process, which she refers to as "flesh waste" to highlight that "extractivism takes from the ocean not just minerals but more-than-human lives, lifeways, and relations, rendering them waste."[73] Under this logic, ocean materialities are abstracted into either "resources," which are deemed valuable in the capitalist scheme, or "waste," which is then dumped back into the sea, wreaking destruction of a second order.[74] This second order of destruction by waste-making is the even darker side of extractivism.

Hydrothermal vents in particular throw into sharp relief the extractive imaginary established by UNCLOS, and this Convention's ostensibly neat juxtapositions of land and water, life and matter, and mobility and immobility. Vents are mineral-rich soups of manganese, copper, iron, nickel, cobalt, gold, and silver that gush out through the cracks and precipitate as sheets, mounds, or chimneys of great heights on the ocean floor, where the earth's plates are constantly shifting.[75] Formed through the circulation of seawater, vents are also home to unique ecosystems, much sought after by biotech and pharmaceutical industries. Discovered too late for detailed attention in the 1982 UNCLOS, the governance of these ecosystems is currently debated under a new mining code, which is examined critically in this volume by both Ranganathan and Reid. De Lucia, too, writes about vents in this volume, suggesting that "[t]heir complexity eludes sovereign legality, obsessively bent on sorting out to which maritime zone hydrothermal vents belong, to which domain (the living or the nonliving) and thus, consequently, which legal regime is applicable."[76]

Language has proven itself essential to the enterprise of extraction. Deploying such terms as the "blue economy" and "blue capital," seabed mining has been lauded as the new and vast ocean frontier. It is in this context that Elizabeth DeLoughrey warns us against endorsing such seemingly innocent language, which Susan Reid similarly criticizes as a form of blue-washing by corporate enterprises.[77] Instead of casting the blame exclusively on corporate powers, though, DeLoughrey points toward academia's neoliberal enterprise in the humanities to show that "we are all complicit."[78] She contends, specifically, that while the blue humanities scholarship is certainly "driven by our environmental crisis and the ecological/multispecies turn in scholarship, [it] is also the product of the neoliberalization of academia and the rebranding of humanities work in an era of intellectual and economic downsizing."[79] Staging an interdisciplinary conversation between recent scholarship about the speculative practices of deep-sea mining and the speculative fiction that imagines techno-utopian futures of human life under the sea, DeLoughrey's contribution to this volume raises questions about the ways in which particular kinds of literary genres and reading practices themselves produce an extractive imaginary, examining the uncomfortable overlap between the concept of innovation as a driver of both the blue economy and the blue humanities.

UNCLOS and BBNJ: Current and Future Ocean Regimes

Human history has seen multiple local and regional attempts to frame law's relationship to the ocean. But one international treaty stands out for its aspiration to define a *global* relationship between ocean and law. UNCLOS is a monumental treaty with 320 articles divided into 17 parts that establish normative concepts, such as the 12-mile territorial sea, the 200-mile exclusive economic zone (EEZ), the high seas, and the seabed and ocean floor beyond national jurisdiction (or "the Area").[80] Under the jurisdictional matrix of UNCLOS, national sovereignty typically diminishes with increasing distance from land.[81] British geographer Stephen Graham characterizes this way of governing as the "classical, modern formulation of Euclidean territorial units jostling for space on contiguous maps,"[82] whereas Elizabeth DeLoughrey describes UNCLOS as "the largest juridical and cartographic change to the globe in human history."[83]

While UNCLOS aspired to heal the fragmented legalities that have permeated the seas ("the fragmentation of ocean space among more than 100 different sovereignties," according to former Malta ambassador Arvid Pardo[84]) and thus cast itself as a comprehensive ocean constitution,[85] it ended up internalizing the fragments and reproducing them from within. Writing in this volume, Elspeth Probyn thus calls UNCLOS a "geopolitical chokepoint," wherein "the smooth connectedness of globalization across liquid expanses"[86] as often as not translates into blockages, constraints, and death.[87] She elaborates in particular on oceans crowded with ships and vessels that hide under the Flags of Convenience regime adopted by UNCLOS, which exposed its particularly ugly face during the Covid pandemic.[88]

UNCLOS figures prominently throughout this volume as a utopian-turned-dystopian legal text, where the world's best intentions were finally put to die. From the dream of a newfound global freedom *for* the sea[89]—*mare nostrum*[90]—and of a "common heritage of mankind," UNCLOS indeed soon revealed itself as the very technology through which capitalist expansion and exploitation would occur. Susan Reid argues in this volume, accordingly, that despite claims to represent "all peoples of the world,"[91] UNCLOS privileges corporations and excludes diverse knowledges, such as those practiced by Indigenous communities of the Pacific for millennia.[92]

This collection's title, *Laws of the Sea*, reveals the broken promise of a constitutional law of the sea and its fragmented, colonial, and racist implementations, at the same time expressing a yearning for pluralistic, and plural, legalities that

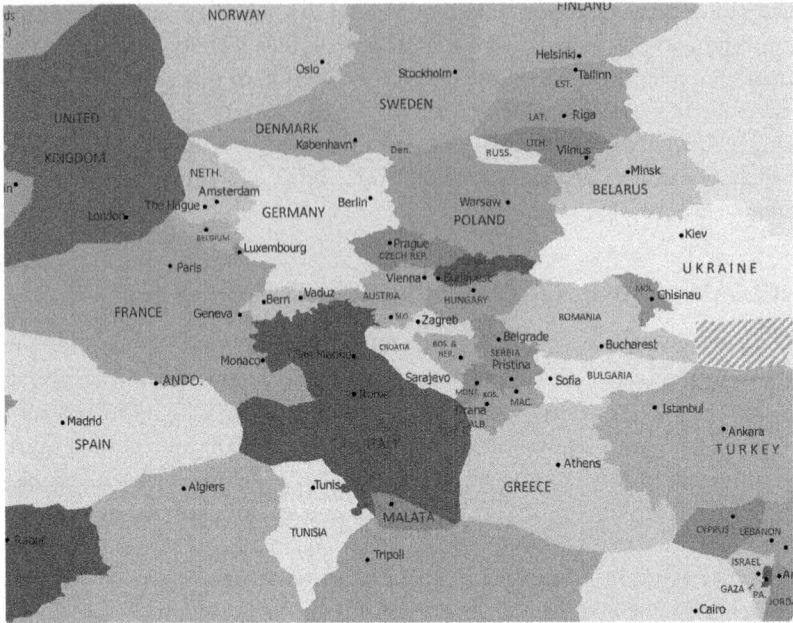

FIGURE I.2 The "Territorial Map of the World" (2013) by Rafi Segal and Yonatan Cohen, depicts the extent of territories, both on land and at sea (submerged lands in EEZs), under the control of nation-states. According to Segal and Cohen, "The rudimentary divide of water and land, a cultural and psychological divide as well as an obsolete mapping artifact, has become irrelevant both to current political and economic concepts of space, and in light of more recent national sovereign claims over sea zones." Used with permission.

correspond with notions of justice.[93] In *Decolonizing International Law*, Sundha Pahuja similarly explores such tensions between a desired and dangerous universality. In her words: "The attempt to use international law was inspired and enabled by international law's promised universality. However, that same promised universality served to constrain, and ultimately to undermine, the radical potential of the Third World demands."[94]

And from one legal utopia to the next: operating under UNCLOS, in 2017 the United Nations General Assembly resolved to convene an international legally binding instrument "on the conservation and sustainable use of marine biological diversity of areas beyond national jurisdiction" (in short, the BBNJ).[95] In the name of biodiversity protection, parties to the negotiation over this "binding instrument" are currently contemplating how best to manage and share the resources of the high seas through a novel regime that would extend national jurisdictions. At the time of writing this, the BBNJ meetings are still underway, with scientists, legal practitioners, and policy makers debating *how*, but not so much *if*, to extract.[96] Thus, an opportunity to promote other-than-humans up the ladder of human priorities might be slipping away—again. It is in this context that Surabhi Ranganathan voices an urgent call on the participants of the BBNJ process to step up to the challenge and "unmake rather than fix" ocean law, quoting from those who mourned the demise of the Okjokull glacier: "We know what is happening and what needs to be done. Only you know if we did it."[97]

During my own participation in some of the discussions of the PrepCom meetings for the BBNJ, I found it quite difficult to follow the nuanced debates over terminology despite my formal legal training. Considering this alongside Probyn's note about international law's frequent use of acronyms—an alphabet soup, as she calls it[98]—I cannot help but think about the ways in which expertise seals itself off from influence by non-experts through using indecipherable jargon. As a direct result, the world's public is largely unaware that such crucial discussions are being held, with profound implications for our shared lives on this blue planet.

Laws on the Edge

If the unifying theme in the first group of chapters in this volume is the extractive governmentalities of the sea, then the second group of chapters dwells on the liminal legal geographies of edges and borders and the potential modes of resistance that ensue. Oscillating between extractive and liminal logics, the chapters at the same time resist these neat classifications even in the collection, performing an amphibious existence in this text, too.

I have argued here that the land/sea divide has been foundational to Eurocentric law. Embodying mythical powers and enshrined in structural realities, this divide can be traced back to the physical site from where it originated: the edge between land and sea. Of course, systems of ocean law have much to say about this edge, which is imagined in the form of a line and abstracted as such.[99] Geographical factors, and coastal geography specifically, are fundamental to international law as it pertains to maritime boundary delimitations. The first legal boundary between land and sea, and the site from which national maritime claims are made, is the "baseline," which is defined in modern international law as consisting of a low-water line and of straight lines joining points on the low-water lines, or of a combination of low-water and straight lines.[100] The myriad doctrinal debates over lines and law in the sea are documented in a range of massive legal casebooks on ocean and coastal law.[101]

Taking a different perspective, Rebecca Solnit defines the seashore as an edge because it "defies the usual idea of borders by being unfixed, fluctuant, and infinitely permeable."[102] Edge communities, and humans as "edge species,"[103] flourish where ecosystems overlap in places known as "ecotones,"[104] thus illuminating the always transient and fluid nature of both land and water. Such "edgeness" resonates with Michael Pearson's studies of littoral societies in the Indian Ocean, which point to the amphibious (human) movement between land and sea.[105] It also resonates with Annet Pauwelussen and Gerard Verschoor's work on coral conservation in Indonesia,[106] which foregrounds the more-than-human components of amphibiousness. Referring to the "hybrid land-water interface" as the "amphibious living environment for both coral and people," their study highlights the generative capacity of ambiguity.[107] For the law to become sea-worthy, they emphasize, it needs to cultivate an amphibious quality that would render it capable of moving in between ways of knowing and being. In their words: "Providing room for ambiguity, thinking with amphibiousness further-more encourages suspension of the (Western) tendency to explain the Other, to fix what does not add up."[108]

The littoral quality of the edge illuminates that the essentialization of land and sea as seemingly natural does not hold water. It therefore makes perfect sense that much of the pushback against land/sea distinctions occurs at the interstitial space of the coastline. Jeffrey Kahn's contribution to this volume about shippers on the Miami River and the urbanization, and gentrification, of the ocean is a case in point. Kahn highlights how Black Haitian seafarers navigating wooden freighters produced riverine and maritime spaces through a host of economic improvisations. The growing presence of Haitian shippers in the heart of Miami triggered a set of regulatory responses, a combination of new safety codes and re-zoning moves that would eventually push many of them out of Miami and, as a result, out of the maritime passages between South Florida and Haiti. Operating in this way, legal interventions in the port have become, by extension, legal interventions at sea. Racializing urban pro-cesses extend to the ocean, Kahn explains, suggesting that the legal dimen-sions of land-focused urban gentrification are, in fact, also amphibious in their ramifications.

As a central theme in legal geography, scale manifests in various ways through-out the volume. Florian Grisel's contribution points to the complex interac-tions between traditional coastal communities, on the one hand, and the global legal impositions of UNCLOS and the French state, on the other hand, offer-ing a mode of "transnational localism." His case study of the private govern-ance by a medieval institution called the *Prud'homie* over fishers in the port of Marseille offers a glimpse into the impact of international law on the everyday life of fishermen. Specifically, he shows how this law reshaped their commu-nity by encouraging conflicts between "grand art" and "small art" fishers. Aron Westholm flips the equation in his chapter by showing how municipal spatial planning in Sweden in fact follows globalized trends of economic growth and development, whereas it is through national law that environmental protections are administered on the local scale. Westholm also points to the perils of scalar fragmentation through law, which results in the structural erasure of ecosystems. From his perspective, marine spatial planning norms should integrate the vari-ous scales of management and administer a more inclusive framework that is not severed along administrative and scalar lines.

Heterolegalities, a term coined by Vito De Lucia, provide yet another poten-tial space of resistance to and through law. In his chapter for this volume, De Lucia specifically draws on Michel Foucault's concept of heterotopia[109] to con-template sites in which space is "othered," including "those counter-sites that resist the dissolution of social space produced by the calculating and abstracting spatiality of modernity that now represents normal space."[110] Heterotopias are currently extending to areas beyond national jurisdiction through the BBNJ negotiations, he offers, pausing to consider, finally, whether such heterotopic ways of organizing space might provide alternative forms of the legal. For him, heterolegalities could either crucially navigate the crises of the Anthropocene, or they could remain forever entangled with such hegemonic legality and even facilitate its reproduction.

One potential site of othering considered by De Lucia is the marine protected area (or MPA)—and it is precisely at this site that ethnographers Annet Pauwelussen and Shannon Switzer Swanson pick up the discussion. Unlike De Lucia, however, they depict how marine protected areas—utilized in Indonesia alongside other national and international regimes for nature protection that define sea turtles as endangered wildlife—often clash with the claims of local communities of Bajau people to their customary rights. Instead of the marine protected area as the site of othering, then, it is the Bajau who are perceived by the Indonesian state and international nongovernmental organizations as outlaws. Pauwelussen and Switzer Swanson sensitively document the ways in which the Bajau have worked around and resisted the criminalization of their kinship with the turtle, utilizing a legal pluralism approach to define such mundane practices of resistance as another form of law.[111] Following this line of inquiry, we might want to consider the role that Indigenous marine law can play in unmaking existing international law.[112]

Amphibious Futures: "Every Law Is Turning Blue"

Every day seems to be bringing along with it a new book about oceans, each proposing a different regional or temporal perspective through which to see the sea. Whether we perceive this immense wave of scholarship on oceans as part of an "oceanic" or "blue" turn, as some have depicted it,[113] or we acknowledge that it is not a turn but a re-turn, as certain Indigenous scholars prefer to see it,[114] and whether we call this body of work "blue" or rather "critical" ocean studies,[115] it seems fair to say that critical legal studies have largely not been part of this wave and thus have yet to fully immerse themselves in the sea.

This is not to say that a solid scholarship on ocean law does not exist, but rather to suggest that this scholarship has often been more doctrinal and less interdisciplinary, which has also made it less accessible for those from outside the legal studies community. Even within the legal world, ocean law is marginalized, as many law schools do not offer related courses and, if they do, the ocean is typically covered in one section within the broader studies of environmental or international law. As a result, the fate of the oceans, especially in areas beyond national jurisdiction, is currently being negotiated and decided mostly behind closed doors, in jargon-laden settings, and with little to no critical legal and interdisciplinary input.

Yet legal studies seem to slowly be picking up on the ocean awakening that has been taking place across academia: a novel field of critical legal ocean studies is emerging, with legal historians (some of whom are contributors to this volume) leading the way.[116] This collection pushes the envelope further by bringing legal scholars, geographers, anthropologists, and environmental humanities scholars on board to promote interdisciplinary insights about the coproduction of space and law in the world of the sea. In a play on UNCLOS's *Law* of the Sea, which imagines an all-encompassing constitutional framework for governing oceans, our collection refers to law in plural ways, applying the insights that are emerging within multiple disciplines and trajectories about law, broadly conceived, to consider the possibilities of a comparable critical ocean approach in legal studies, and in its subfield of legal geography in particular.

Thinking with oceans and their lively creatures through the lens of solidarity and care[117] provides a fresh opportunity to challenge the existing modes of governing oceans, thus bringing about more fluid and adaptable—or, as I call them here, "amphibious"—regulatory modalities. By tracing the interstitial connections and relationships among these regulatory spaces, the symbolic and material binary between land and sea is complicated, illuminating their amphibious legal geographies. Such work is necessary at this precarious time as it provides hope for extant forms of life at sea, who critically depend on legal frameworks to survive within their altering environments.

The central question that the contributors of this volume keep circling back to is whether modern Western law, and international law in particular, is capable of such an amphibious rethinking. Specifically, are alternative legalities of the sea even possible, or is Western law so steeped in colonial histories and legacies, which carry on "in its wake"[118] to this day, that change is impossible?[119] And is international law so infused in such colonial and capitalist frameworks that it indeed cannot be fixed but only unmade? Yet relatedly, could UNCLOS serve as the foundation for change, or would such change require a charting of an entirely new regime, one that would move beyond international law to a planetary regime, even? In answering these questions, we might take our cues from Indigenous systems of law, which suggest that alternative legalities of the sea already exist.

Art could also possibly show us the way. A performative action in art can flesh out the performativity of legal scholarship in unexpected ways, orienting the discussion toward a practice-led amphibiousness. Bringing art practice and academic research together is arguably an ethical necessity at this time. In his contribution to this volume, legal theorist and artist Andreas Philippopoulos-Mihalopoulos contends with these questions through what he calls "wavewriting."[120] Bringing water to law, he reminds us that we are all in fact bodies of water. Legal writing is also wavewriting, he explains. "But more than that. Every text is turning blue. Every law is turning blue." Pouring water onto law, the letters of the law smear and smudge. "Your law is no longer dry. It drips bleeding."[121]

The affective power of Philippopoulos-Mihalopoulos's performative wave-writing calls to mind Caribbean-born Canadian Marlene Nourbese Philip's poem cycle *Zong!*[122] There, Philip revisits the 500-word legal report of the 1783 insurance trial *Gregson v. Gilbert*, the only official documentation of the 1781 voyage of a ship, the *Zong*, which originally carried 470 slaves from the West Coast of Africa toward Jamaica. Late in the voyage, after hundreds of slaves had already perished from lack of water, the captain murdered 150 of the still-living enslaved humans on board. By killing these slaves, the captain reasoned, the ship owners could collect insurance money for their "property."[123] Relying entirely on the legal text produced by the court, Philip's poetry blanks out the court's words and reverses their order, deliberately defacing the official document. By doing so, she subverts the murderous abstractions and categories on which the law is based to highlight the unaccounted for materialities of the 1781 event.[124]

Using the law to subvert itself from within is a powerful form of amphibiousness. Currently performed in the context of the violent policing of the "Black Mediterranean,"[125] the Forensic Oceanography project[126] is an excellent example of such amphibious resistance from within as it is executed on the ground—and

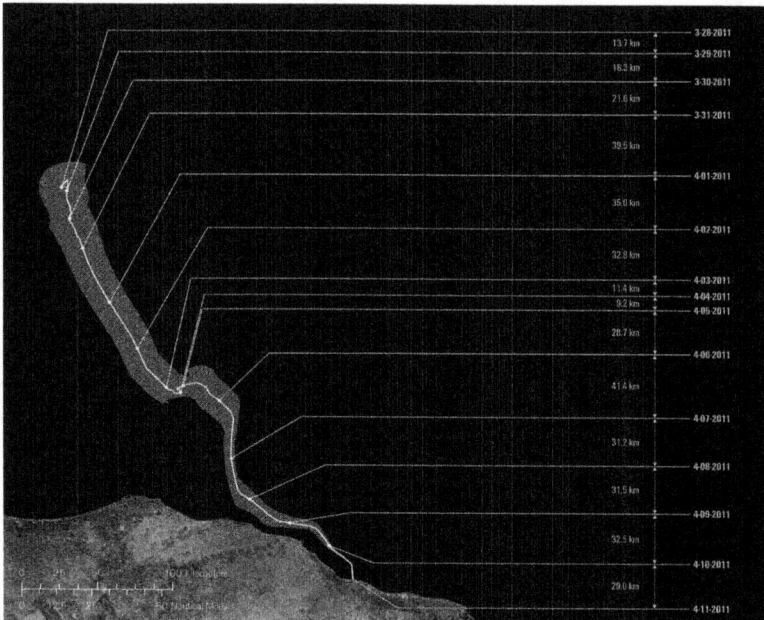

FIGURE I.3 A spatio-temporal mapping of the 2011 Left to Die Boat's time adrift (seen also in Figure I.1). From the Forensic Architecture website: "Our drift model provides day-by-day estimates for the vessels location. The drift trajectory was reconstructed by analyzing data on winds and currents collected by buoys in the Strait of Sicily. Over time, the margin of error in the drifting vessel's track decreases linearly as it is constrained by the known position of landing." Reprinted with permission by SITU Research.

at sea. By utilizing legal technologies such as mapping, survivor testimony, and counternarrative, this project transforms the legal knowledge generated through means of surveillance into evidence of responsibility for the crime of non-assistance. Christina Sharpe describes the Forensic Oceanography project "as another kind of wake work that might counter forgetting, erasure, [and] the monumental."[127] The continued relevance of law as a powerful adjunct of coercion at sea highlights the relevance and urgency of this collection's study of law's amphibious geographies.

Acknowledgements

I am grateful to the participants of the *Laws of the Sea* workshops, and to my colleagues at the National Humanities Center during fall 2021. Special thanks to Jack Schlegel, Annet Pauwelussen, Vito De Lucia, Philip Steinberg, Jeffrey Kahn, Elizabeth DeLoughrey, Barbara Kowalzig, Jane Thrailkill, and Jessica Hurley for their comments on the draft, and to my research assistant Margaret Drzewiecki for her editorial work.

Notes

1 Teresia Teaiwa, "We Sweat and Cry Salt Water, So We Know that the Ocean Is Really in Our Blood," *International Feminist Journal of Politics* 19, no. 2 (2017): 133, 133–136.
2 See also Bernhard Klein and Gesa Machenthun, "Introduction" in *Sea Changes: Historicizing the Ocean*, eds. Bernhard Klein and Gesa Machenthun (New York: Routledge, 2004), 2. Compare with Lauren Benton and Nathan Perl-Rosenthal, eds., *A World at Sea: Maritime Practices and Global History* (Philadelphia: University of Pennsylvania Press, 2020), 187. See also Philip Steinberg, *The Social Construction of the Sea* (Cambridge: Cambridge University Press, 2001), 35. Milun argues that empty space, and the space of the global commons in international law in particular, is a key spatial imaginary of modernity. Kathryn Milun, *The Political Uncommons* (Farnham: Ashgate, 2011), 2. By contrast, Kahn insists that the sea spaces of the northern Caribbean do not fit the Agambenian vision of exception; they cannot be described as "a legal or normative void, so to speak. There is far too much law at work … [in such oceanic jurisdictions] for that." Nonetheless, he adds, certain spaces are sometimes *posited* as exceptional spaces relative to others, a distinction that renders specific actions juridically plausible at sea that would not be elsewhere. Jeffrey S. Kahn, *Islands of Sovereignty: Haitian Migration and the Borders of Empire* (University of Chicago Press, 2019), 13.
3 In recognition of the many ways that this modern and terracentric way of knowing seeps into everyday practices, legal norms, and institutional structures, we label it in this volume with the appellation "Western"—with a capital "W"—not to honor or reinforce it, but to recognize and criticize its power.
4 John R. Gillis, *The Human Shore* (Chicago: University of Chicago Press, 2012), 198. See discussion below.
5 Whereas the term "ecotone" evolved in the environmental sciences since the 1960s, scholars from the environmental humanities have more recently used it to think through relations of power in the social realm. See, e.g., Paul G. Risser, "The Status of the Science Examining Ecotones," *Bioscience* 45, no. 5 (1995): 318–325. In *The Black Shoals*, Tiffany Leffabo King defines the shoal as a type of ecotone in that it

"offers a moment to reassemble the self as an amphibious and terrestrial subjectivity. Not just water (fluid, malleable, and fungible) but also a body landed. A place and time of liminality where one becomes an ecotone, a space of transition between distinct ecological systems and states." Tiffany Leffabo King, *The Black Shoals* (Durham: Duke University Press, 2019), 9.

6 Benton and Perl-Rosenthal, *A World at Sea.*

7 Reid, this volume, 73.

8 Plato, *Phaedo*, trans. David Gallop (Oxford: Clarendon, 1975), 108c.

9 Batsheba Demuth, *Floating Coast: An Environmental History of the Bering Strait* (New York: W.W. Norton, 2019), 11.

10 Irus Braverman, "More-than-Human Legalities," in *The Wiley Handbook of Law and Society*, eds. Patricia Ewick and Austin Sarat (Hoboken: Wiley Press, 2015), 307–321. See also Irus Braverman, "Law's Underdog: A Call for Nonhuman Legalities," *Annual Review of Law and Social Science* 14 (2018): 127–144.

11 As Philip and Steinberg state: "We propose a wet ontology not merely to endorse the perspective of a world of flows, connections, liquidities, and becomings, but also to propose a means by which the sea's material and phenomenological distinctiveness can facilitate the reimagining and re-enlivening of a world ever on the move." Philip E. Steinberg and Kimberley Peters, "Wet Ontologies, Fluid Spaces: Giving Depth to Volume through Oceanic Thinking," *Environment and Planning D: Society and Space* 33, no. 2 (2015): 247–264.

12 King, *The Black Shoals.*

13 This term was coined by Kamau Brathwaite as the rejection of the notion of the dialectic, and taken up by multiple scholars. Stefanie Hessler explains that "*Tidalectics* is an experiment to formulate an oceanic worldview, a different way of engaging with the oceans and the world we inhabit. Unbound by land-based modes of thinking and living." "Tidalectics," Stefanie Hessler, accessed October 20, 2021, http://stefaniehessler.com/entries/tidalectics. See also Elizabeth DeLoughrey and Tatiana Flores, "Submerged Bodies: The Tidalectics of Representability and the Sea in Caribbean Art," *Environmental Humanities* 12, no. 1 (2020): 132–166.

14 Michael Pearson, *The Indian Ocean* (London: Routledge, 2003), 5.

15 Benton and Perl-Rosenthal, *A World at Sea.*

16 Irus Braverman, ed., *Animals, Biopolitics, Law: Lively Legalities* (Abingdon: Routledge, 2015). See also Stacy Alaimo, "Oceanic Origins, Plastic Activism, and New Materialism at Sea," in *Exposed: Environmental Politics and Pleasures in Posthuman Times*, eds. Serenella Iovino and Serpil Oppermann (Bloomington: Indiana University Press, 2014), 186–203; Astrida Neimanis, *Bodies of Water: Posthuman Feminine Phenomenology* (London: Bloomsbury Publishing, 2017); DeLoughrey and Flores, "Submerged Bodies"; Alexis Pauline Gumbs, *Undrowned: Black Feminist Lessons from Marine Mammals* (Chico: AK Press, 2020).

17 Gillis, *The Human Shore*, 198.

18 Michael N. Dawson and William M. Hammer, "A Biophysical Perspective on Dispersal and the Geography of Evolution in Marine and Terrestrial Systems," *Journal of the Royal Society Interface* 5 (2008): 135–150, 136.

19 They explain that: "The chemistry of body fluid carries a historical signature of its marine origin, albeit one that has been greatly modified by the demands of physiology. ... Hypersea consists of the body fluids of land-dwelling (i.e., air and soft-dwelling) eukaryotes that have an intimate symbiosis, ranging from mutualism to parasitism, with other organisms." Mark A.S. McMenamin and Dianna L.S. McMenamin, "Hypersea and the Land Ecosystem," *BioSystems* 31 (1993): 145–153, 145–146.

20 Rachel Carson, *The Sea Around Us* (New York: Open Road Media, 2011).

21 "5 Questions for Dr. Sylvia Earle," Conservation Law Foundation, 2015, https://www.clf.org/conservation-matters-articles/5-questions-for-dr-sylvia-earle/, quoted in DeLoughrey, this volume, 145.

22 On the distinction between private and public international law, see Philip Steinberg, *The Social Construction of the Ocean* (Cambridge: Cambridge University Press, 2001). In his words: "private ocean law (or maritime law) is a set of rules (including national legislation and international agreements) regulating the economic and commercial aspects of marine transport. Public ocean law (or the law of the sea) governs the sea itself, particularly with reference to control of ocean-space and access to its living and nonliving resources. Steinberg ties this juxtaposition to that between land and sea, pointing to its manifestation in the contrasting governance regimes applied to coastal and deep-sea geographies." Ibid., 15. Of course, such neat divisions between public and private international law obscure consequential inner connections. See also Edgar Gold, *Maritime Transport: The Evolution of International Maritime Policy and Shipping Law* (Lexington: Lexington Books, 1981), xx.

23 Benton and Perl-Rosenthal, *A World at Sea*, 188.

24 Ibid., 192. Maori legal expert Moana Jackson writes along these lines that: "For the Maori people, *te tikango o te moana*, or the law of the sea, is predicated on four basic precepts deeply rooted in Maori cultural values. First, the sea is part of a global environment in which all parts are interlinked. Second, the sea, as one of the *taonga*, or treasures of Mother Earth, must be nurtured and protected. Third, the protected sea is *koha*, or gift, which humans may use. Fourth, that use is to be controlled in a way that will sustain its bounty." Moana Jackson, "Indigenous Law and the Sea," in *Freedom for the Seas in the 21st Century: Ocean Governance and Environmental Harmony*, eds. Jon M. Van Dyke et al. (Washington, D.C.: Island Press, 1993), 41–48, 46.

25 Marcus Rediker, *Outlaws of the Atlantic: Sailors, Pirates, and Motley Crews in the Age of Sail* (Boston: Beacon Press, 2014), 2.

26 Gillis, *The Human Shore*, 7.

27 Rediker, *Outlaws of the Atlantic*, 2. See also Elizabeth DeLoughrey, *Routes and Roots: Navigating Caribbean and Pacific Island Literatures* (Honolulu: University of Hawai`i Press, 2007); Peter Linebaugh and Marcus Rediker, *The Many-Headed Hydra* (Brooklyn: Verso Books, 2000); Paul Gilroy, *The Black Atlantic* (Cambridge: Harvard University Press, 1993); Jace Weaver, *The Red Atlantic: American Indigenes and the Making of the Modern World, 1000–1927* (Chapel Hill: University of North Carolina Press, 2014); Vicente M. Diaz, "Voyaging for Anti-Colonial Recovery: Austronesian Seafaring, Archipelagic Rethinking, and the Re-Mapping of Indigeneity," *Pacific Asia Inquiry* 2, no. 1 (2011): 21–32; Fahad Ahmad Bishara, *A Sea of Debt: Law and Economic Life in the Western Indian Ocean, 1780–1950* (Cambridge: Cambridge University Press, 2017); Sunil S. Amrith, *Crossing the Bay of Bengal* (Cambridge: Harvard University Press, 2013); Omisee'eke Natasha Tinsley, "Black Atlantic, queer Atlantic," *GLQ* 14, nos. 2–3 (2008): 191–215.

28 Renisa Mawani, *Across Oceans of Law* (Durham: Duke University Press, 2018), 5, 33. Scholarship in international law has long debated Grotius's legacy. In the words of Martti Koskenniemi: "My purpose, however, is neither to celebrate the 'Grotian tradition' nor to attack it as an unscrupulous instrument for legitimating capitalism, authoritarianism and colonial rule." Martti Koskenniemi, "Imagining the Rule of Law: Rereading the Grotian 'Tradition,'" *European Journal of International Law* 30, no. 1 (2019): 17–52, 24.

29 Jackson, "Indigenous Law and the Sea," 43.

30 Austin Sarat, "'… The Law is All Over': Power, Resistance and the Legal Consciousness of the Welfare Poor," *Yale Journal of Law & the Humanities* 2, no. 2 (1990): 343–379.

31 See, e.g., Nicholas Blomley, *Law, Space, and the Geography of Power* (New York: Guilford, 1994), 7–8. For a different view of the relationship between social and legal norms, see Florian Grisel, this volume.

32 Bishara, *A Sea of Debt*, 9, 12.

33 For a definition of critical legal geography see Irus Braverman et al., eds., *The Expanding Spaces of Law: A Timely Legal Geography* (Stanford: Stanford University Press, 2014).

34 Here, a quick nod to Elspeth Probyn's discussion of "chokepoint" in this volume. The term strangulation might better highlight the agency, the intentionality even, of the perpetrators. See, e.g., Moira Donegan, "Gabby Petito Died of Strangulation: Far Too Many Other Women Have, Too," *The Guardian*, October, 14, 2021, https://www.theguardian.com/commentisfree/2021/oct/14/gabby-petito-wyoming-strangulation-domestic-violence ("But medical experts and domestic violence advocates prefer the word strangulation. ... Using the term strangulation also has greater political valence: it keeps the focus on the perpetrator. Someone can choke by accident. Strangulation is intentional"). Two media articles that illustrate this intentionality in the ocean context were published in a matter of days as I was writing this introduction. First, Patrick Wintour, "Rotting Red Sea Oil Tanker Could Leave 8m People Without Water," *The Guardian*, October 11, 2021, https://www.theguardian.com/world/2021/oct/11/rotting-red-sea-oil-tanker-could-leave-8m-people-without-water; and second, Gabrielle Canon, "Ships Backed Up Outside US Ports Pumping Out Pollutants as They Idle," *The Guardian*, October 15, 2021, https://www.theguardian.com/business/2021/oct/15/us-california-ports-ships-supply-chain-pollution.

35 See, e.g., Alanna Mitchell, *Sea Sick: The Global Ocean in Crisis* (Toronto: McClelland & Stewart, 2009); Sylvia Earle, *The World Is Blue: How Our Fate and the Ocean's Are One* (National Geographic, 2009); and Callum Roberts, *The Unnatural History of the Sea* (Washington, D.C.: Island Press, 2007).

36 Jackson, "Indigenous Law and the Sea," 43.

37 For a discussion of Schmitt's legacy in the context of oceans see Mawani, *Across Oceans of Law*, 5, 232; Kahn, *Islands of Sovereignty*, 9–10.

38 Carl Schmitt, *Land and Sea: A World-Historical Meditation*, trans. Samuel Garrett Zeitlin (Candor: Telos Press Publishing, 2015), 6.

39 Ibid.

40 Ibid., 63.

41 Ibid. But in Schmitt's *Nomos of the Earth*, "air power" would be the determining factor in what he calls a new "spatial ordering" of the world after the Second World War. According to Schmitt, the rise of the United States to global preeminence and the corresponding decline of European power will prompt a new spatial understanding of the world to ground its geopolitics. The Carl Schmitt Reader, "The Distribution of Air Power and the "Nomos" of the Earth," October 4, 2016. https://medium.com/@csreader/the-distribution-of-air-power-and-the-nomos-of-the-earth-fe6eb6f250bd. See also Carl Schmitt, *The Nomos of the Earth in the International Law of the Jus Public European* (New York: Telos Press Publishing: 2006).

42 Benton and Perl-Rosenthal, *A World at Sea*, 186.

43 Similarly, Vicente Diaz's work, as well as that of many other Indigenous scholars, challenge notions of Indigenous "rootedness" in static time and space. See Vicente M. Dias, "No Island Is an Island," in *Native Studies Keywords*, eds. Stephanie Teves et al. (Tucson: University of Arizona Press, 2015), 90–108.

44 This resonates with the "stateless" seafaring communities in Probyn's chapter, this volume. See also Stephen Allen et al., *The Rights of Indigenous Peoples in Marine Areas* (Bloomsbury: Bloomsbury Publishing, 2019).

45 Allen et al., *The Rights of Indigenous Peoples*, 2.

46 Philip E. Steinberg, "Of Other Seas: Metaphors and Materialities in Maritime Regions," *Atlantic Studies* 10, no. 2 (2013): 156–169, 157.

47 Schmitt, *The Nomos of the Earth*, 43.

48 Kahn, *Islands of Sovereignty*, 140. Quoting from his poignant articulation: "the task of unveiling the elements of this architecture requires a painstaking attention to the technical dimensions of law to an extent that may, at times, border on the excruciating for those not initiated into the world of legal draftsmanship. ... [I]t is meant to be taxing, because it is the banality and arcanity of the art of legal manufacture

that abets its invisibility and fosters the type of fetishization that makes such edifices appear monolithic in the first instance." Ibid. See also Mawani's oceans as method in *Across Oceans of Law*, 8.

49 R.P. Anand, *Origin and Development of the Law of the Sea* (Hague: Nijhoff, 1983), 2. Anand suggests that while it casts itself as universal, international law is in fact an imperial promotion of Western values that present themselves as the tallest moral order. See also Milun, *The Political Uncommons*, 78.

50 Janne E. Nijman, "Grotius' 'Rule of Law' and the Human Sense of Justice: An Afterword to Martti Koskenniemi's Foreword," *European Journal of International Law* 30, no. 4 (2019): 1105–1114, 1113.

51 Ibid., 1113–1114. See also Antony Anghie, *Imperialism, Sovereignty, and the Making of International Law* (Cambridge: Cambridge University Press, 2005).

52 Jackson, "Indigenous Law and the Sea," cf. Jon M. Van Dyke et al., eds., *Freedom for the Seas in the 21st Century: Ocean Governance and Environmental Harmony* (Washington, D.C.: Island Press, 1993), 10.

53 Jackson, "Indigenous Law and the Sea," 45.

54 Henry Jones, "Lines in the Ocean: Thinking with the Sea About Territory and International Law," *London Review of International Law* 4, no. 2 (2016): 307–343, 342.

55 Ibid. For a detailed account of the maritime political boundaries of the world, see Victor Prescott and Clive Schofield, *The Maritime Political Boundaries of the World* (Leiden: Nijhoff, 2004).

56 Jones, "Lines in the Ocean," 342.

57 Van Dyke et al., *Freedom for the Seas*, 49.

58 Ibid.

59 Robert Knox, "Marxist Approaches to International Law," in *The Oxford Handbook of the Theory of International Law*, eds. Anne Orford and Florian Hoffman (Oxford: Oxford University Press, 2016), 316. See also Henry Jones, this volume.

60 De Lucia, this volume, 129. Stuart Elden defines territory as a "political technology" aimed at "measuring land and controlling terrain." Stuart Elden, "Land, Terrain, Territory," *Progress in Human Geography* 34, no. 6 (2010): 799–817, 799.

61 Steinberg et al., this volume, 172.

62 Kimberley Peters et al., eds., *Territory Beyond Terra* (Lanham: Rowman & Littlefield International, 2018).

63 Steinberg et al., this volume, 174. See also Jessica Lehman et al., "Turbulent Waters in Three Parts," *Theory & Event* 24, no. 1 (2021): 192–219; Kimberley Peters and Philip E. Steinberg, "The Ocean in Excess: Towards a More-than-Wet Ontology," *Dialogues in Human Geography* 9, no. 3 (2019): 293–307; Steinberg and Peters, "Wet Ontologies."

64 Braverman, this volume. On the sea's microbial life, see Stefan Helmreich, *Alien Ocean: Anthropological Voyages in Microbial Seas* (University of California Press, 2009).

65 Braverman, this volume, 109.

66 Sabina Leonelli, *Data-Centric Biology: A Philosophical Study* (Chicago: University of Chicago Press, 2016).

67 Philip Allott, "Mare Nostrum: A New International Law of the Sea," in *Freedom for the Seas in the 21st Century: Ocean Governance and Environmental Harmony*, eds. Jon M. Van Dyke et al. (Washington, D.C.: Island Press, 1993), 49–71, 51. See also Philip Allot, "Mare Nostrum: A New International Law of the Sea," *American Journal of International Law* 86, no. 4 (1992): 764–787.

68 Surabhi Ranganathan, "Ocean Floor Grab: International Law and the Making of an Extractive Imaginary," *European Journal of International Law* 30, no. 2 (2019): 573–600.

69 Steinberg, *The Social Construction of the Sea*, 39; D.P. O'Connell, *The International Law of the Sea: Volume I*, ed. I.A. Shearer (Oxford: Clarendon Press, 1982), 1; Ranganathan, "Ocean Floor Grab."

70 The common heritage of mankind (CHM) principle applies to the seabed in the high seas beyond national jurisdictions, called the Area. Convention on the Law of the Sea, Article 136, Dec. 10, 1982, 1833 U.N.T.S. 397 [UNCLOS]; UNCLOS, Preamble, ¶ 6. Specifically, it declares that exploiting any solid, liquid, or gaseous mineral resources in this zone must be carried out for the "benefit of mankind as a whole." Among other rules, the CHM prevents a state or legal person (i.e., corporation) from appropriating any parts or resources of the Area. UNCLOS, Article 137(1). For a detailed outline of the common heritage of mankind doctrine, see Kemal Baslar, *The Concept of the Common Heritage of Mankind in International Law* (Hague: Kluwer Law International, 1998).

71 Reid, this volume, 75.

72 Ibid., 72. See also Elspeth Probyn, *Eating the Ocean* (Durham: Duke University Press, 2016); Elspeth Probyn et al., eds., *Sustaining Seas: Oceanic Space and the Politics of Care* (London: Rowman & Littlefield, 2020).

73 Reid, this volume, 79.

74 On the practice of dumping other pollutants, such as radioactive agents, into the deep sea, see, e.g., Astrida Neimanis et al., "Fathoming Chemical Weapons in the Gotland Deep," *Cultural Geographies* 24 no. 4 (2017): 631–638.

75 Ranganathan, this volume, 29.

76 De Lucia, this volume, 132. See also Vito De Lucia, "Ocean Commons and an 'Ethological' Nomos of the Sea," in *International Law and Areas Beyond National Jurisdiction: Reflections on Power, Knowledge, Space and Justice*, eds. Vito De Lucia et al. (Leiden: Brill, in press): 15–44.

77 Reid, this volume. In light of this appropriation of the color blue, I have come to reconsider the term "blue legalities," coined in Irus Braverman and Elizabeth R. Johnson, eds., *Blue Legalities: The Life and Laws of the Sea* (Durham: Duke University Press, 2020).

78 Quoting from the hypnotizing phrase in this volume's Afterword by Andreas Philippopoulos-Mihalopoulos. Susan Reid makes a similar point in her chapter for this volume.

79 DeLoughrey, this volume, 145. DeLoughrey suggests the following source as an early example of the use of the term blue humanities: John R. Gillis, "The Blue Humanities," *Humanities* 34, no. 3 (2013).

80 Braverman and Johnson, *Blue Legalities*.

81 Katherine G. Sammler, "Kauri and the Whale: Oceanic Matter and Meaning in New Zealand," in Braverman and Johnson, *Blue Legalities*, 63–84.

82 Stephen Graham, "Vertical Geopolitics: Baghdad and After," *Antipode* 36, no. 1 (2004): 12–23, 20.

83 DeLoughrey, this volume, 147.

84 Arvid Pardo, "Perspective on Ocean Governance," in *Freedom for the Seas in the 21st Century: Ocean Governance and Environmental Harmony*, eds. Jon M. Van Dyke et al. (Washington, D.C.: Island Press, 1993), 38–48, 39.

85 UNCLOS has often been described by its drafters as a "constitution *for* the ocean." See, e.g., Tullio Treves, "Introductory Note, United Nations Convention on the Law of the Sea," United Nations, 2008, https://legal.un.org/avl/ha/uncls/uncls .html (emphasis added). Cf. Reid, this volume.

86 Probyn, this volume, 193.

87 See Amitav Ghosh's similar use of the term "maritime choke points" in his *The Nutmeg's Curse* (Chicago: University of Chicago Press, 2021), 108.

88 See also Elspeth Probyn, "Doing Cultural Studies in Rough Seas: The Covid-19 Ocean Multiple," *Cultural Studies* 35, nos. 2–3 (2021): 557–571.

89 Van Dyke, *Freedom for the Seas*.

90 Philip Allot, "Mare Nostrum."

91 For example, references to "all peoples of the world," UNCLOS Preamble, First and Seventh Recitals; and "mankind as a whole," UNCLOS Preamble, First Recital.

These references also gesture to the constitutional-based approach to ocean govern-ance developed by one of UNCLOS's co-architects, Elisabeth Mann Borgese as well as others. Betsy Baker, "Uncommon Heritage: Elisabeth Mann Borgese, Pacem in Maribus, the International Ocean Institute and Preparations for UNCLOS III," *Ocean Yearbook Online* 26, no. 1 (2011): 11–34.

92 Reid, this volume, 75.
93 For an ethnographic discussion of legal pluralism see Pauwelussen and Switzer Swanson, this volume.
94 Sundha Pahuja, *Decolonising International Law: Development, Economic Growth and the Politics of Universality* (Cambridge: Cambridge University Press, 2011), 2. See also Anand's critique of international law in *Origin and Development of the Law of the Sea*.
95 "Intergovernmental Conference on Marine Biodiversity of Areas Beyond National Jurisdiction," United Nations, accessed October 16, 2021, https://www.un.org/bbnj/.
96 Braverman, Ranganathan, and De Lucia, this volume.
97 Ranganathan, this volume, 42. See also Reid, De Lucia, and Braverman, this volume.
98 Probyn, this volume, 189.
99 But see Kahn's different account of sovereign boundaries at sea, in Kahn, *Islands of Sovereignty*, Chapter 6.
100 Prescott and Schofield, *The Maritime Political Boundaries of the World*, 2.
101 Ibid. See also Alison Rieser and Donna R. Christie, eds., *Ocean and Coastal Law, Cases and Materials* (St. Paul: West Academic Publishing, 2019); Randall S. Abate, ed., *Climate Change Impacts on Ocean and Coastal Law* (Oxford: Oxford University Press, 2015); Donald C. Baur et al., *Ocean and Coastal Law and Policy* (Chicago: ABA Book Publishing, 2015); Louis B. John et al., *Cases and Materials on the Law of the Sea* (Leiden: Brill, 2014).
102 Rebecca Solnit, "Seashell to Ear," in *Unravelling the Ripple*, ed. Helen Douglas (Edinburgh: Pocketbooks, 2001), n.p.
103 Gillis remarks in this context: "Why does the story of the human edge species remain so obscure? An obvious reason lies in the absence of the kind of evidence that is so abundant at inland sites. … Humans occupied many places that have long since disappeared, either washed away by rising ties or drowned so deep beneath the waves that only now … have we been able to explore them." Gillis, *The Human Shore*, 24.
104 Ibid., 19.
105 "There has to be attention to land areas bordering the ocean, that is the littoral. A history of the oceans needs to be amphibious, moving easily between land and sea." Michael Pearson, *The Indian Ocean* (London: Routledge, 2003), 5. See also Laleh Khalili, *Sinews of War and Trade* (London: Verso, 2020), 5. Renisa Mawani also briefly alludes to "amphibian legalities" in her review of law and settler colo-nialism. In her words: "The global exigencies that arise from the past, organize the present, and impinge on the future demand a shift away from terra firma toward the aqueous and amphibian legalities through which settler colonial power contin-ues to expand and flourish." Renisa Mawani, "Law, Settler Colonialism, and 'the Forgotten Space' of Maritime Worlds," *Annual Review of Law and Social Science* 12 (2016): 107–131, 126.
106 Annet Pauwelussen and Gerard Verschoor, "Amphibious Encounters: Corals and People in Conservation Outreach in Indonesia," *Engaging Science, Technology and Society* 3 (2017): 292–314. See also Annet Pauwelussen, "Visits from Octopus and Crocodile Kin: Rethinking Human-Sea Relations Through Amphibious Twinship in Indonesia," in *Environmental Alteritie*, eds. C. Bonelli and A. Walford (Manchester: Mattering Press, 2021); and Annet Pauwelussen, "Leaky Bodies: Masculinities and Risk in the Practice of Dive Fishing in Indonesia," *Gender Place & Culture* (2021), https://doi.org/10.1080/0966369X.2021.1950642. From Pauwelussen's PhD thesis: "The concept of 'amphibiousness' is mobilised to refer to

living in and moving between different worlds that can intermingle but that cannot be reduced to each other. The concept is used to describe the human capacity to live in different worlds at the same time. This amphibious capacity is further elaborated 1) in terms of living in a hybrid land-water interface, 2) in terms of being able to move along with different understandings of the world, of reality, and 3) It refers to the methodology of the anthropologist who also needs to move in these worlds bodily and cognitively, to develop a sensitivity to and understanding of these different worlds. Amphibiousness captures the anthropological engagement with flow, multiplicity and otherness by way of moving between worlds in order to explore the moving interface between worlds, realities or ways of life that partly interact." Annet Pauwelussen, "Amphibious Anthropology: Engaging with Maritime Worlds in Indonesia" (PhD dissertation, Wageningen University, 2017).

107 Pauwelussen and Verschoor, "Amphibious Encounters," 295. See also Zoe Todd, "Fish Pluralities: Human-Animal Relations and Sites of Engagement in Paulatuuq, Arctic Canada," *Inuit Studies* 38, nos. 1–2 (2014): 217–238.

108 Pauwelussen and Verschoor, "Amphibious Encounters," abstract.

109 Michel Foucault, "Of Other Spaces: Utopias and Heterotopias," *Architecture/ Mouvement/ Continuité*, trans. Jay Miskowiec (1984).

110 De Lucia, this volume, 122.

111 Their contribution shows that "sea turtles are particularly 'troubling' legal objects because of their mobile and amphibious nature. They transgress administrative boundaries between land and sea and their corresponding governmental departments and law enforcement agencies. Moreover, [they] also express a certain 'amphibiousness' as plural objects (and subjects), engaging different yet intersecting ontological and normative systems. This also troubles any neat division between 'Bajau' and 'conservationist' legalities and normative systems by pluralizing perspectives in and between." This volume, 272. See also Pauwelussen, "Amphibious Anthropology"; Pauwelussen and Verschoor, "Amphibious Encounters." This resonates with Todd, "Fish Pluralities," 218, which discusses the "slipperiness" of fish as beings, existing as "simultaneously different entities" that "challenge existing articulations of human-environment relationships."

112 See, e.g., Allen et al., *The Rights of Indigenous Peoples*.

113 DeLoughrey, this volume. See also Elizabeth DeLoughrey, "Toward a Critical Ocean Studies for the Anthropocene," *English Language Notes* 57, no. 1 (2019): 22–36; Stacy Alaimo, "Introduction: Science Studies and the Blue Humanities," *Configurations* 27, no. 4 (2019): 429–432.

114 Certain Indigenous scholars have pointed out that the various ocean "turns" that have been eclipsing across the academic world are in fact indicative of a myopic historical vision that discounts the experiences by those Indigenous peoples who never left the oceans in the first place. See Epeli Hau'ofa, "Our Sea of Islands," *The Contemporary Pacific* 6, no. 1 (1994): 148–161.

115 DeLoughrey, "Toward a Critical Ocean Studies for the Anthropocene." See also Karin E. Ingersoll, *Waves of Knowing* (Durham: Duke University Press, 2016); Steven Mentz, "Towards a Blue Cultural Studies: The Sea, Maritime Culture and Early Modern Literature," *Literature Compass* 6, no. 5 (2009): 997–1013.

116 Kahn, *Islands of Sovereignty*; Mawani, *Across Oceans of Law*; Benton and Perl-Rosenthal, *A World at Sea*.

117 See, e.g., Probyn, *Eating the Ocean*; Probyn et al., *Sustaining Seas*.

118 Christina Sharpe, *In the Wake: On Blackness and Being* (Durham, Duke University Press, 2016).

119 As Khalili writes in this context: "Law mattered: to sovereignty, to capital accumulation, and to maritime businesses. … [C]olonial legal regimes left their traces in much of maritime law. Legal infrastructures conceived in the North Atlantic continue to be hegemonic, even as centres of capital accumulation move across the seas. The law acts as the powerful adjunct of coercion, according to imperial civilisation

hierarchies." Khalili, *Sinews of War and Trade*, 91–92. See also Macarena Gómez-Barris, *The Extractive Zone* (Durham: Duke University Press, 2017).

120 See also Stefan Helmreich, "Nature/Culture/Seawater," *American Anthropologist* 113, no. 1 (2011): 132–144; Stefan Helmreich, "Waves: An Anthropology of Scientific Things—Transcript of the Lewis Henry Morgan Lecture given on October 22, 2014," *HAU: Journal of Ethnographic Theory* 4 (2014): 265–284.

121 Philippopoulos-Mihalopoulos, this volume, 287, 288.

122 M. NourbeSe Philip, *Zong!* (Toronto: Mercury Press, 2008).

123 The description is from Audrey Golden, "*Gregson v. Gilbert*," *American Book Review* 36, no. 4 (2015): 6–7.

124 Anita Rupprecht, "'A Limited Sort of Property': History, Memory and the Slave Ship *Zong*," *Journal of Slave and Post-Slave Studies* 29, no. 2 (2008): 265–277.

125 Utilizing this term, Sharpe explains that, "The Mediterranean has a long history in relation to slavery." Sharpe, *In the Wake*, 58. See also Naor Ben-Yehoyada, "'Follow Me, and I Will Make You Fishers of Men': The Moral and Political Scales of Migration in the Central Mediterranean," *Journal of the Royal Anthropological Institute* 22, no. 1 (2015): 183–202; Naor Ben-Yehoyada, *The Mediterranean Incarnate: Region Formation between Sicily and Tunisia Since World War II* (Chicago: University of Chicago Press, 2017); Itamar Mann, "Maritime Legal Black Holes: Migration and Rightlessness in International Law," *European Journal of International Law* 29, no. 2 (2018): 347–372; Itamar Mann, "The Right to Perform Rescue at Sea: Jurisprudence and Drowning," *German Law Journal* 21, no. 3 (2020): 598–619.

126 This project "investigat[es] and address[es] social, legal and scientific challenges—sometimes all three—through an architectural lens." Lorenzo Pezzani and Charles Heller, "Forensic Oceanography Report," SITU Research, 2012, https://situ.nyc/research/projects/forensic-oceanography-report. See, relatedly, the Forensic Architecture Project, based at the University of London, which investigates violence committed by states, police forces, militaries, and corporations. "Agency," Forensic Architecture, accessed October 19, 2021. https://forensic-architecture.org/about/agency.

127 Sharpe, *In the Wake*, 59.

1

THE VEXED LIMINALITY OF HYDROTHERMAL VENTS

An Opportunity to Unmake the Law of the Sea

Surabhi Ranganathan

Introduction

On the evening of February 17, 1977, having returned to the R/V Knorr after a 2000-meter dive into the Galapagos Rift in the Pacific Ocean, Tjeerd Hendrik van Andel wrote his diary:

> The landscape was both gorgeous and geologically fascinating … a gently undulating terrain of smooth broad lava domes, glittering in the lights … And in the middle of this stark and barren vastness, hard, prickly, new and untamed, a small oasis, perhaps an acre, sharply defined, with coral gardens, pink and gold anemones, white crabs in great variety and profusion, yellow, brown, liver-colored fish, medusoid large clumps of some kind of mussel ten inches long, crevices filled with their huge bleached shells looking from afar like rims of snow in a boulder field. It was like an aquarium, huge but carefully arranged with elegance and taste and grace … What produced this little paradise in the big and unusual deep sea floor desert? I have no idea.[1]

As the historian of science Naomi Oreskes notes, the discovery of this deep-sea ecosystem caught the cruise by surprise. While the emergence and general acceptance of the theory of plate tectonics in the previous decade had prepared scientists for the discovery of vents in the seafloor at points where two plates were moving apart, they had not expected that such sites would be *loci* for the

FIGURE 1.1 A dense bed of hydrothermal mussels covers the slope of the Northwest Eifuku volcano in the Mariana Arc region, Western Pacific Ocean, near the seafloor hot spring Champagne vent. Other vent animals living among the mussels include shrimp, limpets, and Galatheid crab. Credit: Pacific Ring of Fire 2004 Expedition. NOAA Office of Ocean Exploration; Dr. Bob Embley, NOAA PMEL, Chief Scientist.

DOI: 10.4324/9781003205173-2

"profusion of life."[2] Their amazement is nicely captured in an account of the cruise by the Woods Hole Institute:

> Observing the scene from Alvin's viewports, [Jack] Corliss talked by acoustic telephone to his graduate student Debra Stakes, who was aboard R/V Lulu.
>
> "Isn't the deep ocean supposed to be like a desert?" Corliss asked. When Stakes answered, "Yes," Corliss replied: "Well, there's all these animals down here."[3]

The cruise, in fact, had so little expected to find marine life that it had included no biologists at all in its team of scientists. Nor had it taken along apparatus to preserve the specimens found. The team had to use their stock of vodka as a preservative.[4] The cruise would return properly equipped in 1979, its discoveries then opening the way to an understanding of the deep seafloor as hosting multiple, thriving, and highly unusual vent communities.[5]

Vent communities differ radically from other forms of life. They are "powered not by sunlight but by the planet's inner heats and energies."[6] The starting point of the so-called "dark" food chain of a vent community are chemosynthetic bacteria, autotrophs that produce energy by oxidizing compounds such as hydrogen sulfide or methane that emanate from vents. Creatures such as tube worms and clams draw sustenance from these microbes and are in turn fed upon

FIGURE 1.2 Hermit crab from a cold seep site sampled during seeps voyage on RV Tangaroa in 2006 NOAA Expedition. Note the seep-associated bacterial filaments on its claws. Image courtesy of NOAA/NIWA.

by other species. Robert Ballard, leader of the 1977 and 1979 expeditions, states that while scientists had been aware of the chemosynthesis process, "it wasn't until the discovery of the Galapagos Rift hydrothermal vents [that they] realized it could form the basis of an entire ecosystem."[7]

Vents, however, also offered up other secrets in the form of new types of mineral ores. David Perlman, the science correspondent on board the 1977 cruise, noted in his dispatches:

> [H]ere on the Galapagos rift zone—and presumably elsewhere around the world where vast plates of the Earth's crust are grinding against each other and moving like ice floes—this expedition has found solid evidence of abundant manganese, copper, nickel, sulphur and iron, leached out of the lava by ocean brine.[8]

Unlike the poorly crystalline manganese nodules found on abyssal plains, and until then the key known potential source of marine minerals, the manganese spires found at vents were "perfectly formed."[9] In time, vent fields and marine volcanic regions were found to contain two new types of ores: polymetallic sulfides—massive chimneys containing rich metal lodes; and cobalt-rich ferromanganese crusts coating the sides of volcanic islands and seamounts. Vent ores "were quite potent compared to those of the icy manganese nodules, being 'approximately one thousand times' as concentrated ... like fistfuls of gold versus tiny flakes."[10]

The discoveries of vent communities and minerals coincided with the Third United Nations Conference on the Law of the Sea (UNCLOS III, 1973–1982), which adopted the United Nations Convention on the Law of the Sea (UNCLOS). However, the discoveries came too late to be accommodated in the discussions. UNCLOS's text comprehends vents only in oblique ways. Vent *minerals* fall within the definition of the resources of the international seabed: Article 133(a) provides, "'resources' means all solid, liquid or gaseous mineral resources *in situ* in the Area at or beneath the seabed, including polymetallic nodules." But their specific properties are not addressed in the international seabed regime, which is built around the properties of manganese nodules and their mining. It has been left to the International Seabed Authority (ISA), also established by UNCLOS, to tailor the application of the regime to these other minerals. ISA is at work on a "mining code."[11] Vent *communities* are enfolded within references to the marine environment of the international seabed area. In the "deepest of ironies," vent *microbes*, the focus of much present interest for their pharmaceutical and biotechnology applications as marine genetic resources (MGRs), are not identified as resources at all.[12] An ongoing law-making exercise toward the adoption of a treaty on biodiversity beyond national jurisdiction (BBNJ) seeks to repair the omission.[13]

In this chapter, I bring into focus these ongoing law-making exercises, which, viewed in terms of their attempts to address vents, reveal the unsatisfactory foundations of the law of the sea. I argue, first, that the law of the sea, represented pre-eminently by UNCLOS, establishes an extractive imaginary of the ocean. It does so, firstly, by a series of complex divisions of ocean space which have the paradoxical effect of simplifying it into places of feasible economic activity; and, secondly, by linking exploitation, especially of seabed minerals, with a greater utopian vision of world order. Next, I discuss the ways in which vents confound the divisions and assertions that underpin UNCLOS. I then turn to the point that both the BBNJ negotiations and ISA's work on the mining code operate as "fixes"—relying here on David Harvey's discussion of this term[14]—that seek to hold in place UNCLOS's extractive imaginary. The chapter concludes with some reflections on what it might mean to "unmake," rather than fix, the law of the sea.

An Extractive Imaginary

What does it mean to suggest that the law of the sea, pre-eminently represented by UNCLOS, establishes an extractive imaginary? The ocean was of course viewed as an economic resource even prior to this treaty, and the pre-UNCLOS law of the sea likewise sought to unlock its resource value. The instruments adopted in the 19th and early 20th centuries on shipping, navigation, fishing, submarine cables, piracy, neutrality, and interdiction were directed toward max-imizing the ocean's resources of connection (and hence "free" trade and specu-lation) and fisheries.[15] This law of the sea, pitched in the language of universal truth, consolidated a view of the ocean as the hinterland of human societies, simultaneously connected and distanced—eclipsing both worldviews that would imagine it otherwise,[16] and ways of thinking that would "racinate" the ocean, bringing forth the embodied impacts of its utilization.[17]

Even so, the law which developed from 1945, culminating in UNCLOS, was different in several respects from prior regimes. For one, UNCLOS was cast as a "constitution," claiming comprehensiveness and "covering every aspect of the uses and resources of the sea."[18] It was a far cry from the limited (and unsuccess-ful) codification effort of the interwar period, and even from the early post-war efforts of the International Law Commission (ILC), which gingerly undertook to draft rules relating to the high seas and territorial waters. The ILC's work led to the 1958 UN Conference on the Law of the Sea, and the chopped-up four Geneva Conventions on the Law of the Sea (GCs). While the Geneva process can be placed within UNCLOS's genealogy, and it took up novel issues such as rights over the continental shelf, at no point did it envisage a comprehen-sive outcome. The four GCs cover some of the same ground as UNCLOS, but they do not exhaust the topics that it contains. As well as superseding the GCs, UNCLOS is also the filter of many other legal developments of the same period, and the original instrument dealing with topics such as deep seabed mining. It is the fulcrum of the law of the sea, providing "the framework within which most uses of the seas are located."[19]

UNCLOS's constitutional standing does not suffer from the non-participation of some states, most prominently the United States. Not only does the United States formally accept (as custom) most substantive UNCLOS rules, except those on the international seabed area, it is also in regular attendance at ISA meetings and active in the BBNJ negotiations. It has also participated in side-agreements that complement the UNCLOS seabed regime, had provisionally signed the 1994 "Implementing Agreement" on seabed issues, and has long abandoned its once strenuous efforts to articulate an alternative multilateral arrangement. UNCLOS, then, for all that it inscribes particular visions and interests, is not a particular legality of limited application, in contestation with other international laws of the sea. Of course, it may still be fleshed out, expanded, and sometimes even revised, by implementing agreements, regional treaties, and other particular arrangements.

It is not only in formal respects that the "new" law of the sea differs from the "old"; more significant are the changes it effects on the ocean. A characteristic of the old law, summed up by Hugo Grotius, and unchallenged in the main by his interlocutors, was that treaties relating to the sea could only have certain, limited effects. They could obligate persons (those under the jurisdiction of the princes who made the treaty, or pirates), but they could not affect the thing itself.[20] UNCLOS, however, does affect the thing itself, ordering the ocean by its spatial divisions, resource classifications, and allocations *in rem* of rights and responsibilities. And in doing so it consolidates an extractive imaginary of the ocean in two steps.

Firstly, UNCLOS culminates divisions of the ocean that effectively parcel it into discrete sites of economic activity. The ocean is divided into zones enclosed within national jurisdiction (territorial seas, exclusive economic zones (EEZs), archipelagic waters, continental shelf) and beyond national jurisdiction (high seas and deep seabed). While such enclosures were often championed on the basis that they gave expression to the "natural" affinities of the ocean (such as underwater prolongation of land, adjacency to coasts, connection of fragmented territorial unities), the specific rules depart from these criteria. The law constructs a geography of the sea, even as the claim is that it simply reflects it.[21]

The same is true *a fortiori* of further divisions, effected by medium. Only in the zones closest to coasts are regimes of land and water unified. Further out, the continental shelf is governed by separate provisions from the rest of the EEZ,[22] and the high seas are likewise distinguished from the international seabed area. While the *conceit* informing these divisions was that actions exploiting seabed resources did not implicate the above waters, the *concern* was to ensure that seabed claims did not affect claims to the waters.

UNCLOS also classifies and allocates the contents of the ocean into the national and international, and land and sea regimes, enacting fictions as to what is found where, or—in the case of "sedentary species"—what does not move (crustaceans). For the high seas and international seabed area, the classification is sharply binary: marine life is placed in the high seas, and minerals under the seabed regime. The effect of all this complicated legal work is, paradoxically, to simplify the ocean. The law transforms complex and tangled ecosystems into dedicated and insular resource areas. This goes to convey that, far from being the great unknown, the ocean (and what is in it) is entirely comprehended through legal categories.

Secondly, UNCLOS makes the extraction of resources a normative obligation: not just something permitted, or to be regulated, but also something to be promoted. We can see this from the provisions on fisheries—for example, in the injunction to coastal states to "promote the objective of optimum utilization of the living resources" in their EEZs.[23] States are required not only to set the total allowable catch within their EEZs by calculating maximum sustainable yield,[24] but also to allocate fishing rights to other states if they cannot themselves fish to the full extent of the set limit. Although UNCLOS recognizes overexploitation as a concern,[25] the details make clear that *under*-exploitation is also considered undesirable.

The normative championship of exploitation is at work in much stronger terms in relation to the international seabed. Here, the language of "common heritage of mankind," "benefit of mankind as a whole," and "consideration [of] needs of developing states [and non-self-governing peoples]" is tied to the "development of the resources of the area," that is, to seabed mining.[26] While UNCLOS represents an effort to construct a more equitable political economy of mining (facilitating the participation of developing states, and providing that the financial and economic benefits of seabed mining must be shared), and highlights the need for the protection of the marine environment, it also casts mining as desirable and necessary.

In fact, the development of the law with regard to seabed minerals much preceded the development of mining itself. Moreover, at various points, the law has shifted in response to the demands of mining corporations and their representative states. Over the 1970s, UNCLOS III moved away from the idea of limiting mining to an international public enterprise to a "parallel system," in which private and state-owned entities could also mine the seabed. In the 1980s, further alterations were made to the UNCLOS regime to accommodate the interests of those, like the United States, who were unhappy with it; and then the 1994 agreement squarely foregrounded commercial principles, stripping the common heritage principle of much of its redistributive and participatory potential. ISA today acts as an advocate, and not just regulator, of seabed mining.[27]

This slant of the law as it developed from 1945 to UNCLOS is not surprising. In this period, as we will also see later, both North and South (especially states) perceived the ocean as the answer to multiple challenges. To the Global North, it offered solutions for food scarcity, as "population explosion" became a concern;[28] "supply security," as anticolonial struggles made southern minerals less easily accessible, at least for a time; and financing development in the South, as they began to look for alternatives to international aid. For southern states, the ocean was key to realizing a New International Economic Order (NIEO), through the construction of cooperative and redistributive political economies of resource extraction, and elimination of zero-sum dependencies (including nuclear testing and waste dumping). The speech by Maltese Ambassador, Arvid Pardo, often cited as a catalyst of UNCLOS III, is a good snapshot of the hopes reposed in the ocean. Pardo made his case for the common heritage principle by emphasizing that the ocean, as being made accessible through rapidly evolving technologies, enabled the overcoming of the structural limits that usually made the pursuit of common benefit less likely than conflict. He spoke of quantities of minerals that could support human consumption for thousands of years; billions of barrels of oil and natural gas; the vast potential for farming and fish husbandry; and fish protein concentrate, which could provide "adequate animal protein to meet daily requirements of one child at less than 1 cent of US money."[29] This abundance, in short, could grease the wheels of just and cooperative international relations.

UNCLOS's jurisdictional classifications and conceptual associations were transformative and constitutive of the ocean itself. They constructed what they claimed merely to be describing: differences between land and sea, life and non-life, fixity and mobility, and so on. Into this scheme of purportedly understood and legally ordered space, enter messy hydrothermal vents.

The Vexed Liminality of Hydrothermal Vents

In the time since the historic cruise of 1977, vents have become the object of fascinated study among scientists of all specializations. Hundreds of research expeditions have discovered vents in a variety of locations: mid-ocean ridges, volcanic arcs, and back-arc spreading centers; in the northern hemisphere and the southern; in all the oceans. Moreover, the findings have revealed a great diversity in vent structures, species, and ecosystems. Vents can be "black smokers" or "white smokers," formed in areas of slow or fast spreading, and hosted in a range of geologic formations.[30] They may be dominated by particular species, such as tubeworms or clams, or contain a profusion of biodiversity. As oceanographers Andrew Thaler and Diva Amon note, "[j]ust as 'forest' can describe ecosystems ranging from boreal forests to tropical rain forests, 'hydrothermal vent' describes a suite of deep-ocean ecosystems united by a shared dependence on chemosynthetically derived primary production and above-ambient temperatures but diverse in their composition and connection to one another."[31]

Research on vents has revealed much—new species (roughly two described each month of the last 40 years),[32] possibilities of new biotechnological and pharmaceutical applications based on vent MGRs (in fact, the test to diagnose Covid-19 uses an enzyme from a vent microbe),[33] and new mineral reserves. Vents, then, appear to invite the application of UNCLOS's extractive imaginary. Unfortunately (for UNCLOS), what has been learned so far calls into question its jurisdictional divisions and allocations. For vents, it turns out, sit right on the boundary lines of UNCLOS's classifications.

Take for instance the land/water distinction. Vents are formed by the circulation of seawater. Seawater percolates through the cracks in the ocean crust into the hot subsurface, heats up, becomes enriched with chemicals and volatile gases, then rises through the cracks. Upon mixing with colder seawater, it precipitates the chemicals into tall chimneys, through which yet more water precipitates and emerges; the whole process is aided by chemosynthetic microbes.[34] At vent sites, oceanic mediums meet and merge in a dynamic process that makes a land/water distinction an impossible one.[35]

Vents also throw into relief the inadequacy of UNCLOS's principles of classifying oceanic resources. Recall that in areas beyond national jurisdiction the Convention places all "solid, liquid, or gaseous mineral resources" in the regime of the seabed, and all living resources in the high seas regime. In doing so, it ignores their attachments, as many deep-sea creatures depend on mineral substrates to live and grow. Vent communities are built around the life that is encrusted in the mineral ores of hydrothermal vents. Indeed, the very question of what is "living"—and what is a discrete life form—is particularly complicated for marine microbes.[36] Consider only the mutually constitutive relationship between vent microbes and minerals, evidence of intense lateral gene transfers between microbes (twenty million billion times *per second*), flourishing vent communities in conditions once regarded as impossible for living, and of course the dubious claims of some microbes to a definition of what it means to be alive.[37]

But even with respect to larger creatures, the idea of separability of seabed and water regimes is fraught with difficulty. The logic of UNCLOS would suggest that, in areas beyond national jurisdiction, vent chimneys are part of the seabed regime but, say, tubeworms attached to those chimneys are part of the high seas regime. These are rather specious distinctions.

The situation is not much better as regards vents within zones of national jurisdiction. In fact, that UNCLOS principles differ for those only makes matters more complicated. For, recall that in zones within national jurisdiction, the regime of the continental shelf includes minerals as well as sedentary species. This might on the face of it solve problems—clarifying that both vent minerals and microbes and certain sea creatures (or creatures at certain life stages, like larvae) are part of the shelf regime. But where do we draw the line? Not only is there a lack of sufficient information to make distinctions between fixed and mobile forms of life in vent communities, but also little is known about the interactions between vent and non-vent communities, or transfers that take place between different vent communities.[38] It would be inappropriate to fix the characteristics of vent communities in the absence of actual knowledge—a rehearsal of older UNCLOS classifications.

Vents, then, also challenge the lines that UNCLOS seeks to draw between extraction and the protection of the marine environment. Here, the fact that UNCLOS was developed particularly with manganese nodules in mind is significant. This is because the nature of that resource suggests (although this too is challenged)[39] that recovery of minerals *can* take place at sufficient remove from (large sections of) the living creatures in the area, so that threats to marine biodiversity can be mitigated. ISA relies on that in proposing a scheme of no-mining zones ("Areas of Particular Environmental Interest" or APEIs) around exploitation areas, to ensure that the distinctive biodiversity of each part of the ocean is preserved. Thus, the nodule-rich Clarion Clipperton Zone has nine APEIs scattered around the mining sites. However, it is not clear how this approach translates to vents, especially those with tightly integrated communities. Not much can be assumed about their rates of recovery post mining operations. In particular, scientists caution against assumptions that data can be extrapolated from the handful of vent sites that have been more extensively studied, given significant variations in vent ecosystems both between and within regions.[40]

More significant even than what we have learned about vents, is what those findings reveal of what we do not know, and might discover with more research. Despite the many cruises, the study of vents remains patchy. Some have received detailed attention; others barely sighted. Thaler and Amon note especially a North/South imbalance. The southern hemisphere has more vent fields (353 vent fields to 300 in the northern) and is the subject of more exploration activity towards mineral extraction (36 vent fields to 9 in the northern). But vents in the northern hemisphere have been studied to a greater degree (189 cruises to 72 in the southern), although even there, only a few have been sampled extensively.[41] The capacity for new discovery remains at par with that associated with an older age of oceanography.

Vent research is also opening vast new fields of thought and speculation. Vents likely shape global biogeochemistry, create sinks for atmospheric carbon dioxide, reveal how photosynthesis emerged, push our understanding of the boundaries of life, and might be key to the origins of life on earth (and the possibility of life on other planets).[42] Even setting aside non-instrumental considerations, and purely from the perspective of their utility to humans, it is almost impossible to quantify the potential scientific, ecological, societal, and cultural value of vents.[43] In such case, how can we conclude that extraction of minerals is the best way to engage with them?

The work that UNCLOS does to simplify the ocean and construct a political economy of resource development that makes extraction appear desirable comes undone at vents. The UNCLOS imaginary is unsatisfactory vis-à-vis other resources too, including nodules. But vents throw its inadequacies into sharp relief. Even for those concerned only with the recovery of resources, the imbrications of vent minerals and MGRs raise questions about the suitability of jurisdictional categories that place the two in different regimes, and about the ways in which extraction of the one might take place without destroying the possibility of the other. For its adherents and critics alike then, vents reveal cracks in the UNCLOS regime. The difference is that for the adherents the need is that of a legal fix, whereas for critics, the messiness of vents and their vexed liminality hold the key to a deeper rethinking of the law of the sea.

Vents in Ongoing Law-Making Processes

That vents do not sit easily within UNCLOS categories has not escaped states, organizations, or corporations concerned with the business, and regulation, of extraction. With respect to vent minerals, ISA has been working out rules tailored to their specific characteristics, adjusting the UNCLOS regime for this purpose. In 2010, it issued regulations on prospecting and exploring polymetallic sulfides, the ores found at vents. It has since awarded seven exploration contracts for vent sites in the mid-Atlantic and Southern Indian Ocean ridges, and is preparing exploitation regulations for all types of seabed minerals. It is now under pressure to complete this process by June 29, 2023, following Nauru's invocation of the "two-year rule."[44] Nauru is the sponsor of NORI, or Nauru Ocean Resources Inc., a wholly owned subsidiary of the Canadian The Metals Co., formerly DeepGreen Metals.

The ISA regulations do adapt the UNCLOS regime to the particular characteristics, patterns of occurrence, and removal methods associated with vent minerals in several respects: for example, the size of the licensed area and the terms upon which explorations licenses are granted.[45] However—and though it is fair to acknowledge ISA's efforts to develop UNCLOS's provisions on the marine environment—its regulations do not explain how it approaches the different ecological considerations presented by vent sites. The sulfide regulations simply replicate the provisions on the protection of marine environment adopted in the case of nodules.

ISA is of course studying the issues, and clearly *will* tailor its general approaches, such as environmental impact assessments, APEIs, and regional environmental management plans (REMPs) to vents. However, despite many workshops and meetings—many organized by the ISA Secretariat—and despite assurances, its practices have raised questions about the extent to which its organs recognize the complexities of vents, as also about its priorities in a choice between extraction and protection. In 2017, for example, ISA's Council granted a 15-year exploration contract to Poland for an area of the Mid-Atlantic Ridge that includes, among other vent fields, the "Lost City" vents. These vents are important for a number of reasons: significantly, they offer "a contemporary analogue of conditions where life may have originated."[46] The Lost City has been recommended for the award of World Heritage status by UNESCO; and ISA's decision to allow activities that could threaten it is regarded as both cavalier and the too-likely outcome of a non-transparent governance process.[47] Moreover, it is three years *after* granting that contract that ISA is discussing a REMP for the Mid-Atlantic Ridge.[48]

Meanwhile, the BBNJ process is considering the question of vent communities as part of its work on access to and sharing the benefits of MGRs. The outcome of these negotiations, which also cover area-based management tools, environment impact assessments, and capacity building and technology transfer issues in addition to MGRs, will take the form of an Implementing Agreement to UNCLOS, similar to the 1994 Agreement, and will be "interpreted and applied in the context of and in a manner consistent with [it]."[49]

In the course of the BBNJ process, there has been much discussion about the appropriate way to qualify vent MGRs. A key question is whether it is possible, despite UNCLOS classifications, to regard them as part of the common heritage. For those who favor the principle, the case for doing so rests on the language of Article 136 of UNCLOS, which states: "The Area and its resources are the common heritage of mankind." The argument is that unlike the restrictive definition of "resources," "the Area," meaning the seabed beyond national jurisdiction, is not a restrictive category. So long as they are *of* the seabed then, vent communities, and therefore MGRs, would count as part of the common heritage. However, and setting aside difficulties of placing MGRs in this way, this argument has also been rebutted on a doctrinal assessment of the treaty provisions. A leading analysis is by Konrad Marciniak, who points out that, read as a whole, UNCLOS provisions do not support treating non-mineral resources as the common heritage.[50] For instance, all references to "activities in the Area" refer solely to the extraction of minerals. Moreover, ISA's role vis-à-vis vent communities is limited to protecting the marine environment, and promoting and disseminating scientific research; it is not given the responsibility to regulate their extraction.[51] UNCLOS's drafting history also reveals a narrowing of the focus of the common heritage regime to minerals alone. To then interpret common heritage as including vent MGRs is to read against the grain of the ordinary rules of interpretation.

However, some states and scholars have argued that the common heritage principle should apply to MGRs as a matter of *lex ferenda*—the law as it should be(come)—even if it does not apply *lex lata*.[52] Thus, current UNCLOS categories should be revised to bring vent MGRs within the more normatively conditioned regime of the seabed. For, as seen, while the common heritage principle is oriented toward extraction, it makes the redistribution of benefits, environmental protection, and—in some understandings—intergenerational equity, all part of the calculus. In contrast, fewer conditions restrict extraction in the high seas areas governed by the principle of freedom. Guided by this, some developing states have even recommended a more radical revision of categories bringing *all* MGRs, not just those of vents, within the common heritage principle. The North/South distinction is stark here as northern states, especially the United States and Russia, strongly oppose extending the principle's scope in any way.[53] The common heritage of mankind has consequently jumped in and out of BBNJ drafts. It was dropped from the text prepared for the session in August 2019, but restored in the revised draft prepared for the (postponed) April 2020 session.[54] Even so, it features in a limited way: listed among a number of generally applicable principles, but not mentioned directly in relation to MGRs. Instead, the draft presented options for conditioning the exploitation of MGRs on some of the factors associated with the principle, such as benefit-sharing.[55]

It is worth thinking about the inadequacy of all approaches to solving the problem of vent MGRs. Given the political differences over the common heritage principle, it is unlikely that the BBNJ agreement will place all MGRs under it. It is only marginally more probable that states may agree to place vent MGRs under it. And if they do, then a line will have to be drawn between the MGRs of the seabed, and those that are not of it; a process that will once again be arbitrary, and based on legal fictions and scanty knowledge. In view of these difficulties, some scholars suggest that the best approach is to subject all MGRs to their own special, integrated, non-common-heritage regime, which also seems to be the direction of the current BBNJ draft.[56] However, this approach also raises many questions, in fact rehearsing those which follow from the current classification placing vent minerals and living resources under separate regimes. Notably, who will administer vents? On what principles will institutional conflicts between ISA, and some other institution administering the new special regime, be resolved? The BBNJ looks set to elide the potential for conflicts between exploiting MGRs, exploiting minerals, and protecting biodiversity and ecosystems, by deflecting these questions into the future and onto the terrain of managerialism and technical decision—leaving it to ISA and other bodies charged with jurisdiction over vents to find ways to cooperate with each other.

From Fixing to Unmaking

One possible way to read UNCLOS is as enacting a legal fix to various crises of capitalism arising from the conjuncture of the 1950s to 1970s. As Harvey explains, the word "fix" can be understood in several ways, adding up to an explanation of how space is (re)made for capitalism.[57] Firstly, fix may be understood as "fixity," the fixing in place of infrastructures like transportation and communications that enable capitalist activity. Such infrastructures are established and maintained until the point in time when it becomes more profitable to destroy them to make way for new ones. Secondly, fix as a "solution" to problems that impede or disrupt production and sales, including shortages of raw materials, or labor, saturation or closure of markets, political instabilities, or other difficulties. Related to this, a further way of understanding "fix" is in terms of its colloquial use, when it describes something marketed and consumed as meeting a desire—e.g., your "morning fix."

UNCLOS acted in all these senses of the word. Its regimes opened up the ocean for new types of extractive activities, and expanded and deepened the prospects for others, and provided the legal scaffold for various marine technologies, from shipping to cables. Especially for northern states, it offered the promise of solving many crises: resource scrambles and raw materials insecurity in the wake of decolonization and the Cold War, insecurity of tenure over oceanic sites, threats to critical infrastructures. It was also a space to park the emancipatory desires of the time, offering simultaneously, a "constitutional" experiment in *reforming* the global political economy in the face of demands for an NIEO, and smoothening the path for commercial actors. In this sense, the common heritage regime especially served as a "makeshift geographic solution to a mainstream economic crisis."[58] And when some aspects of that crisis dropped away with the end of the Cold War and the spread of liberalization, and the seabed regime of 1982 was found wanting in the direction of smoothening the path for commercial actors, it was fixed by way of further modifications in 1994. Subsequent developments, such as the UN Fish Stocks Agreement of 1995 or the expansion of marine protected areas to address the problem of dwindling marine life, have offered further fixes to the UNCLOS regime.

On this reading it is a given that the BBNJ and ISA processes will continue the trend, providing the framework for access and use of new resources and fixing problems in resource classification and conservation. Moreover, as such, these processes can accept and make visible only some types of crises: the need for the raw materials of the fourth, "planet-saving" industrial revolution (as mining companies like to present seabed mining); not enough science and inadequate environmental impact assessment standards (the version of the problem that the ISA accepts and addresses); no clear administrative body, not enough marine spatial planning, need for accountability and clearer rules (problems that the BBNJ process recognizes and is seeking to resolve). In short, on offer in both processes are possible improvements to the existing law. However, neither presents a venue for revisiting its underlying assumptions of extraction, "blue growth," and commercial profitability. Indeed, such questions are purported to be off the table, either because decided in the past, or because present needs admit no alternative.[59] Unless, that is, other spatial fixes become available--and both outer space, and, as is more conventional, immiserated areas of the world are under pressure.[60]

If all this sounds rather bleak, then it may be worth thinking about another way of reading UNCLOS. Although in outcome everything that this chapter has described in terms of establishing an extractive imaginary, UNCLOS was also, at least at the inception of UNCLOS III, a bid for something more ambitious, as a world (re)making experiment, to replace colonial structures of the exploitation of lands, resources, and peoples, with regimes foregrounding equality and distributive justice between states, and to alter the protocols of international law-making itself.[61] It did not succeed in those terms, for a number of reasons. The inexorable logic of extractivism, modernization, and neoliberalism capitalism, on the one hand, and constraints of state-making in the Global South (which generated both expansive claims of national jurisdiction over oceanic resources, and narrowed recognition of the participation and rights of non-state/non-self-governing entities), on the other hand, were among the factors that limited more radical possibilities. The determined undoing, by the Global North, of the NIEO movement also had its ramifications on UNCLOS III, which by the late 1970s was overtaken by the "ill-controlled appetites" of northern states.[62] Important critiques of the conference, international law more broadly, the logics of extractivism and neoliberalism, and the ways in which racial capitalism and imperialism were being reinscribed into the new world order, all remained on the margins. But, even amidst all these limitations and failures, UNCLOS III saw the articulation of novel ideas for engaging in, administering, and distributing the spoils of, the extraction of ocean resources; and, despite retrenchments, its seabed regime remained somewhat normatively conditioned. And so, although UNCLOS III ended up "fixing" crises of capitalism in the West, it did not begin with just that ambition.

And this is the difference with the ongoing law-making processes, which do not reveal any similar ambition. Both the BBNJ negotiations and ISA's work are explicitly designed as fixes only, operating as they do under injunctions not to "undermine" existing UNCLOS frameworks.[63] At neither venue has there been much thought about what it might mean to realize UNCLOS's inaugural ambitions. ISA takes for granted that its purpose is to enable seabed mining even though the history of UNCLOS makes clear that the purpose was equality and distributive justice, with seabed mining seen as a conduit for this, according to the scientific and economic information available at the time. BBNJ takes for granted that the purpose is to fill UNCLOS gaps vis-à-vis newer resources without upsetting existing frameworks, even though UNCLOS III was an effort to go beyond precisely such an exercise of placing the square pegs of new information into the round holes of existing law.

That, for many of its participants, UNCLOS III was an exercise in sincerity as much as anything else—applied toward turning the political, economic, scientific, and technological churn of the time into new law—is perhaps attributable to its timing. In the words of one scholar, the NIEO took shape in a "narrow and specific window of geopolitical opportunity, a 'moment of disjunction and openness,' when wildly divergent political possibilities appeared suddenly plausible."[64]

The question we need to ask of the lawmaking of the moment is why that sense of disjunction and openness is not available today. For if anything the present moment should be one of compounded disjunction and openness. To paraphrase David Chandler, we live in a time of ontological and epistemological uncertainty about how to know and how to govern the world.[65] This is of course due to the recognition of the Anthropocene and of the deep flaws in long-standing assumptions "central to the modernist constructions of governance." With incontrovertible evidence of the destructive effects of human activity on global ecosystems, and of the corresponding disasters of increasing intensity that have followed the same, it is no longer plausible to build governance frameworks on understandings of human separation from and mastery over nature. It is not only marginal(ized) critical voices (of post-humanists, new materialists, and, prominently, Indigenous, Black, and postcolonial scholars) but also scientific ones that are pressing a rethinking of the standard Enlightenment classifications, of subject and object, living and nonliving, agent and matter.[66]

With serendipitous timing, hydrothermal vents have been discovered and (are being) studied in this context of uncertainty. And to this context they afford an almost ludicrously appropriate microcosm. Depending on who is thinking about them, vents are clues to the beginning of life and to solutions to stave off the end of the world through bio- and nano-technologies; resources to master the climatologically changing world anew, and homes to some of the worst threats (zoonotic diseases, methane pollution) that can follow in the wake. But in their vexed liminality, they are also invitations to reflect on the boundings and categories that underpin the modern world, and how it has defined rights of access to and use of the oceans. Vents, then, offer an opportunity to reflect on the history of the law of the sea, and the ways in which its doctrines—of freedom, plenty, conservation, remoteness, and of appropriate classification principles—have furthered imperialism, inequality, exploitation, and the destruction of forms of life. As Kathryn Yusoff has put it pointedly, "[t]he Anthropocene might seem to offer a dystopic future that laments the end of the world, but imperialism and ongoing (settler) colonialisms have been ending worlds for as long as they have been in existence."[67] The sea and the law of the sea have been part of this. Indeed, it was recognition of the destructive consequences of the older law of the sea (albeit not in the Anthropocene context) that had provided the impetus for many of those participating in UNCLOS III, and for the ideas they had advanced for a new law. And yet, the venues currently shaping the law of the sea are largely disconnected from this arguably fertile period for critical reflection and action.

The thrust of the BBNJ and ISA processes to fix the law of the sea enacts an erasure: of all voices that speak to its unequal historical and ongoing operations, of the experiences that these voices seek to articulate, and also of the world remaking ambition that had guided the negotiation of UNCLOS. But it would be wrong to attribute the erasure to ignorance, either of the law and its operations, or its critiques. The people who regularly gather for ISA and BBNJ meetings in Kingston and New York, and at workshops around the world, have much expertise about the oceans, the details of the law of the sea, its politico-economic effects, and its history. As individual conversations regularly reveal, delegates at these venues are rarely confused about the potency of the case for a moratorium on seabed mining (something that several indigenous groups, scientists, some states, and several civil society organizations are pressing for); or for stronger mechanisms for the protection and equitable use and distribution of high seas biodiversity. While there is not of course an exact sense of the ideal outcome, most—and especially third world—participants understand readily that the issues confronting them are not just about the crises of the ocean and of modernity today, but of reckoning with the much longer history of crises—of centuries of ocean and world endings, as Yusoff might put it—that modernity has been catalyzing, and of listening to the voices that articulate these histories and ongoing experiences.[68] Many recognize the implausibility of fixes. Yet, in what we might perhaps label as a performance of "cynical reason," following Peter Sloterdijk, the collective common sense articulated through these venues does not meet that which is individually understood.[69]

In his paper discussing the challenges of governance in the Anthropocene, Chandler quotes Bruno Latour's observation of the inversion of geological and "human or cultural" time: now "glaciers appear to slide quicker, ice to melt faster, species to disappear at a greater speed, than the slow, gigantic, majestic, inertial pace of politics, consciousness, and sensibilities."[70] But it is not just that political processes have yet to catch up with the ontological and epistemological uncertainty unleashed by the recognition of the Anthropocene. It is that they seek to bypass this uncertainly altogether, using the old trick of valorizing superficially presented "epochs" in time—from the Enlightenment, to UNCLOS III—which are turned into a simple story of ever-expanding, and universally promising, capitalist modernity. In this story, all problems *can be* fixed (and only fixable problems are officially acknowledged); and to seek radical unmaking is irresponsible, a negation of all the ideals which too were part of the assemblage of these times.[71] This story is a time warp, an alternate history in which the assimilation of the ideal to corporate interests is maintained against the weight of critical and scientific knowledge. In the oceanic context—and borrowing a trick from JJ Abrams's *Star Trek*—we might think of this story as the NORI/DeepGreen timeline.

Conclusion: The Demands of Loyalty (to UNCLOS)

How, then, to think about calls not to "undermine" UNCLOS? Sloterdijk's critique made clear the falsity of warped valorizations to maintain the status quo against much-needed unmakings:

> There can be no healthy relation of modern-day enlightenment to its own history without sarcasm. We have to choose between a pessimism that remains "loyal" to its origins and reminds one of decadence and a lighthearted disrespect in the continuation of the original tasks. As things stand, the only loyalty to enlightenment consists in disloyalty.[72]

In the oceanic context, loyalty to the UNCLOS III moment, and the goals of equality and distributive justice that marked its outset would demand matching its world-remaking ambition over preserving its flawed classifications. Again, this is well known to those participating in the ISA and BBNJ processes. The challenge is of bringing their individual knowledge into the collective law-making spaces; and it is one of mounting urgency, for the ocean and for all the communities dependent on it. In the words of the "letter to the future" from the scientists and locals who mourned the demise of the Okjokull glacier: "we know what is happening and what needs to be done. Only you know if we did it."[73]

Notes

1 Naomi Oreskes, "A Context of Motivation: US Navy Oceanographic Research and the Discovery of Sea-Floor Hydrothermal Vents," *Social Studies of Science* 33, no. 5 (2003): 697–742, 722.
2 Ibid.
3 "Discovering Hydrothermal Vents," Woods Hole Oceanographic Institution, accessed August 11, 2021, https://www.whoi.edu/feature/history-hydrothermal -vents/discovery/1977.html.
4 Robert Ballard, "The History of Woods Hole's Deep Submergence Program," in *50 Years of Ocean Discovery: National Science Foundation 1950–2000*, ed. National Research Council (Washington, D.C.: National Academy Press, 2000), 76.
5 Richard T. Barber and Anna Hilting, "Achievements in Biological Oceanography," in *50 Years of Ocean Discovery*, ed. National Research Council (Washington, D.C.: National Academy Press, 2000), 11–21.
6 William Broad, *The Universe Below: Discovering the Secrets of the Deep Sea* (New York: Touchstone, 1997), 109.
7 Ballard, "History of Woods Hole," 77.
8 David Perlman, "Astounding Undersea Discoveries," *San Francisco Chronicle* (San Francisco, CA), March 9, 1977.
9 David Perlman, "Deep Questions from the Sea," *San Francisco Chronicle* (San Francisco, CA), March 25, 1977.
10 Broad, *The Universe Below*, 261.
11 "The Mining Code," International Seabed Authority, 2021, https://www.isa.org.jm /mining-code.
12 Lyle Glowka, "The Deepest of Ironies: Genetic Resources, Marine Scientific Research, and the Area," *Ocean Yearbook* 12, no. 1 (1996): 154.
13 "Intergovernmental Conference on Marine Biodiversity of Areas Beyond National Jurisdiction," United Nations, accessed August 11, 2021, https://www.un.org/bbnj/. See also Braverman, this volume.
14 David Harvey, "Globalization and the Spatial Fix," *Geographische Revue* 2, no. 3 (2001): 23–30.
15 Philip Steinberg, *The Social Construction of the Ocean* (Cambridge: Cambridge University Press, 2001).
16 Epeli Hau'ofa, "Our Sea of Islands," *Contemporary Pacific* 6 (1994): 47.
17 Tamara Fernando, "Death at the Pearl Fishery," *Hypocrite Reader* 95 (2020), http://hypocritereader.com/95/tamara-fernando-mannar-pearls-cholera;   Isabel Hofmeyr, "Oceans as Empty Spaces? Redrafting Our Knowledge by Dropping the Colonial Lens," *The Conversation*, September 6, 2018, https://theconversation.com /oceans-as-empty-spaces-redrafting-our-knowledge-by-dropping-the-colonial -lens-102778.
18 Tommy Koh, "A Constitution for the Oceans," in *United Nations Convention on the Law of the Sea, with Index and Final Act of the Third United Nations Conference on the Law of the Sea* (Tokyo: United Nations, 1983), xxxiii.
19 Robin Churchill and Vaughan Lowe, *The Law of the Sea* (Manchester: Manchester University Press, 1999), 24.
20 Hugo Grotius, "Defense of Chapter V of the *Mare Liberum*," (ca. 1615), trans. Herbert Wright (1928), repr. *The Free Sea*, ed. David Armitage (Indianapolis: Liberty Fund, 2004), 112.
21 More detail in Surabhi Ranganathan, "Ocean Floor Grab: International Law and the Making of an Extractive Imaginary," *European Journal of International Law* 30, no. 2 (2019): 573–600. See also Henry Jones's chapter in this volume.
22 See UN Convention on the Law of the Sea Articles 56(3), 68, 77(3), Dec. 10, 1982, 1833 U.N.T.S. 397.
23 Ibid., Article 62.

24 Carmel Finley and Naomi Oreskes, "Maximum Sustainable Yield: A Policy Disguised as Science," *ICES Journal of Marine Science* 70, no. 2 (2013): 245–250.

25 UNCLOS Article 61.

26 UNCLOS Preamble; UNCLOS Articles 140, 148, 150; Isabel Feichtner, "Sharing the Riches of the Sea: The Redistributive and Fiscal Dimension of Deep Seabed Exploitation," *European Journal of International Law* 30, no. 2 (2019): 601.

27 See Michael Lodge and Philomene Verlaan, "Deep Sea Mining: International and Regulatory Challenges and Responses," *Elements* 14 (2018): 331; Michael Lodge, "How to Mine the Oceans Sustainably," *Scientific American*, August 11, 2020, https://www.scientificamerican.com/article/how-to-mine-the-oceans-sustainably/?print=true. Lodge is the Secretary General of the ISA.

28 Paul Ehrlich, *The Population Bomb* (New York: Ballantine Books, 1968).

29 UNGA, First Committee Debate, U.N. Docs. A/C.1/PV.1515-1516 (Nov. 1, 1967).

30 Andrew Thaler and Diva Amon, "262 Voyages Beneath the Sea: A Global Assessment of Macro- and Megafaunal Biodiversity and Research Effort at Deep-Sea Hydrothermal Vents," *PeerJ* 7 (2019): e7397.

31 Ibid., 2.

32 Ibid., 13; Eva Ramirez-Llodra et al., "Biodiversity and Biogeography of Hydrothermal Vent Species: Thirty Years of Discovery and Investigations," *Oceanography* 20, no. 1 (2007): 30–41.

33 "COVID-19: The Ocean, an Ally Against the Virus," UNESCO, April 16, 2020, https://en.unesco.org/news/covid-19-ocean-ally-against-virus.

34 Gregory Dick, "The Microbiomes of Deep-Sea Hydrothermal Vents: Distributed Globally, Shaped Locally," *Nature Reviews: Microbiology* 17, no. 5 (2019): 271–283.

35 Of course, strict land/water separations are questionable even in other parts of the ocean.

36 Stefan Helmreich, "What was Life? Answers from Three Limit Biologies," *Critical Inquiries* 37, no. 4 (2011): 671–696.

37 See inter alia, ibid., 685; Frederic Bushman, *Lateral DNA Transfer: Mechanisms and Consequences* (Cold Spring Harbor: Cold Spring Harbor Laboratory Press, 2002); Luis P. Villarreal, "Are Viruses Alive?" *Scientific American*, August 8, 2008, https://www.scientificamerican.com/article/are-viruses-alive-2004/; Lynn Rothschild and Rocco Mancinelli, "Life in Extreme Environments," *Nature* 409 (2001): 1092–1101.

38 Dick, "Microbiomes"; Lauren Mullineaux et al., "Exploring the Ecology of Deep-Sea Hydrothermal Vents in a Metacommunity Framework," *Frontiers of Marine Science* 5, no. 49 (2018): 1.

39 Steven Haddock and Anela Choy, "Treasure and Turmoil in the Deep Sea," *New York Times* (New York, NY), August 14, 2020.

40 Thaler and Amon, "262 Voyages," 19.

41 Ibid., 10.

42 Mathiu Ardyna et al., "Hydrothermal Vents Trigger Massive Phytoplankton Blooms in the Southern Ocean," *Nature Communications* 10 (2019): 2451; Elisabetta Menini and Cindy van Dover, "An Atlas of Protected Hydrothermal Vents," *Marine Policy* 108 (2019): 103654; Dick, "Microbiomes."

43 Menini and van Dover, "An Atlas," 1; Amon and Thaler, "262 Voyages," 2.

44 "Nauru Requests the President of ISA Council to Complete the Adoption of Rules, Regulations and Procedures Necessary to Facilitate the Approval of Plans of Work for Exploitation in the Area," International Seabed Authority, June 29, 2021, https://isa.org.jm/news/nauru-requests-president-isa-council-complete-adoption-rules-regulations-and-procedures. The two-year rule refers to Annex S.1 paragraph 15 of the 1994 agreement, which allows a state to request the ISA Council to complete the adoption of exploitation regulations within two years. If ISA fails to do so, then it must consider any plan of work sponsored by Nauru on its merits.

45 See Surabhi Ranganathan, "The Law of the Sea and Natural Resources," in *Community Interests Across International Law,* eds. Eyal Benvenisti and Georg Nolte (Oxford: Oxford University Press, 2018), 121, 133–134.

46 David Johnson, "Protecting the Lost City Hydrothermal Vent System: All Is Not Lost, or Is It?" *Marine Policy* 107 (2019): 103593.

47 Olive Heffernan, "Seabed Mining Is Coming—Bringing Mineral Riches and Fears of Epic Extinctions," *Nature: News Feature,* July 24, 2019, https://www.nature.com/articles/d41586-019-02242-y.

48 "Workshop on the Development of a REMP for the Area of the Northern Mid-Atlantic Ridge with a Focus on Polymetallic Sulphides Deposits," International Seabed Authority, accessed August 11, 2021, https://www.isa.org.jm/event/workshop-remp-area-northern-mid-atlantic-ridge.

49 Article 4, Revised BBNJ Draft.

50 Konrad Marciniak, "Marine Genetic Resources: Do They Form Part of the Common Heritage of Mankind Principle?" in *Natural Resources and the Law of the Sea,* eds. Lawrence Martin et al. (Huntington: Juris, 2017), 373–405.

51 UNCLOS Articles 143, 145, 147.

52 Dire Tladi, "The Common Heritage of Mankind and the Proposed Treaty on Biodiversity in Areas beyond National Jurisdiction: The Choice Between Pragmatism and Sustainability," *Yearbook of International Environmental Law* 25 (2014): 113–132.

53 Alice Vadrot et al., "Who Owns Marine Biodiversity? Contesting the World Order through the 'Common Heritage of Humankind' Principle," *Environmental Politics* (2021) DOI: 10.1080/09644016.2021.1911442.

54 Revised draft text, A/CONF.232/2020/3 (Nov. 18, 2019). The session finally took place in March 2022.

55 See Braverman, this volume for a broader discussion of the definitions of MGRs.

56 Joanna Mossop, "Towards a Practical Approach to Regulating Marine Genetic Resources," *ESIL Reflection* 8, no. 3 (2019): 1–11.

57 Harvey, "Globalization," 24–25.

58 Madhuri Ramesh and Nitin Rai, "Trading on Conservation: A Marine Protected Area as an Ecological Fix," *Marine Policy* 82 (2017): 25–31, 26.

59 On this narrative, see for instance, Wil S. Hylton, "History's Largest Mining Operation is About to Begin. It's Underwater—and the Consequences are Unimaginable," *Atlantic,* January/February 2020, https://www.wrongkindofgreen.org/tag/international-seabed-authority/.

60 Unsurprisingly, these tie together, including in the visions and enterprises of particular corporations. See for example, on Elon Musk: Simon Hannah, "Elon Musk: The Prince of Capitalism," *Anticapitalist Resistance,* May 9, 2021, https://anticapitalistresistance.org/elon-musk-the-prince-of-capitalism/.

61 I examine UNCLOS III in these terms in Surabhi Ranganathan, "Decolonization and International Law: Putting the Ocean on the Map," *Journal of the History of International Law* 23, no. 1 (2021): 161–183.

62 Mohammed Bedjaoui, *Towards a New International Economic Order* (New York: Holmes and Meier, 1979), 224–227. On the deliberate undoing of the NIEO see also Mark Mazower, *Governing the World: The History of an Idea* (London: Penguin, 2012), Chapter 12.

63 On the BBNJ, see G.A. Res. 72/249 (Jan. 19, 2018). That the ISA operates under a similar restriction is seen from the writings of its Secretary General, e.g., Lodge and Verlaan, "Deep Sea Mining."

64 Nils Gilman, "The New International Economic Order: A Reintroduction," *Humanity* 6, no. 1 (2015): 1–16.

65 David Chandler, "Planetary Boundaries and the Challenge to Governance in the Anthropocence," *REPATS* n.01 (2018): 21–41, 26. The quotes in the paragraph are from this article, at 22.

66 Ibid.
67 Kathryn Yusoff, *A Billion Black Anthropocenes or None* (Minneapolis: University of Minnesota Press, 2018), xiii.
68 Yusuff, *A Billion Black Anthropocenes.*
69 See Peter Sloterdijk, *The Critique of Cynical Reason* (Minneapolis: University of Minnesota Press, 1988), 5–6.
70 Chandler, "Planetary Boundaries," 23.
71 In pondering the slippages and elisions in this narrative, it is useful to think with Ellen Meiksins Wood, "Capitalism or Enlightenment," *History of Political Thought* 21, no. 3 (2000): 405–426.
72 Sloterdijk, *The Critique of Cynical Reason.*
73 Jon Henley, "Icelandic Memorial Warns Future: 'Only You Know if We Saved Glaciers,'" *The Guardian* (London, UK), July 22, 2019.

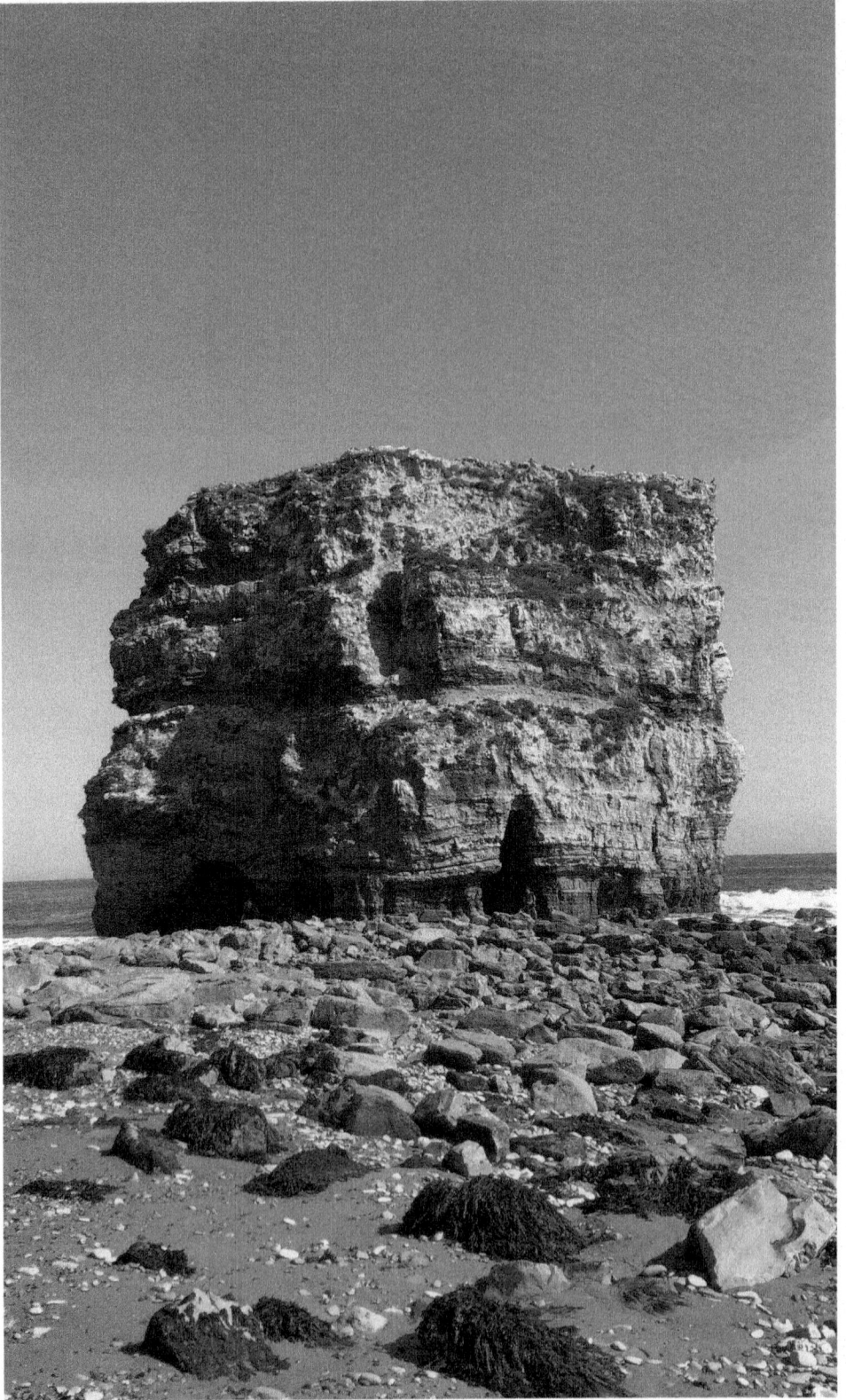

2

COMMODIFYING THE OCEANS

The *North Sea Continental Shelf Cases* Revisited

Henry Jones

Introduction

In 1969 the International Court of Justice (ICJ), the World Court at The Hague, established under the Charter of the United Nations, issued its judgment in the dispute between the Netherlands and Denmark, on the one hand, and the Federal Republic of Germany on the other.[1] The dispute was over how to draw the boundaries allocating rights over the shallow portion of seabed off the coast of these states, under the North Sea, known as the continental shelf. This case happened during a period where the law of the continental shelf was still being made and argued over. The Court faced a novel challenge of law, geography, and technology. In this chapter, I return to those judgments to find how the law turns the oceans into discrete, exploitable objects, how it separates and fixes this fluid space, and how the materiality of the seas nevertheless always pushes back against this process.

The standard teleology of doctrinal legal history explains legal change by reference to social change and portrays law as a rational, functional response to the problem of organizing society.[2] However, when we look at the history of the law of the sea, we often see the law leading the way. Whether it is Grotius declaring a free sea to justify Dutch privateering, or the United Nations Convention

FIGURE 2.1 The North Sea with Marsden Rock in the foreground, as viewed from Marsden Bay, South Tyneside, UK. On this crumbling limestone coast, the distinction between land and sea is particularly unstable. It is also here where an 18th-century coal miner, known as Jack the Blaster, moved into a cave, to live "free from impost." Here Thomas Spence found him, and wrote in chalk on the wall "Ye landlords vile, whose man's peace mar, Come levy rents here if you can; Your stewards and lawyers I defy, And live with all the RIGHTS OF MAN." As recounted in Alastair Bonnett, "The Other Rights of Man: The Revolutionary Plan of Thomas Spence," *History Today* 57(9) (2007).

DOI: 10.4324/9781003205173-3

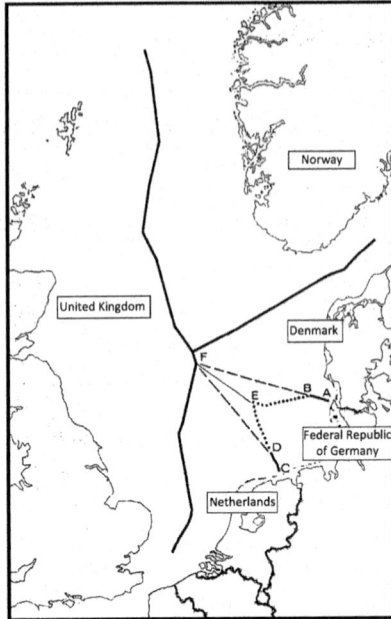

FIGURE 2.2 Produced by the International Court of Justice, this map illustrates the positions of each side. The lines drawn represent the various claims made. Public domain.

on the Law of the Sea (UNCLOS) regulating the then non-existent industry of deep seabed mining, the law is a productive force in the world. Productive in the sense that it is connected to the dominant mode of production, Henri Lefebvre's production of space as a social form. Lefebvre distinguishes social space from abstract, mental space and from given, natural physical space.[3] Legal geographers have further highlighted that this process is co-constitutive, that as law produces space so too does the materiality of the world feedback into the law.[4] These social phenomena inform the way we think about law, and how it is possible to think about law. To contribute to understanding the ways the law shapes the world and vice versa is the purpose of this chapter.

James Crawford and Thomas Viles attempted something similar in an essay entitled "International Law on a Given Day," in 1994, presumably prompted by the pending UNCLOS Implementation Agreement.[5] In this essay they ask what international law was on September 29, 1945, the day after the Truman Proclamations on the continental shelf and coastal fisheries.[6] This essay looks at the development of the law of the continental shelf before and after the Proclamations, and the effect of the proclamations themselves, as a case study of customary law formation, custom being a key source of international law. While some reference is made to material changes and the use of the oceans, such as the benefit to American oil companies of clarity over the continental shelf regime, the focus is almost exclusively on ideas. While they find the answer to their question to be indeterminate, "a question that cannot be answered, of conduct that was neither lawful nor unlawful (or perhaps contingently both)," they locate this indeterminacy in legal thought, not in material social relations.[7] Their answer to this indeterminacy is that custom "is an *ex post facto* construct … international law has to be brewed." International law on a given day is indeterminate, but they give no answer to what Susan Marks calls the false contingency question, that is why given indeterminacy, how are arguments nonetheless resolved?[8]

It is surely much more useful to start from the other end, the material, and ask how it came to be. Therefore, in seeking to answer again what was international law on a given day, and looking instead at how the judges of the ICJ did it in *North Sea Continental Shelf*, my starting point is what were the material conditions, who was exploiting the continental shelf, how and why? What was at stake for the parties to the case, and what were the implications which the Court was, or should have been, aware of?[9]

I accept Crawford and Viles's argument that the continental shelf is a legal product, but this is how the indeterminacy question should be asked, with a focus on material relations not ideas. How was the continental shelf produced, and what did its legal character confer upon it? Ultimately, I suggest that this is a pure example of commodification, and the law as key to producing an abstract and universal commodity from the physical continental shelf. It is then the work of a critical oceans account of the law to undo this abstraction, to allow for the material to push back. My aim is to simultaneously explain the origin of the continental shelf regime in historical materialist terms, and to rid the concept of this abstraction. Further, by bringing in the historical and colonial context of the decision, the temporality as well the materiality of the law is highlighted.

To do this, I will first explain the historical materialist approach to the study of international law, then think about what the ocean is, drawing on scholarship from across the humanities and social sciences, that can be loosely called "critical ocean studies."[10] Taken together these methodologies allow for a re-understanding of international law and space and create an opportunity to see and make use of the spatial feeding back into the legal. After this theoretical section attention will turn to reading *North Sea Continental Shelf*, before, in the final part, a study of the history and material conditions around the case.

Law and the Production of Commodities

Humanity burns about 40 gigatons of fossil carbon a year. It has been calculated that we can burn about 500 more gigatons before the average global temperature rises over two degrees. There remains at least 3,000 gigatons of fossil carbon in the ground.[11] Finding and extracting these resources requires a large investment before it is profitable. A clear and certain system of ownership has been developed. If we are to prevent the burning of fossil fuels, essential for the survival of life on this planet, then one small part of this will be to change how we value and control these resources and spaces, how they are constructed and produced.

What does it mean to say the ocean is legally produced? It means to connect legal innovation to the dominant mode of production, which is capitalism.[12] The understanding of the sea and seabed change due to a change in the imperatives of economic exploitation. These spaces are remade, as commodities, through law. What is a commodity? That is the question with which Marx began *Capital*. "A commodity is, first of all, an external object, a thing which through its qualities satisfies human needs."[13] Through an analysis of use value and exchange value, Marx quickly comes to a different answer: "analysis brings out that it is a very strange thing, abounding in metaphysical subtleties and theological niceties."[14] It is exchange value which provides these effects, the abstraction away from the thing itself to an exchangeable commodity.

The role of law then becomes to guarantee these commodities. In order for commodities to be exchanged, their owners must "recognize each other as owners of private property."[15] According to the commodity form theory of law, as developed by Bolshevik legal theorist Evgeny Pashukanis, it is law which allows for this recognition. Law "is that which regulates disputes between formally equal, abstract individuals," over equal and abstract things.[16] This theory holds that law only becomes a universal system under capitalism. It is capitalism which turns specific goods into generalizable commodities, with an abstract value. Everything has a value and everything can be exchanged. Law both creates and secures this abstract value.

International law in this understanding is structurally connected to imperialism, first because the international legal form is bound up with the spread of international capitalism, and second because only imperial violence can enforce international law. As Robert Knox explains, there is a structural connection between international law and imperialism. The violence of imperialism is the enforcement mechanism of international law. As capitalism spread internationally and became global, so too did international law, to the point where international law, constituted by imperialism and violence, comes to structure the world.[17]

If all international law is tied up with imperialism, how does this play out in a dispute apparently limited to northern Europe? Prior to the case, the Netherlands and Denmark are concerned for their colonial territories' potential continental shelves. The preference for negotiation and equity hands over the enforcement to the formally equal states. But states are only formally equal, as China Miéville explains in his reading of Pashukanis,[18] and so the force behind the states is the actual enforcement. It will be seen that this judgment reiterates a fundamental feature of international law—strong states win over weak states, imperialism is baked in, even in something as strict and worked out as the continental shelf would seem to be after its codification in the UNCLOS.[19]

The Marxist understanding of international law illuminates what is happening in the development of the continental shelf regime. In terms of changing it, that is tactical engagement with the international legal system, a focus on oceans offers solutions. One of the contributions of a legal focus on ocean geography is that it demands a systematic legal geographic engagement with international law, something which to date has been piecemeal. What is needed is an emphasis on the materiality of the seas, a reconnection of these disconnected zones and ideas, a refusal of environmental protection as a discrete legal project but instead as the basis of all action, this will get us closer to understanding the relationship between law and social change, "*the* question of Marxist legal theory."[20]

When we relate this back to the ocean, the process is actually more visible than more routine commodities, or even land, because the thing itself, whether the continental shelf, the deep seabed, or exclusive economic zones, is already abstract. It is on this basis that I argue that the continental shelf and the deep seabed are legally produced. As geographical facts they do not exist in anything like the form they are regulated. The continental shelf would not exist without the law. And this is where it is essential to bring in more direct attention to the oceans.

Critical Ocean Studies

In critical ocean studies, the work of Kimberley Peters and Phil Steinberg and that of Elizabeth DeLoughrey get us started on how to think with and about the ocean. Steinberg and Peters have developed the concept of wet ontologies to capture the potential for thinking with and about the seas.[21] What they mean by wet ontology is to understand the ocean in all its complexity, "as forces, as vectors, as assemblages of molecules and meanings, as spaces of periodicity, randomness, instability and transformation, and as volumes (depths) and areas (surfaces)," gives rise to an oceanic politics and an understanding of space as unstable, transforming, voluminous.[22] This approach puts emphasis on the materiality of the oceans, and tries to move beyond accounts of the ocean which treat it as flat or inert, as a stage for human history, but rather as a space with its own history. It draws attention to the fabrication and instability of line drawing as a governance technique in general, by first understanding it as completely unsuitable to ocean geographies, before then questioning the practice in general. In Marxist terms, we see here very clearly a clash between the classes of governance and the governed, and how open to struggle and contestation lines, space, and therefore law really are.

DeLoughrey has developed complementary thinking from a literary discipline: "unlike terrestrial space—where one might memorialize a space into place—the perpetual circulation of ocean currents means that the sea dissolves phenomenological experience and diffracts the accumulation of narrative."[23] Where Steinberg and Peters seek the more-than-wet, to use the sea to think geopolitically in general, DeLoughrey uses the absence of the human at sea to access thinking about the nonhuman, and to de-center the human from cultural and political thought. Of course, the specific human usually centered is the Western, masculine, capitalist subject, and DeLoughrey highlights other perspectives, experiences, and ontologies of the ocean. For example, using Indigenous Pacific poetry to undo the US military spatial construction of the ocean,[24] or using Caribbean art to emphasize the depths and currents that create ocean space, in a form of thinking DeLoughrey and Flores call "Tidalectics."[25]

The *North Sea Continental Shelf* Cases

In the North Sea, gas was found from 1964, and oil was discovered in December 1969.[26] During this period, Denmark, the Netherlands, and Germany tried and failed to negotiate their overlapping claims before referring their dispute to the ICJ in February 1967.[27] Alex Oude Elferink provides a masterful and exhaustive historical account of the cases.[28] Before the cases were referred to the ICJ he reveals the different interests and anxieties of the parties. Interestingly, all parties had an eye on claims to resources beyond the North Sea, meaning that some of the claims were not connected to the specifics of the North Sea. Denmark had concerns over its colonies of Greenland and the Faroe Islands; the Dutch had interests in resources connected to their territories of Suriname, the Antillies, and New Guinea in particular; Germany had resisted the continental shelf regime in general, fearful that it would miss out, and also had concerns of the claims to be made for the German Democratic Republic.[29] Then again, some concerns were directly related to the geographical context, such as the choice to ignore the Norwegian Trough as it was in nobody's interest to drag Norway into the dispute.[30] Denmark had given its first concession to prospect for hydrocarbons in its territorial sea in 1962, including an option to extend the search if Danish sovereignty was extended.[31] In the Netherlands, gas had been discovered in Gröningen in 1959.[32] Interestingly, given that the case is famous for setting out the rule that such disputes should be negotiated equitably, when negotiations began between the states both the Netherlands and Denmark were caught by surprise when Germany suggested splitting the area equally.[33]

Also of interest is the position of other North Sea states not party to the dispute. The UK North Sea Continental Shelf Act of 1964 stated in its introduction that it was to give effect to the Geneva Convention of 1958, while never offering its own definition or limit on the continental shelf, nor making any claims.[34] All three parties to the case also made efforts to keep the United Kingdom out of the dispute, having accepted equidistance agreements there which suited the Netherlands and Denmark.[35] In the submissions from all parties to the ICJ the North Sea is described as all being at a depth of less than 200 meters, with the exception of the Norwegian Trough running along the Norwegian coast, which is much deeper. This was a considered choice, but it remains striking that this geographical fact could so easily be ignored without even explaining why it does not have legal effect. While it may make sense not to give legal effect to it in this specific situation, by ignoring it the Court instantly detached the legal geography from the physical geography.

The case remains interesting as an example of how to wrangle with international law sources, not as a source of law on the continental shelf. On that front it was overtaken by UNCLOS. However, it is incredibly instructive as an example of how international law deals with a new geographic space, a new resource. As Judge Tanaka saw it in his dissent:

An originally geological and geographical concept ... by reason of its intrinsic economic interests which have become susceptible of exploration and exploitation as the result of recent technological development, has been vested with legal interest and presents itself as a subject matter of rights and duties subject to the rule of law and constituting an institution belonging to international law.[36]

Interesting features from the main judgment include that the majority limit themselves to delimitation, not apportionment, explained as:

Delimitation is a process which involves establishing the boundaries of an area already, in principle, appertaining to the coastal State and not the determination *de novo* of such an area. Delimitation in an equitable manner is one thing, but not the same thing as awarding a just and equitable share of a previously undelimited area, even though in a number of cases the results may be comparable, or even identical.[37]

The significance of this is that the territorial claims here are not new, the Court is not granting territory, it sees itself as only clarifying the means by which to agree the boundaries. The Court here is insisting on a lack of novelty in what it is doing, it is simply clarifying the rules for allocating the continental shelf, not making new ones. The Court may think that is what it is doing, but in the very next paragraph it says:

rights of the coastal state in respect of the area of continental shelf that constitutes a *natural prolongation of its land territory* ... exist *ipso facto* and *ab initio*, by virtue of its *sovereignty* over the land ... an *inherent right* ... Its existence can be *declared* but does not need to be *constituted* ... [T]he right does not depend on its being exercised.[38]

Suddenly this still very new concept is natural, inherent, and automatic; something which can be, but does not need to be, declared, and does not need to be constituted. The continental shelf by 1969 just *is*. This is an extreme statement of how sovereignty over territory works, and I struggle to find any comparable example before this. Settler colonialism still needed something like discovery, occupation, or use. The early development of the continental shelf regime was clearly weighed down by issues around extension of territory and questions of discovery, of symbolic vs. actual occupation, of *terra nullius* and more.[39] The ICJ breaks free of all of this, saying that a state has a continental shelf simply by having a coastline. The law is producing territory out of nothing and producing and guaranteeing property in the seabed. Even if the state doesn't know it, hasn't explored or made any attempt to claim it. That states have this huge extent of underwater territory, and always have, is quite a thing for the ICJ to discover in 1969.

This reasoning continues, where the continental shelf is described as "*actually part of the territory* over which the coastal State *already has dominion*" and "a prolongation of continuation of the territory." The Court follows the parties in just ignoring the geographic fact of a massive trench far deeper than 200 meters just off Norway. Prolongation obviously makes no sense if the trench was taken into account. The curvature of the coastline of Germany is described as "an incidental special feature from which an unjustifiable difference of treatment could result."[40] No explanation of what makes this special is given, and as noted by Elferink this was a surprise to both Denmark and the Netherlands when Germany first raised it. The judges close by calling the continental shelf "submerged land,"[41] finally betraying the understanding that has informed their entire judgment. Only by seeing the continental shelf as land can such strange things be said seriously at the start of the judgment. If the shelf is land, then it doesn't matter about depth or shape. Discovery and occupation are not needed because this land was always there. By understanding the seabed as land then the problem just becomes a question of clarifying the borders. In my view this is more than just a convenient legal fiction, it is an ontological choice that the lawyers and judges in the case understand the sea as if it were land. Either way, the outcome is to transform the near coastal seas into commodities, abstract, certain and fungible, ready for exploitation.

What becomes clear re-reading the judgment is that for the most part the Court is satisfied to abstract entirely from material reality. Whether it is ignoring a deep trench, finding something special about a concave coastline, or saying that the continental shelf existed before it was ever explored or named, there is denial of the material. This process commodifies the seas, fixes them with lines and definitions that bear little relation to geography. The seabed can then more easily be packaged up to be exploited and exchanged. In the next section I tell a materialist history of the North Sea, to try and bring some of the flow and churn back to the case, and to unsettle international law's commodifying effect.

A Materialist History of the North Sea

In this section I trace the history of the North Sea, with a focus on the materiality of this ocean space, how it had been constructed and understood, and how this further enhances our understanding of the ICJ case. The case took place as the North Sea was first being explored for oil, and the decision constructs the seabed in a way which is optimal for this type of exploitation. Interestingly, while lawyers argued about distance from shore, in the history of offshore oil it is depth which is the key consideration. This makes intuitive sense, as the challenges of drilling increase in line with depth, ever increasing pressure, turbulence, stresses on materials, et cetera. The difficulties of being far from shore are not much more if it is 100 nautical miles or 200 nautical miles, simply the complexity of supply and transport. The distance from shore is more prominent for non-industry perspectives. Close to shore operations pose a greater environmental threat to the coastline, and have a bigger effect on the local economy, such as the influence of North Sea oil on the development of the city of Aberdeen.[42] The law's preference for understanding this regime in terms of distance from shore betrays a terracentric ontology, while oil platform builders and operators have a more fluid ontology.[43] As the use and understanding of the geographical area has changed, so too has the way it is regulated.

Archaeologists have suggested that the North Sea was a large prehistoric plains area until the end of the last ice age, and Stone Age artifacts have been found.[44] In ancient history, Pliny called it the Northern Ocean,[45] the Celts knew it as *Morinaru*, the Dead Sea, and Germanic peoples as *Morimarusa*.[46] This naming convention, which lasted into the Middle Ages, referred to still water patches on the sea, a name based on the materiality of the sea itself rather than its relationship with land. The North Sea was also known this way in Dutch—*lebermer* or *libersee*.[47] By the late Middle Ages, its name as the German Ocean was common in English.[48]

The North Sea was central to the late Viking Empire, and was primarily important as a means of transport.[49] With the Norman conquest of England, the North Sea lost its prominence as a travel route, with attention shifting to the Baltic Sea, dominated by the Hanseatic League.[50] Bruges's deep port made it the center of trade between the Hanseatic League and London and therefore the rest of southern Europe.[51] The Danish Sand Toll, first recorded in 1461, was a tax specifically on use of the beach, or more generally on launching and landing boats and fishing equipment.[52] Denmark dominated herring fishing in the North and Baltic seas in the high Middle Ages, but by the late Middle Ages this position was already dwindling, with Dutch ascendancy.[53] The Hanseatic deal had prohibited Dutch herring fishing in the 14th century, but this restriction had led to Dutch fishermen developing other herring fisheries in the 15th century. As the Hanseatic League broke down in the early 15th century, the Netherlands became the center of the North Sea economy.[54]

As European exploration and colonialism spread out into America and the East Indies, the North Sea remained important for connecting Dutch spices and Spanish silver. It also became a key economic area for fishing and whaling. Norway, Denmark, and Scotland all made claims to territory in North Sea herring fisheries. Grotius's argument in *De Jure Praedae* is more generally associated with the East Indies, and this was of course a top priority, but the North Sea was also a major concern. Alison Rieser argues persuasively that the Battle of the Books, and Grotius's debates with Welwood in particular, was primarily about herring fishing.[55]

Potentially the first legal construction of the North Sea came with the English Navigation Acts of 1651 and 1660. The 1651 Act required all trade between England and its colonies to be carried out on English vessels and tried to impose a 30-mile exclusive fishing zone.[56] This led to the first Anglo-Dutch war. These Acts created tensions between England and the Netherlands both in the North Sea, and in North America where trade between English and Dutch colonies was prohibited. Anglo-Dutch wars were fought at least partly over herring fishing and shipping in the North Sea in 1652–1654, 1665–1667, and 1672–1674, each time resulting in Dutch victory.[57] This understanding of the North Sea as important for transit, trade, and fishing remains well into the 19th century, when we start to see a change in the understanding of the sea due to a change in the relationship of production.

The mercantilism of the early modern period required the ever-greater exploitation of fisheries for trade, and Dutch dominance of herring fisheries was the key reason for its growth and dominance as the trade hub for Baltic grain and timber with French and Iberian salt, oil, and wine. Spanish and Portuguese gold and spices changed this dynamic again, and the involvement of the Dutch in imperialism in the east demanded a new assertion of the freedom of both fisheries and seas.[58] As we see the emergence of capitalism, the law becomes more generalized as a tool of social organization. Freedom is not enough, and property must be secured. On land this is the key innovation of English imperialism. However, this change comes more slowly to the seas. The commodification of the seas really arises in the 20th century, and the possibility of the exploitation of the resources of the continental shelf. It is at this point that the North Sea stops being understood and constructed in its specificity, as a place for transit and fishing, and becomes abstracted into a space for exploitation of commodities. As such, the focus of the history changes to the continental shelf.

A Materialist History of the Continental Shelf

The growth of oil and gas as an alternative to coal changed both the labor market around fossil fuels and the geography of energy production. Commercially viable oil wells had been drilled in the United States since the middle of the 19th century. Offshore drilling began in 1896, on a Santa Barbara beach in California. Connected to the land by a 300 foot wooden pier, Henry L Williams was the first person to drill for oil under the sea.[59] At that time nobody argued that the United States did not have sovereignty over the land below the water. In 1911, Shell built a well on Caddo Lake, Louisiana, ending the reliance on piers for drilling under water. The year 1938 saw the successful establishment of the first oil rig in the Gulf of Mexico, about a mile from the shore. By 1947, there was a well 10 miles out. Today, the world's most isolated oil platform is Shell's *Perdido*, in the Gulf of Mexico, nearly 200 nautical miles from land, 8,000 feet deep.[60]

The term continental shelf itself slowly emerges over the first half of the 20th century, and its usage tracks the legal history, peaking with the negotiation of UNCLOS.[61] The earliest use of the term I have so far found is from 1888, in a paper on fish habitats published in the *Scottish Geographical Magazine*.[62] This paper defines the term as meaning "applied to the shallow portion of the continental slope, lying within the 100-fathom line, which is usually terminated seawards by a very abrupt descent to abyssal soundings."[63] The paper cites as authority one from the previous year, on soundings required to lay underwater cables, but in that paper the feature is at all times referred to as the continental slope, without differentiating different parts. The term appears sporadically in other geographical meetings in the 1890s and begins to appear more widely in scientific literature in the early 20th century, in a variety of places but mostly in geological surveys, particularly in the United States, the Arctic, and the Antarctic. At this time, it certainly hasn't established a specific technical meaning, still being interchangeable with continental slope.

The initial interest in the continental shelf is related to fishing and to the laying of submarine cables. But the use of the continental shelf for energy also has a history going back to a similar point. The legal history of the continental shelf starts with the Cornwall Submarine Mines Act of 1858, declaring that ownership of minerals and workings from mines below the low tide mark adjacent to the coast of Cornwall belonged to the Crown.[64] The deeper Cornish mines went out under the sea. This drove various technological developments such as the Cornish steam engine to pump water out efficiently, and developments in law as the Duchy of Cornwall and the Crown clashed over ownership of the land beyond low tide. This Act was a result of this clash, as described in the judgment of Lord Coleridge in *R. v. Keyn*.[65]

The Duchy of Cornwall, the estate belonging to the Prince of Wales, owned and operated mines which extended out underground beyond the low-water mark. It was found by the arbitrator in that case, Sir John Patteson, that the Crown owned the land beyond the low-water mark. The Duchy argued first that the seabed which adjoined the county of Cornwall was passed to the Duchy under the original grant, and second and in the alternative that the seabed was unowned, and thus belonged to the Prince of Wales as first occupier. At this time the argument that the seabed was *res nullius* was not successful, and the decision of the arbitrator is reflected in the language of the Act. Section 2 declared and enacted that:

> All mines and minerals lying below low-water mark under the open sea adjacent to but not being part of the County of Cornwall are, as between the Queen's Majesty, in the right of her Crown, on the one hand, and His Royal Highness Albert Edward Prince of Wales and Duke of Cornwall, in right of his Duchy on Cornwall, on the other hand, vested in Her Majesty the Queen in right of her Crown as part of the soil and territorial possessions of the Crown.

This was raised and discussed in *R. v. Keyn* as the dispute was an appeal against a criminal conviction of the captain of a German ship which had collided with a British vessel within three nautical miles of the coast. In the case we see English High Court judges grappling with all manner of authority on international law. Lord Coleridge concludes based on this exercise of sovereignty over mines which extend below the low tide mark that "the realm does not end with [the] low-water mark, but that the open sea and the bed of it are part of the realm and territory of the sovereign."[66]

However, Coleridge was in the minority. In his majority judgment Lord Chief Justice Cockburn proceeds along a very different line of reasoning. He traces the development of the jurisdiction of the Admiralty, and case law where jurisdiction was claimed at sea since Edward I. The judges who find jurisdiction rely on the opinions of international legal scholars, while the majority who do not find jurisdiction rely on domestic case law. Cockburn dismisses Selden and Hale's extensive claims, alongside others, as "vain and extravagant pretensions." Where he relies on jurists, he traces a more modest line from Grotius of qualified jurisdiction at sea.[67]

Crucially, while writers on international law gradually accepted a three-mile territorial sea, English lawyers were ignorant of this. Cockburn finds no reference in English legal history to claims of territory over the sea. Enough inconsistency is shown in the international legal authorities as to doubt the obviousness of criminal jurisdiction extending out to sea. Furthermore, claims that the bed of the sea is the territory of the state are all found to start from claims of ownership over all the seas. If that claim falls away as outdated, then so does the accompanying claim of ownership of the seabed.

In relation to the Cornish mines, he finds the territory where the mine starts belongs to the Crown, and that presumably the seabed is capable of being appropriated by first occupier: "I should not have thought that the carrying one or two mines into the bed of the sea beyond low water mark could have any real bearing on a question of international law."[68] The Act itself only conveys rights "in right of her Crown," not because of any ownership of the soil. That a carefully limited piece of legislation, in response to one very specific dispute, should be the basis of "a parliamentary recognition of the universal right of the Crown to the ownership of the bed of the sea below low water mark" is, as Cockburn says, surprising.[69] In short, the majority is not convinced that international law can convey rights to a state without that state actively claiming them, it doesn't find anything like a claim to criminal jurisdiction up to three miles in the legislation over mines, and as such finds that the German ship captain was not subject to English law.

This case is best remembered as being about the rule of law,[70] but it also illustrates the relationship between domestic and international law at this time in fascinating ways. To just focus on the legal doctrine is to miss the way the sea is being constructed and commodified through law, and the way the materiality of the ocean is driving the development of the law. A few Cornish mines might not have represented a claim to ownership of the seabed at the time, but as ownership of the seabed became a more pressing matter, this history was reinterpreted.

In 1923 Cecil Hurst claimed in the *British Yearbook of International Law* that this English mining legislation from 1858 was the starting point for the authority for a state to claim ownership of the seabed.[71] He was not convinced by Cockburn, finding that the only basis for this legislation could be a belief that the Crown had territorial rights over the bed of the sea. His reading of the common law authority separates the question of territorial waters from the question of ownership of the bed of the sea. Where Cockburn had found no necessary connection between property in a couple of mines and the extension of criminal jurisdiction, Hurst reads it the other way around: "the property in the bed of the sea and not merely sovereignty and jurisdiction over it was vested in the Crown."[72] Property in the bed of the sea has existed since people claimed exclusive fisheries beyond the low-water mark, and as such "the rights of the Crown in the bed of the sea must have been fixed at least as early as the thirteenth century."[73] That is quite a change in interpretation of the same piece of legislation.

In the interim between *R. v. Keyn* and Hurst's article, there had been several decisions which accepted property in the seabed, mostly Privy Council decisions concerned with British colonies. He also extends the claim of property beyond three miles where the concern is sedentary fisheries. The right to these fisheries is a property right, and the ownership of the benefit is "based on their being a produce of the soil."[74] So having dismissed *res nullius* arguments for the seabed beyond the low-water mark, it comes back in here in the language of settlor colonialism, with title in property being derived from occupation, usage, and enjoyment of the benefits. The areas in question are also largely off the coast of colonies. Furthermore, this is a distinctly terracentric understanding of the bed of the sea.[75]

The 1930 Hague Codification Conference reached no outcome on the continental shelf, although preparatory materials noted that there was unanimity about territory over at least three miles. The Truman Proclamations in 1945 gave new impetus. The most relevant and best-known Proclamation over the continental shelf has several interesting features. First, it situates the declaration in the context of the need to secure and exploit petroleum resources.[76] By the middle of the 20th century oil had decisively overtaken coal as the most important hydrocarbon in the global economy. Second, it uses the term continental shelf, with no limits. The origin of title here is the seabed being contiguous. The Proclamation also claimed "jurisdiction and control," but only over the resources of the seabed and subsoil. The second Proclamation, on fisheries, is notable for the assertion of the power to regulate fishing activities on the high seas, whittling down the freedom of the seas to navigation alone.

Reaction to the Proclamations is interesting. Commentators were skeptical, but states were either silent or followed the US practice.[77] Panama made a declaration in 1946, as did Argentina, which went a big step further in claiming sovereignty over the continental shelf. Also in 1946, the United Kingdom negotiated with Venezuela to divide the seabed between Trinidad and Venezuela, again basing the title on occupation and exploitation. However as quickly as 1951 the International Law Commission found that "the seabed and subsoil were subject to the exercise, by the littoral states, of control and jurisdiction for the purposes of exploration and exploitation. The exercise of such control and jurisdiction was independent of the concept of occupation."[78] The Geneva Convention in 1958 defines the continental shelf based on adjacency, confers sovereignty over resources, with the only limit being 200 meters depth, or up to the depth that "admits of the exploitation of the natural resources."[79] Whether there is significance in the difference between near synonyms contiguity and adjacency is hard to say. Depth as a limit, much as it was for the oil industry, is more interesting, again part of a long line of extractivism driving the framing of the law in this area.[80]

That brings us back to the cases, and again the question of what the law is, but now this is not a metaphysical question but a materialist one. In 1967, Maltese Ambassador Arvid Pardo gave a famous speech at the United Nations to argue for the deep seabed and its resources to be protected as the common heritage of mankind.[81] As made clear by R. P. Anand in his account of these years, the ICJ is responding to these developments, albeit indirectly, in its assertions of sovereignty for these European states. This other major tension, between First and Third World, would go on to be central to the UNCLOS negotiations. How the law would continue to be derived from and shaped by the materiality of the seabed, the demands of the dominant mode of production, and the ideological effect of the law, is beyond the scope of this chapter. The judgment of the Court is not consciously commodifying the seabed and producing something abstract and fungible, it is responding to the demands of the material conditions. The final abstractness of the continental shelf is achieved as it is disassociated from the water, conceptualized as land which does not need occupying, and after UNCLOS is disconnected from any material definition when it is granted to every coastal state up to 200 nautical miles.

Conclusion: Freeing the Sea

> For the chains of the sea
> Will have busted in the night
> And will be buried at the bottom of the ocean.
> —Bob Dylan, "When the Ship Comes In"[82]

The history I tell here has revealed that as the use of the North Sea changed, so too did the way it was conceptualized. This happened incredibly quickly, with the reorganization of the North Sea based on the seabed rather than surface and water column activities preceding the first commercial exploitation of the seabed resources. As demonstrated here, in different times the sea was understood differently. In the pre-modern, it was primarily a way of travelling. In the early modern, a fishing resource, and then very quickly in late modernity, the seabed became all important. The hydrocarbons contained under the continental shelf became the resource which dominated the understanding of the sea.

What should be clear from this chapter is that thinking about law with the sea makes international law central to legal geography. Legal geography's central contribution has been to demonstrate that law and space are co-constitutive, and that legal justice and spatial justice rely on each other. On land, this can often mean a focus on property law and local legal constructions, but to understand the legal constitution of the oceans is to understand the legal co-constitution of the world.

North Sea Continental Shelf is particularly suitable for this sort of analysis, treating the ocean as flat, empty, and easily divided into different, separate, zones. It doesn't take much ontological questioning to see that to separate the seabed from the water above it, to prioritize its connection to the land beside it, is a peculiar way to understand the sea. But putting this in the context of critical oceans thinking, we can see this as using law to respond to a specific oceanic anxiety, the very fluidity, the smooth and de-territorializing effect of the ocean. Law's very abstractness and abstracting force makes it the perfect, necessary tool for rendering the oceans comprehensible for exploitation, not just as flat surface for movement, but also the seabed for mining, life forms for biotech research, et cetera.

North Sea oil and gas reserves went on to be exploited over the next five decades, peaking around 1999. North Sea oil is now nearing depletion. The oil infrastructure is being decommissioned. Some argue for decommissioned oil rigs to be left in place, as nature reserves,[83] sea-steading bases,[84] or diving hotels.[85] The Oslo-Paris Commission has instead demanded that all non-natural infrastructure should be removed.[86] The North Sea today faces being re-constructed a commodified anew, as the commercial viability of oil and gas under the sea falls, and the promise of sea wind power rises. The innovation of the North Sea Windpower Grid, for example, would see a whole new understanding of the sea and energy which connects seabed anchoring and cables with the air currents above.[87]

New uses of space, as demonstrated here, demand new legal constructions. The law is central to how the space is constituted, and by understanding how a space has been made we can try to understand how it can be remade. The oceans are a generative space for law as commodity producer and commodity guarantor. The ocean as commodity and the law as commodifier are co-constitutive, and as I have sought to demonstrate elsewhere, original and generative.[88] Thus, the challenge of Peters and Steinberg, to do *more* than just take account of the oceans' fluidity, depth, volume, et cetera, but to understand the more-than-wet, the ice, the mist, the winds, currents, atmosphere, dissolving, and precipitation, will be key in re-imagining not just the continental shelf, not just ocean space in general, but the international legal constitution of the world as a whole.

Notes

1 *North Sea Continental Shelf* Judgment, *International Court of Justice Reports* (1969), 3.
2 The critical legal historian Robert Gordon called this "evolutionary functionalism" in Robert W. Gordon, "Critical Legal Histories," *Stanford Law Review* 36 (1984): 57–125.
3 Henri Lefebvre, *The Production of Space*, trans. Donal Nicholson-Smith (Oxford: Blackwell, 1992), 27.
4 Andreas Philippopoulos-Mihalopoulos, "And for Law: Why Space Cannot Be Understood Without Law," *Law, Culture and the Humanities* 17, no. 3 (2018): 620–639.

5 James Crawford and Thomas Viles, "International Law on a Given Day," in *International Law as an Open System: Selected Essays*, ed. James Crawford (London: Cameron May, 2002), 69–94.

6 For the text of the Truman Proclamations see *American Journal of International Law* 40, supp. 45 & 46 (1946).

7 This is China Miéville's critique of Martti Koskenniemi's indeterminacy thesis in China Miéville, *Between Equal Rights* (Leiden: Brill, 2005), 54.

8 Susan Marks, "False Contingency," *Current Legal Problems* 62, no. 1 (2009): 1–21.

9 I have attempted something similar, in collaboration with others, with the *SS Lotus* case, C Chinkin et al., "Bozkurt Case, aka the Lotus Case (France v Turkey): Ships That Go Bump in the Night," in *Feminist Judgments in International Law*, eds. Loveday Hodson and Troy Lavers (Oxford: Hart Publishing, 2019).

10 The term "critical ocean studies" is most associated with the work of Elizabeth DeLoughrey, see Elizabeth DeLoughrey, "Submarine Futures of the Anthropocene," *Comparative Literature* 69, no. 1 (2017): 33 and Elizabeth DeLoughrey, "Toward a Critical Ocean Studies for the Anthropocene," *English Language Notes* 57, no. 1 (2019): 21–36.

11 All statistics taken from "Energy and Carbon Tracker 2020," International Environment Agency, accessed January 9, 2021, https://www.iea.org/data-and-statistics/data-product/iea-energy-and-carbon-tracker-2020.

12 On the question of modes of production and production of space more generally, see Henri Lefebvre, *The Production of Space*, trans. Donald Nicholson-Smith (Oxford: Blackwell, 1991).

13 Karl Marx, *Capital Volume 1*, trans. B. Fowkes (Middlesex: Penguin Books, 1990), 125.

14 Ibid., 163.

15 Ibid., 178.

16 Robert Knox, "Marxist Approaches to International Law," in *The Oxford Handbook of the Theory of International Law*, eds. Anne Orford and Florian Hoffman (Oxford: Oxford University Press, 2016), 316.

17 Ibid., 319.

18 Miéville, *Between Equal Rights*.

19 United Nations Convention on the Law of the Sea, December 10, 1982, Montenegro Bay Jamaica, UN Treaty Series 31363.

20 Knox, "Marxist Approaches," 326.

21 The concept is developed through a trilogy of important articles: Philip Steinberg and Kimberley Peters, "Volume and Vision: Toward a Wet Ontology," *Harvard Design Magazine* 39 (2014): 124; Philip Steinberg and Kimberley Peters, "Wet Ontologies, Fluid Spaces: Giving Depth to Volume Through Oceanic Thinking," *Environment and Planning D: Society and Space* 33, no. 2 (2015): 247–264; Philip Steinberg and Kimberley Peters, "The Ocean in Excess: Towards a More-Than-Wet Ontology," *Dialogues in Human Geography* 9, no. 3 (2019): 293–307.

22 Steinberg and Peters, "Wet Ontologies," 261.

23 Elizabeth DeLoughrey, "Submarine Futures of the Anthropocene," *Comparative Literature* 69, no. 1 (2017): 32–44, 33.

24 Elizabeth DeLoughrey, "Toward a Critical Ocean Studies for the Anthropocene," *English Language Notes* 57, no. 1 (2019): 21–36.

25 Elizabeth DeLoughrey and Tatiana Flores, "Submerged Bodies: The Tidalectics of Representability and the Sea in Caribbean Art," *Environmental Humanities* 12, no. 1 (2020): 132–166.

26 James Bamberg, *British Petroleum and Global Oil, 1950–1975* (Cambridge: Cambridge University Press, 2000), 201–203.

27 Alex G. Oude Elferink, *The Delimitation of the Continental Shelf between Denmark, Germany and the Netherlands* (Cambridge: Cambridge University Press, 2013), 95.

28 Ibid.

29 Ibid., 20-31.
30 Ibid., 18.
31 Ibid., 70.
32 Ibid., 99.
33 Ibid., 104–107.
34 Ibid., 36.
35 Ibid., 153–155.
36 *North Sea Continental Shelf* (F.R.G. v. Den.; F.R.G. v. Neth.), 1969 I.C.J. 3 (February 20, 1969) (Tanaka, J., dissenting).
37 Ibid., 18.
38 Ibid., 19 (emphasis added).
39 Reading articles in leading international law journals from the time before the case demonstrates just how debated the concept was. See, for example, Shigeru Oda, "A Reconsideration of the Continental Shelf Doctrine," *Tulane Law Review* 32 (1957): 21–36; J.A.C. Gutteridge, "The 1958 Geneva Convention on the Continental Shelf," *British Yearbook of International Law* 35 (1959): 102–123.
40 *North Sea Continental Shelf* at [91].
41 Ibid. [96]
42 Colin Jones & Duncan Maclennan, "The impact of North Sea oil development on the Aberdeen housing market", *Land Development Studies* 3, no. 2 (1986): 113–126.
43 However, it should be noted that it was a concerted effort from Third World states acting collectively which ultimately secured distance as the criteria. They saw this as giving them equal standing regardless of geographic specifics. See R.P. Anand, *Origin and Development of the Law of the Sea* (Hingham: Nartinus Nijhoff, 1982).
44 Simon Fitch et al., "The Archeaology of the North Sea Palaeolandscapes," in *Mapping Doggerland: The Mesolithic Landscapes of the Southern North Sea*, eds. Simon Finch et al. (Oxford: Archeopress, 2007): 105–118.
45 Pliny the Elder, *Natural History Book 2* (Cambridge MA: Loeb Classical Library, 1938) 303.
46 Oskar Bandle et al., *The Nordic Languages: An International Handbook of the History of the Northern Germanic Languages* (Leiden: Walter de Gruyter, 2002), 596.
47 Ibid.
48 This naming continued in to the 19th century, for example in the 6th edition of the *Encyclopedia Brittanica* (Edinburgh: Archibald Constable, 1823).
49 James Graham-Campbell, *The Viking World* (London: F Lincoln, 2001), 10.
50 Arthur Boyd Hibbert, "Hanseatic League" in *Encyclopedia Britannica*, 21 Oct. 2019, https://www.britannica.com/topic/Hanseatic-League, accessed April 5, 2022.
51 Ibid.
52 P. Holm, "South Scandinavian Fisheries in the Sixteenth Century" in Juliette Roding et al., *The North Sea and Culture (1550–1800)* (Leiden: Uitgeverij Verloren, 1996), 110–112.
53 Ibid.
54 Alison Rieser, "*Clupea Liberum*: Hugo Grotius, Free Seas, and the Political Biology of Herring," in *Blue Legalities*, eds. Irus Braverman and Elizabeth R. Johnson (Durham: Duke University Press, 2020), 206–207.
55 Ibid.
56 "October 1651: An Act for Increase of Shipping, and Encouragement of the Navigation of this Nation", in *Acts and Ordinances of the Interregnum, 1642–1660*, ed. C.H. Firth and R.S. Rait (London, 1911), 559–562.
57 James R. Jones, *The Anglo-Dutch Wars of the Seventeenth Century* (London: Longman, 1996).
58 Rieser, "*Clupea Liberum*."
59 "History of Offshore Oil and Gas in the United States," National Commission on the BP Deepwater Horizon Oil Spill and Offshore Drilling, https://www.iadc.org

/archived-2014-osc-report/history/history-of-offshore-oil.html, accessed August 12, 2021.

60 Ibid.

61 Google Books n-gram search for "continental shelf", available at https://books .google.com/ngrams/graph?content=continental+shelf&year_start=1800&year _end=2019&corpus=26&smoothing=0, accessed September 1, 2021.

62 H.R. Mill, "Sea-Temperatures on the Continental Shelf," *Scottish Geographical Magazine* 4 (1888): 544–549. Mill cites a paper from the previous year in support of his definition, but this paper, on laying submarine cables, does not use the term continental shelf.

63 Ibid., 544.

64 Cornwall Submarine Mines Act 1858 c.109.

65 *R. v. Keyn* (1876) *UK Law Reports Exchequer Division*, 2, at 155–157.

66 Ibid., 158.

67 Ibid., 176.

68 Ibid., 199.

69 Ibid., 201.

70 A.W. Brian Simpson, *Leading Cases in the Common Law* (Oxford: Oxford University Press, 1996), Chapter 9, 227–258.

71 Cecil J.B. Hurst, "Whose Is the Bed of the Sea?" *British Yearbook of International Law* 4 (1924): 34–43.

72 Ibid., 36.

73 Ibid., 37.

74 Ibid., 40.

75 For more on territory and terrain in the seas, see Phil Steinberg et al. in this volume.

76 For how this need leads US maritime policy see DeLoughrey, "Critical Ocean Studies."

77 Cecil J.B. Hurst, "The Continental Shelf," *Transactions of the Grotius Society* 34 (1948): 153–169; C.H.M. Waldock, "The Legal Basis of Claims to the Continental Shelf," *Transactions of the Grotius Society* 36 (1950): 115; Sir Hersch Lauterpacht, "Sovereignty over Submarine Areas," *British Yearbook of International Law* 27 (1950): 376; J.L. Kunz, "Continental Shelf and International Law: Confusion and Abuse," *American Journal of International Law* 50 (1956): 828; Oda, "Reconsideration of the Continental Shelf Doctrine."

78 International Law Commission Yearbook, Volume II (1951), 384–385.

79 United Nations Convention on the Continental Shelf, Geneva, April 29, 1958, *United Nations Treaty Series*, vol. 499, p. 311.

80 On the extractivism of UNCLOS in general, and the deep seabed regime in particular, see Surabhi Ranganathan, "Ocean Floor Grab: International Law and the Making of an Extractive Imaginary," *European Journal of International Law* 30, no. 2 (2019): 573–600.

81 For more on Pardo's speech see Ranganathan, "Ocean Floor Grab."

82 Bob Dylan, "When the Ship Comes In," in *The Lyrics* (London: Simon & Schuster, 2012).

83 Known generally as rigs-to-reefs programs, see, for example, "Rigs to Reef Programs Create Valuable Fish Habitat," American Petroleum Institute, 2021, https://www.api .org/oil-and-natural-gas/environment/environmental-performance/environmental -stewardship/rigs-to-reef-programs.

84 "Reimagining Civilization with Floating Cities," Seasteading Institute, 2021, www.seasteading.org. On seasteading more generally, see Surabhi Ranganathan, "Seasteads, Landgrabs, and International Law," *Leiden Journal of International Law* 32, no. 2 (2019): 205–214.

85 See the Seaventures Dive Rig Hotel, off Borneo, Malaysia (https://seaventuresdive .com/).

86 OSPAR Convention for the Protection of the Marine Environment of the North-East Atlantic, Ministerial Meeting of the OSPAR Commission, July 22–23, 1998 Decision 98/3.

87 "Political Declaration on Energy Cooperation Between North Sea Countries," European Commission, 2016, https://ec.europa.eu/energy/sites/default/files/documents/Political%20Declaration%20on%20Energy%20Cooperation%20between%20the%20North%20Seas%20Countries%20FINAL.pdf.

88 Henry Jones, "Lines in the Ocean: Thinking with the Sea About Territory and International Law," *London Review of International Law* 4, no. 2 (2016): 307–343.

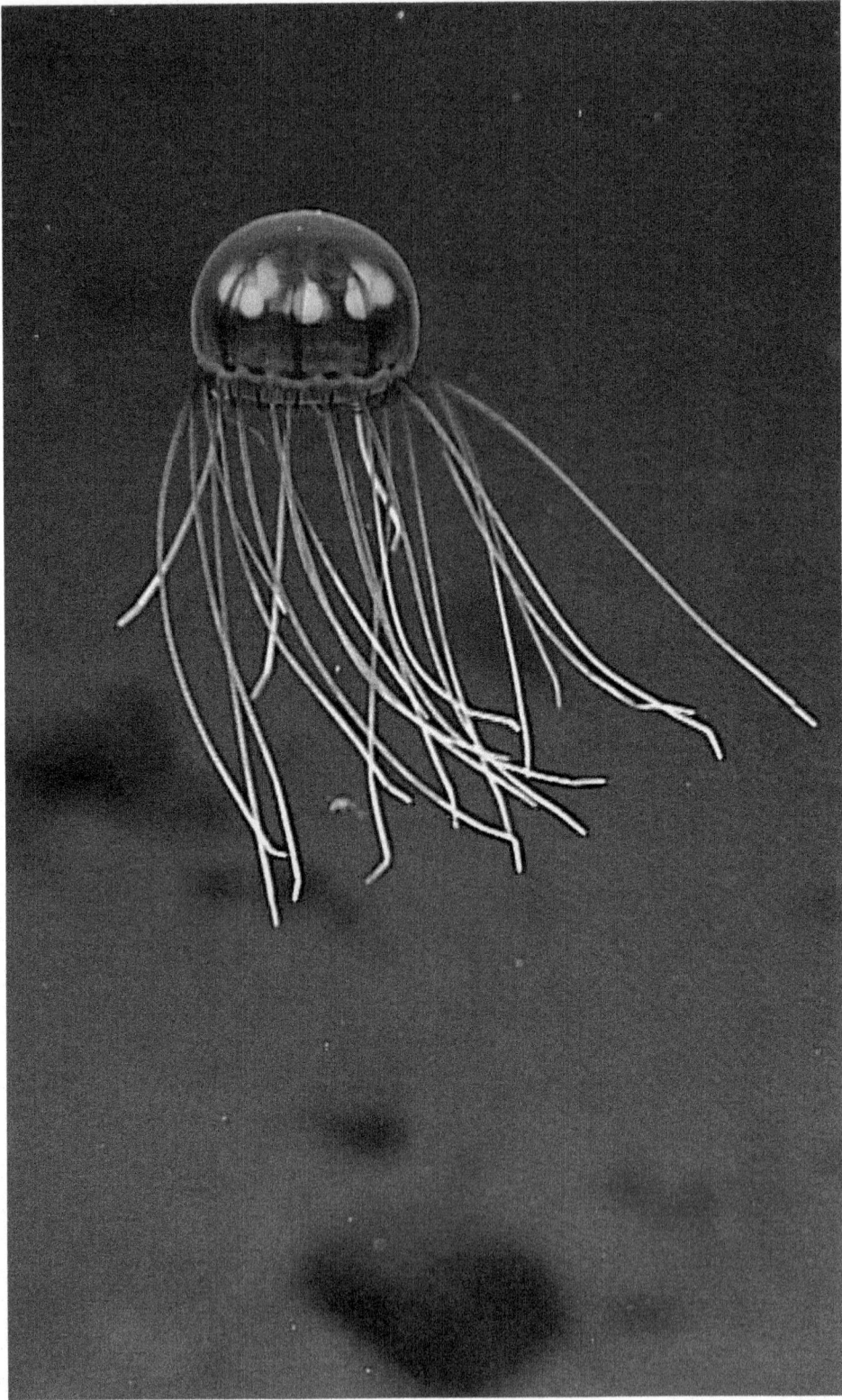

3

IMAGINING JUSTICE WITH THE ABYSSAL OCEAN

Susan Reid

The Abyssal Ocean

The abyssal ocean is entirely without light, except for the luminous signals emitted by the deep dwelling beings gifted with chemiluminescence.[1] Within abundant darkness lie cold, dark plains punctuated by seamounts, knolls, island arcs, and incised by trenches and valleys. Abyssal sonics bend and channel in correspondence with these topographies. Two known energy communities live here: the chemosynthetic ones that feed at the hydrothermal vents; and the detrivores, who are entirely dependent on pulses of nutrition from above. Food packages arrive in the form of carcasses, which take several days to sink through the water column before landing in a stir of sediment. Nutrition also falls in seasonal rains of detritus that comprise spent phytoplankton and zooplankton from the surface.[2] Falls of organic matter mix with other particulates to form sedimentary compositions of diatom and radiolarian oozes, sponge spicules, and clay that vary basin to basin. Beneath dense atmospheres of water, several kilometers deep, lives and lifeways move and transition epochally. Currents creep the abyssal plains, erosion is low, and sedimentation is slow. Some of the deep-sea fauna here have direct provenance to the Miocene era, 13 million years ago.[3] Little is understood of this watery archive of material relations and their near and far temporalities and indeterminant futures. Neither is it easy for humans to feel implicatedness in the conditions of the remote abyssal ocean.

By contrast, this deep realm experiences humans at intimate and planetary-wide material scales.[4] Slow old currents trace submarine cables, wrecks, and

FIGURE 3.1 A gorgeous jellyfish (*hydromedusa*) seen in the Marianas Trench Marine National Monument at a depth of 3,700 meters (12,000 feet). Image courtesy of NOAA Office of Ocean Exploration and Research, 2016 Deepwater Exploration of the Marianas.

DOI: 10.4324/9781003205173-4

abandoned oil rigs, projecting their paths into waters above. Fossil fuel-related heat, carbon, and plastic wastes mix deep into the ocean's heart; plastics knot cetacean bellies and settle in sediments. Human connectivity is there too in the fewer large carcasses that reach the seafloor because of over-fishing; or in the rust-encrusted drums of wartime and industrial toxic wastes. This is not to say that abyssal depths are immune to nonhuman ruptures. Amongst other forces, winter cooling and increases in salinity can trigger cascades of shelf water through the canyons and slopes to the abyssal depths.[5] Unlike such events though, anthropogenic impacts ripple multidimensionally and multi-temporally through the ocean's interconnected ecosystems.[6]

One particularly significant impact will result from commercial deep seabed mining activities. The industry intends to mine the manganese nodule fields in a nine million km² region of the Eastern Pacific seafloor, known as the Clarion Clipperton Zone (CCZ). Some of earth's most diverse ecosystems are associated with nodule fields,[7] such as those found in the CCZ. Deep-living beings of the CCZ are particularly vulnerable to disturbances due to their slow growth rates, maturation at a relatively old age, long life expectancies, and low or unpredictable recruitment.[8] The effects of 24/7 mining activities in this region would likely be severe and last well beyond human time scales. Civil society groups and conservation organizations are understandably concerned about the potential impacts of mining on these seabed ecologies. Historic and ongoing violences,

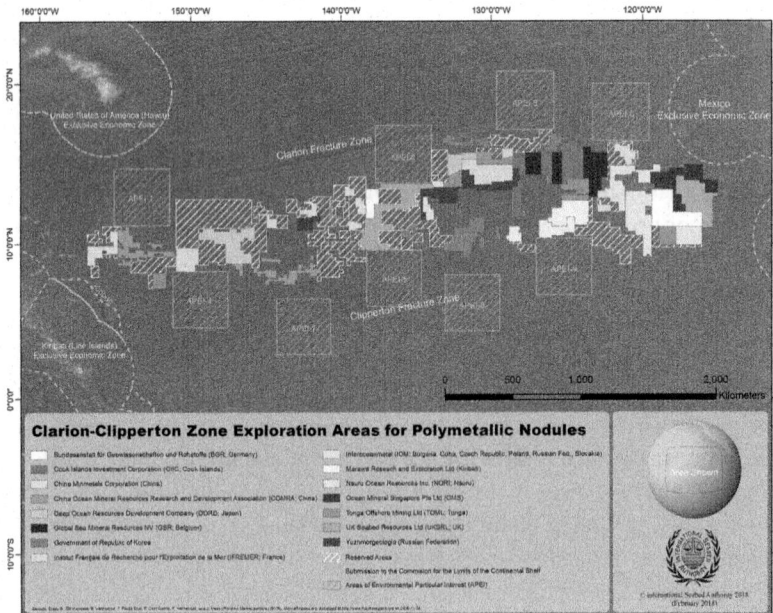

FIGURE 3.2 A map of the Clarion Clipperton Zone in the central Pacific Ocean. The colored areas are those licensed for mining and shaded squares are areas currently protected from mining. Image adapted from the International Seabed Authority, 2018.

inflicted by mining corporations on terrestrial environments, indicate that this concern is warranted.[9]

Deep seabed mining would be another assault on the ocean. Though seas are transitional by nature, the scale and cumulative effects of industrialization, such as heat and plastic pollution, manifest oceanic changes of an entirely different order and consequence for planetary habitability. Across the seas and into their depths, for example, anthropogenic climate change already affects circulation, hydrodynamics, temperature levels, and acidity. Humans are also vulnerable to the changes that industrialization forces on the abyssal ocean—what happens in this zone influences whether rains come, plants thrive, temperatures are liveable; or whether the ocean provides sufficient food or enough oxygen for humans and other terrestrial dwellers to breathe. Given human interdependence with the abyssal ocean, our relations with these worlds matter.

This chapter emphasizes relationships: the co-constitution of beings, lifeways, and materialities of the deep ocean as well as human interconnectivity with them such as the specter of deep seabed mining underscores. As well, it highlights how, by controlling the dominant human relation with the ocean, the United Nations Convention on the Law of the Sea (UNCLOS)[10] intervenes as an eco-logical force. It begins with an analysis of the seabed mining regime as the poster industry for blue capitalism; and how the regime's enabling structures privilege resource corporations. UNCLOS's representational strategies are then examined to understand how diverse seabed worlds and constituent materialities are appropriated into regimes of global capital. The ecological violences of mining are both cloaked within discursive legal strategies and justified on the grounds of economic accumulation. The effect of these strategies is to diminish recognition of both the ocean's agency and vulnerabilities, and prioritize commercial interests—key factors contributing to the deepening ocean crisis.

Legal and economic discourses of global extractivism obscure the nature of extraction as a form of "predation." Neither are the ontological dimensions of predation sufficiently factored within considerations of individual material consumption. Configuring extraction as a form of predation brings to the fore the agency and lifeworlds of that which is extracted and enables a fuller, albeit contingent, reckoning of what (if anything at all) constitutes a just need for their destruction and consumption. It is by dint of being materially embodied that humans are vulnerable to needing materials from the world, sourced through acts such as mining. I describe this as "material predation"—that is, killing animals and lifeways to obtain the minerals and other materials that feed the embodied prosthetics, such as communication equipment and technologies, through which we extend into and participate in the world. Acknowledging material predation as a dimension of ecological subjectivity is a key tenet within my concept of ocean justice, in which I reflexively consider human material vulnerabilities alongside relational approaches with more-than-human ocean worlds.

Material vulnerability and predation implicate us all, in some way, with extractive industries. If, as material predators, humans do need to secure more minerals and other materials needed to live well, how do we do this ethically? How, for example, do we extend hospitality to the worlds of our prey, as philosopher Jacques Derrida, whose politics of eating thesis is referenced later in this chapter, invokes us to do? I propose a figuration of ethical predation of ocean realms by thinking with, and extending Derrida's concept of eating well—or, more precisely, being a better predator.

Blue Extractivism: Deep Seabed Mining Frictions

Having depleted most of the easy to extract terrestrial supplies of minerals, mining corporations are seeking more profitable sources, and turning their attention to the high grade of minerals at the deep seabed.[11] Though the mining industry has its sights on these "blue" riches, its frontier ambitions are meeting with resistance and concern from multiple directions.[12] Proponents claim that seabed mining would benefit countries of the economic south, framing their extractive development ambitions as within terms such as "blue economy" or "blue capital"[13]—as if "blue" makes seabed mining somehow benign.[14] Taking up blue capitalism's agenda, the International Seabed Authority (ISA), an organization responsible for overseeing seabed development,[15] asserts that seabed mining can expand the resource base for Pacific Island nations and enable growth of their "sustainable Blue Economy."[16] Responding to the ISA's assertions, Pacific Island Association of Non-government Organisations, Deputy Executive Director, Emeline Ilolahia argues that, "If mining was the panacea to the economic issues of the Pacific, we'd have solved all our problems long ago. Instead the environmental and social impacts of mining have made our peoples poorer."[17]

Mining proponents also argue that exploitation of the deep seabed is necessary to source minerals for humanity's renewable energy future.[18] However, a 2016 report from the Institute for Sustainable Futures challenges the view that current renewable energy markets rely on seabed minerals.[19] It finds that, even with the most ambitious energy scenarios and growth in demand, a transition to 100 percent renewables is possible without recourse to deep-sea mining.[20] Along these lines, and exploiting the growing public resistance against seabed mining, large corporations such as Google, BMW, and Volvo recently pledged not to use seabed minerals, and publicly distanced themselves from the industry.[21]

And yet the seabed mining regime appears to be pressing forward despite the concerns of marine scientists about how little is known of deep-ocean ecologies.[22] Avoiding losses from 24/7 deep-sea mining operating over 30 years is unlikely.[23] The ISA's Secretary General, Michael Lodge, acknowledges seabed mining will involve "the crushing of living organisms, the removal of substrate habitat and the creation of sediment plumes," as well as the risk of "environmental damage through malfunctions in the riser and transportation system, hydraulic leaks, and noise and light pollution."[24] Neither is remediation for seabed mining feasible given the likely material, temporal, and spatial scales of the losses.[25] Against this background, claims that seabed mining will save the planet are unconvincing. In a recent interview, University of Hawaii oceanographer Jeff Drazen said: "We're about to make one of the biggest transformations that humans have ever made to the surface of the planet. We're going to strip-mine a massive habitat, and once it's gone, it isn't coming back."[26] Asked about the potential environmental impacts of deep seabed mining, ISA's Michael Lodge effectively gaslighted the widespread concern of marine biologists, conservationists, and concerned publics by responding: "I don't believe people should worry that much."[27]

The Seabed Mining Regime and its Architecture of Exploitation

The seabed mining regime is a mode of extractivism in which economic accumulation is pursued through the maximum yield of profitable materials from the ocean's seafloor.[28] It is facilitated through the legislative framework of UNCLOS.[29] In this section, I add to the growing body of literature that critiques UNCLOS's extractive agenda,[30] with a perspective that emphasizes how UNCLOS, and broader ocean governance systems, form an architecture of exploitation that privileges corporations and economic accumulation to the detriment of the ocean. UNCLOS normalizes extractivist exploitation as the default human relation with the ocean and assumes to bind us all on these same terms. In the context of the deep seabed mining regime, UNCLOS instrumentalizes the seabed for the benefit of the global economic order. It ensures that marine protection and conservation provisions convey an impression of care for the deep ocean but in reality, exist to shore-up raw material supplies for present and future human users. This anthropocentric, supply-depot approach forecloses relations of ethical responsibility toward the oceans themselves, as well as the living and nonliving entities that depend on them.

Mining the seafloor entails violences that extend beyond "crushing living organisms"[31] to the extinguishment of networks of embodied relations and lifeways. These violences are accentuated by UNCLOS's denial of human and more-than-human interconnectivities with deep-ocean worlds. Failure to recognize these relational factors can be attributed, in part, to the Enlightenment imaginary of mastery that operates within the undertow of law itself—and the particular Western concept of human that it envisages.[32] Insights from feminist legal scholars reveal that an imaginary of mastery imbues law's ideal person with qualities that include disembodiment and autonomy, and a metaphysical detachment from nonhuman natures. As Naffine and Grear assert, these are qualities that only corporations can really achieve[33]—they are unattainable for materially embodied, ecologically interdependent beings.

The concept of human, around which law pivots, is a corporation—the paradigmatic legal person.[34] From this perspective it seems inevitable that corporate privileging flows from UNCLOS's foundational imaginary of mastery into explicit provisions and regulations that favor economic accumulation and, further, enable "regulatory capture"[35] of institutions and implementation practices, by powerful corporations. Through this architecture of exploitation, the seabed mining regime advances without adequate accounting for, or ethically responding to, the relational implications of its violences.

UNCLOS mandates oversight for the seabed mining regime in the international seabed jurisdiction (the Area) to the ISA.[36] It is a jurisdiction encompassing over 40 percent of the earth's surface. The ISA's explicit mandate to advance seabed mining[37] legitimates the industry's expansion and subjugates obligations to protect and conserve natural resources in the Area.[38] The privileging of economic interests continues through the regulatory capture of the ISA, by corporations. A recent investigation by Casson et al. documents cases where contractors finance and sit on the Authority's 30-member Legal and Technical Commission (LTC).[39] Additionally, representatives of the contractors are employed by the ISA.[40] Further, key decisions and discussions are conducted behind closed doors between only selected contractors and experts.[41] In these ways, corporations have the opportunity to influence commercial outcomes, favorable legal and policy formulations, and key decisions, and to minimize costs related to environmental obligations.[42] The absence of a strict regime of independent audit of mining exploration activities, or plans to implement this for exploitation activities,[43] adds further opacity to ISA operations and to the mining practices of contractors.

This architecture of exploitation potentially undermines the Authority's ability and willingness to adequately prevent harm to the marine environment, or to ensure that the harms resulting from mining activities are attended to thoroughly and ethically. It continues with the regime's economic focus that monetizes and constrains environmental protection measures. For example, the Draft Regulations on Exploitation of Mineral Resources in the Area (herein, the Draft Seabed Regulations) provide for the establishment of an Environmental Compensation Fund (ECF) that will finance measures to "prevent, limit or remediate damage" to the seabed, "the costs of which cannot be recovered from a Contractor or sponsoring State."[44] To consider a financial remedy an adequate response to the extinguishment of seabed lives is consistent with an extractivist imaginary that perceives nonhuman natures as fungible. However, even before exploitation activities commence the regulations appear to anticipate that corporations will not be required to fully finance measures to prevent or remediate environmental harms associated with their commercial mining activities. In the case of remediation, the ECF's obligations will also only finance such activity where it is "economically feasible" and supported by "Best Available Scientific Evidence."[45] This scientific evidence too is capped by "economic constraints."[46] While conveying the look of concern for the long-range damage associated with seabed mining, the seabed mining regime only tolerates its environmental responsibilities provided they do not interfere with profits.

Although states have primary responsibility for the implementation of UNCLOS,[47] corporations are the key actors of the regime at the seabed. The heavily reinforced machinery of extractive corporations will operate in remote, sunless lease areas kilometers below the surface, removing the seafloor and their multiplicities of unknowable life-forms, relations, and materialities over the terms of potentially multi-decadal licenses. This has particular implications for deep-ocean worlds, for example: sediment plumes arising from mining activity and returned waste water will deny detrivore communities of critical nutritional falls and deplete oxygen resources from already oxygen-poor zones; and the sonic pollution of mining machinery will impact deep-ocean dwellers for whom sound is vital to communication and orientation in the absence of light. As well, abyssal beings are dependent on depth for habitable conditions of pressure and temperature, which narrows their options for escaping the miners. The remoteness of depth also advantages corporations by veiling how and what they do at the seafloor under cubic kilometers of seawater. Corporations will conduct their commercial business extracting seabed materials and worlds relatively free of scrutiny by independent auditors or general publics, in closed ISA meetings, and quite literally out of sight in the hinterseas.

Despite structurally and operationally privileging corporations, UNCLOS claims to represent the interests of "all peoples of the world."[48] Such assertions fail to acknowledge UNCLOS's exclusion of diverse other human relations of stewardship and kinship with the ocean, including those practiced by Indigenous communities of the Pacific for millennia.[49] Mandated as the institutional manager to oversee UNCLOS's international seabed development goals, the ISA also claims to act on behalf of all of us.[50] However, given the ISA's current regulatory, law-making, and institutional privileging of economic and corporate interests, this is structurally impossible. UNCLOS further deems that the common heritage of mankind (CHM) principle applies to the international seabed jurisdiction, the Area,[51] and that the principle applies to us all.[52] Specifically, it declares that exploiting any solid, liquid, or gaseous mineral resources in this zone must be carried out for the "benefit of mankind as a whole."[53] If it is accepted that humans are materially embodied, ecologically interconnected, and reliant on the ocean for wellbeing, how could the environmental violences that result from multi-decadal commercial seabed mining activity benefit humanity "as a whole"?

The Area: More-than-Seabed, More-than-Metal

Though described as the "constitution *for* the ocean,"[54] UNCLOS utterly misrepresents the dynamic, embodied lifeways and relations of the seas and their seabeds.[55] Guided by imaginaries of mastery and with a discursive sleight of hand, UNCLOS represents the biologically and geologically diverse seabed realm that lies beyond national jurisdiction as "The Area."[56] By so doing, it discursively empties and generalizes 57 percent of the total area of the earth's ocean,[57] severs the connection of the surface to the water column, and through its abstractions, UNCLOS renders this deep-living realm as a quarry. This is the familiar biopolitical force that operates within environmental law in particular, whereby human and more-than-human natures are bifurcated and the nonhuman ones are represented in ways that suit the interests of global economic systems.[58]

Similar strategies are deployed by corporations, such as DeepGreen, with their pro-mining videos depicting the ocean as an empty, featureless space into which industrial machinery seamlessly conducts its extractive operations without any discernible ecological impacts.[59]

These fictions deny the agency of the more-than-human lives in which we are all connected, obscure mining's immediate and processual violences to deep-sea ecologies and relations, and distract from the ecological force of law itself.[60] That is, the power of the law, through the exploitative relations that it prescribes and legitimates, to intervene in human and more-than-human worlds in ways that destroy, or render more vulnerable, the material and social relations and lifeways that create the possibility of liveable worlds. Through its redactions, UNCLOS has already written-off the deep seabed ecological relations as collateral to economic growth. Once the Draft Seabed Regulations are finalized, exploitation licenses may be approved for mining lease areas of up to 75,000 kilometers^2 in size.[61] But these leases refer to UNCLOS's quarry realms—abstracted zones of non-agentic, biological matter—not the actual seabed that will bear the brunt of multi-decadal mining activity.

Under the seabed mining regime of the Area, realms that are continually mixing and mingling are denied their interconnectivity due to the jurisdictional partition between the seabed and the water column above.[62] Provisions further identify "mineral resources" as those that are "*in situ* ... at or beneath the seabed."[63] This broad/vague definition serves to assimilate mineral resources, such as nodules, into extractive regimes of value.[64] In reality, the mineralized nodules resting on the seabed are formed through mineral kinships and accretion of materials circulating in the water column and cycled through sediments. Partitioning happens as well under the Draft Seabed Regulations, where polymetallic nodules are defined as "any deposit or accretion of nodules, on or below the surface of the deep seabed, which contain metals."[65] Nodules are identified exclusively by the value of their constituent minerals—the multiplicity of other materials and living communities with which they are co-constituted having been bracketed out.

While very little is known or understood of the deep ocean, the scientific and cultural material that is available is selectively harvested by UNCLOS. Applying a transdisciplinary approach that I conceive as "seatruthing," I read and imagine this material back into the text of UNCLOS to reveal the injustice of its legal fictions and exclusions. Seatruthing makes no claims for singular, rarefied notions of truth but rather is concerned with noticing and interrogating what particular speech acts, words, and representations of the ocean do when they are brought into relation with actual ocean milieu.[66] Such encounters can create potentially generative "conceptual displacements" that highlight the unseeing of different beings and material relations that can arise from different biases or opportunities for perspectival changes.[67] For example, the little that we know of deep-living beings, such as octopuses, is enough to trouble the jurisdictional boundaries between the regimes of the Area and the high seas or the very narrow concept of mineral resources. How octopuses live in the ocean challenges the legal view that beings "belong to the maritime area in which they live."[68] Octopuses have been recorded, at depths of four kilometers adhering their eggs to the stem of micro-sponges that, in turn, are fixed to the hard substrate of manganese nodules.[69] The octopuses stay nearby guarding the eggs and foraging in surrounding sediments. Given their intimate connection with the seabed, it is evident that octopuses do not belong entirely to the legal regime of the high seas where they also spend significant periods of their life. Additionally, far from being just a manganese resource, the nodules form a vital role within the ecosystem.

The denial of deep-ocean materialities and social relations continues within the CHM regime where representations of nodule assemblages as "mineral resource"[70] or a "deposit or accretion" containing metal[71] exclude multiplicities of seabed kinships. These descriptions omit any sense of nodule fields' diverse ecosystems; or that the communities of these realms are in lively relations with one another and with the materials and nodules that they co-become with. Octopuses are a part of the seabed and nodule fields by virtue of their eggs fixed to seabed substrate; and whereby the seabed and nodules provide a vital crèche, feeding place, and end of life location. Neither is the seafloor sediment, which is slated for mining, a heterogeneous or inert substance. They have agency and are alive with material relations, providing nurturing ooze for the eggs and larvae of deep-living beings and affording conditions of livability for free-swimming adults.

Seatruthing insists that the seafloor, sediments, and nodules be acknowledged as "more-than" worlds that exceed their appropriation into UNCLOS's extractivist regime of value. The Area is more than a seabed jurisdiction for the extraction of minerals. In this sense, "mineral resources" too can be

FIGURE 3.3 Ghost octopus, among the 90 percent of unknown/yet to be described marine beings, is found at 4,290 meters depth. Image courtesy of the NOAA Office of Ocean Exploration and Research, Hohonu Moana, 2016.

FIGURE 3.4 A ghost octopus, possibly spooked by the approach of a massive hard-shelled, multi-tentacular alien (aka the ROV Deep Discoverer). Image courtesy of the NOAA Office of Ocean Exploration and Research, Hohonu Moana, 2016.

recognized as more-than-mineral and in diverse kinship relations across the sea-bed. Xenophyophores, brittle stars, crustaceans, mollusks, and the "hedgehog-skinned" echinoderms, live in these worlds, in relation to one another and the materialities of the abyssal depths. Starfish and sea cumbers feed on different ectomorphs of Xenophyophores that are abundant on the abyssal plains—most as epifauna living atop the seabed but some known to be infaunal, buried deep into the sediment.[72]

Xenophyophores come into lumpy and decorative being by drawing on their mineralized surroundings to create delicate exoskeletons.[73] Their reticulated, heavily folded or fan-shaped forms can also function as nurseries for snailfish.[74] Nodule assemblages also either host or enable marine beings with which they share mineral kinships. For example, the Xenophyophores come into being through the same mineralized solution of deep-ocean waters relied on by man-ganese nodules. In this sense, both complicate the boundaries delineating min-erality with that of biological being. Their mineral kinship blurs distinctions between what could reasonably be understood as nonliving or living resources for the purposes of the CHM principle.[75] Foregrounding these kinships reminds us that the manganese, copper, and lithium that constitute batteries, household wiring, and computers all come to us with their worlds.

Thinking with the entangled relations of seabed sediments, water column, manganese nodules, and other beings, such as the Xenophyophores, reveals just how inalienable they are from their constituent minerals. The formation and material relations of nodules exceed their extractivist representations as inani-mate, potato-shaped rocks and fungible units rich in mineral wealth.[76] Nodules are lifeways in progress in the deep time of abyssal worlds. They are indivisible from the sediments on which they rest and the watery atmosphere of the ocean through which their materiality slowly layers into being. Most nodules form through precipitations of different minerals gathered both from sediment pore waters (diagenetic) and cold seawaters (hydrogenetic), growing around ten millimeters per million years. Defying the pressure from several kilometers of watery atmosphere above, the yielding ooze below, and their incredibly slow formation, mysteriously the nodules remain at the sediment surface. Nodules intimately and materially connect with the lifeways and beings of their ecological communities—as their abundance diminishes so too does ecological diversity.

Re-Imagining the Lively Matter and Relations of Extraction

The extractive regime of seabed mining doesn't just abstract the ocean into artificial zones, it also performs the same disappearing act on the material and social substances that are extracted. Seatruthing reveals the multiple lives and kinships that are bracketed out in order to appropriate minerals resources into discourses and commodity regimes of value. Animating the term "mineral resource" to reconnect minerals with biological life and networks of lifeways brings into view their potential extinguishment by mining activities, and that the act of mining is better understood as material predation. The visceral nature of this act entails stripping ancient and agentic sediment from the earth's surface along with unknowable multiplicities and tatters of crab, fish, worms, and others who haven't been able to escape. I evoke this unknowable, unseen biological material through the figurations "flesh waste," rubbled "bio-ore," and "kin-waste water." It is rubbled bio-ore and flesh waste that will be sucked up riser pipes to processing vessels at the surface, before being dumped back into the water column in plumes of "kin-waste water." These more proximate representations evoke what mining proponents and legal frameworks omit—which is that extractivism takes from the ocean not just minerals but more-than-human lives, lifeways, and relations, rendering them "waste."

Through seatruthing, the implications of the economic motivation of seabed mining can be more closely recognized. The CHM principle subjects vast, planet-spanning seabed worlds to exploitation by private and state corporations and sustains this extractive relationship doctrinally for economic gains. The regime's promissory offer to share these economic benefits is used as a key justificatory tool to advance seabed mining.[77] In other words, the ecological extinguishments, flesh waste, and rubbled bio-ore, rendered as background collateral, are justified for the trophy of profit, without either acknowledging the worlds from which these profits were derived or ensuring that they can endure.

Material Embodiment, Vulnerability, and Predation

The previous section explored how international ocean governance facilitates mining corporations and the global regime of extractive capitalism in their expansions to the deep seabed. Ruder and Sanniti describe the motivating drive of these regimes as arising from a "predatory ontology,"[78] which they characterize as an extreme pursuit of economic accumulation achieved through the externalization of environmental harms. Relatedly, the market strategies of extractive capitalism dematerialize the connections between the material and ecological origins of the computers, cars, and other prosthetics—as if these materials arrive through some form of "spontaneous genesis."[79] Just as processes of dematerialization obscure material connections between commodity goods and more-than-human worlds, so too the relations of violence associated with material provisioning are also obscured. Whether through ignorance, denial, or ontological oversight, the violences associated with material extraction (which I explore in the context of predation in the next section) are out of sight, out of mind.

Given the sheer scale of biodiversity loss, and plastic and heat wastes caused by ocean industrialization, criticism of the exploitations of resource corporations are justified. However, blaming declining oceanic conditions exclusively on corporations and extractive regimes risks a "politics of purity"[80] that neglects to account for our individual implicatedness. There are many ways, for example, that material embodiment implicates us in ecological harms. Being embodied necessitates everyday predation on other worlds to ensure that we are biologically fueled, informed, sheltered, and techno-socially creative and connected with close and distant human communities.[81] Embodiment also tethers us to other bodies and lifeways through inalienable relations of violence.[82] These relations are not often brought into light because their ontological darkness seems as unfathomable as the ocean. Outsourcing our individual material provisioning to mining corporations doesn't diminish individual accountability for our predator interventions into lively worlds, it merely commissions corporations as our proxies.

Extending feminist inquiries about *who* we think we are[83] to also ask *how* we think we are affords a way of fathoming material predation as a dimension of ecological subjectivity. Thinking ecologically through both inquiries reveals multiple ways that humans come to be co-constituted materially and relationally with other worlds and "wes." Humans come into continual being through food, minerals, and other materials without which we risk heightened vulnerability. Our material vulnerabilities extend to needing minerals for the everyday "prosthetics," as Haraway envisages them, that connect embodied subjects to their worlds.[84] Prosthetics enjoin us to other worlds and, whether battery, household wiring, or train carriage, they are constituted in some way by bio-ore and kin waste from uncounted multiplicities of lively beings and the worlds from which they were taken.

Just as animality requires biologically derived food, prosthetics also need food, such as metal. Each of us "eats" metal, in the metaphoric sense that Derrida (whose work is invoked later in this chapter) used the term "eat" to refer to the assimilation of physical worlds and relations.[85] Provisioning kills other beings. With similar bluntness, Shotwell declares that "living our lives relies on the suffering and death of others."[86] This is the case. By acknowledging that relations of violence are inherent to shoring up our material exposures and livability needs, we might at least be more sensible to the character and scale of our violences and predations and the depth of ethical responses they require.[87]

The condition of material embodiment reminds us how much our being is contingent on multiple other bodies. Reflecting on our collectivity with microbial communities, for example, Haraway offers that "to be one is always to *become with* many."[88] As we multiply so we also extend into the world and, thinking with Haraway again, do not "end at [our] skin,"[89] or scales or feathers or jellies or shells. Neimanis writes further that we are always all "becoming in webs of mutual implication."[90] Human bodies extend into the world via biological dependencies and prosthetics that keep our heart beating, allow us to work and communicate in certain ways, or provide mobility and transport. The prosthetics that enjoin us in these ways are constituted by bio-ore and kin waste and the worlds from which they were taken. Our being, therefore, is contingent on these multiple other bodies and sutured materially with their worlds.

Given these mutual implications, how we qualify and justly exercise the ontological need to materially predate within shared worlds are matters of profound dimensions. Scale is a significant factor—it is hard to imagine how much damage will be wreaked on the abyssal seabed within mining leases that are collectively the size of small countries, and where, once the lights and machinery get going, they will not be switched off for potentially 30 years. Further to scale, other factors significant to understanding the nature of material predation and its implications, include the context of an indeterminately changing ocean; the shifting material relations and needs of near and future human and more-than-human communities; and the need for continual revisioning of what ethical obligations toward the worlds of our prey ought to entail.

Extractive Development: Predation and the Ethics of Hospitality

In this section, I draw on Derrida's "how to eat well" to conceptualize an ocean justice approach that responds to the conflict inherent in material predation on worlds that we care about and urgently need to protect. As well, offering a way of approaching ethical practices of care and reciprocity toward the worlds of those we kill and harm, Derrida's thesis gestures to a way of understanding material predation and its violences.

Given that we need to eat, writes Derrida, it matters not what or how we "eat this and not that, the living or the non-living, man or animal"; the key question is how to eat well.[91] This is the implied politics of food that underlies the question of how to eat.[92] I reformulate Derrida's concept as a potential politics and ontology of predation. As embodied beings we need to predate on other worlds for material provisioning, the key question is, *how to be a more ethical predator*. As a

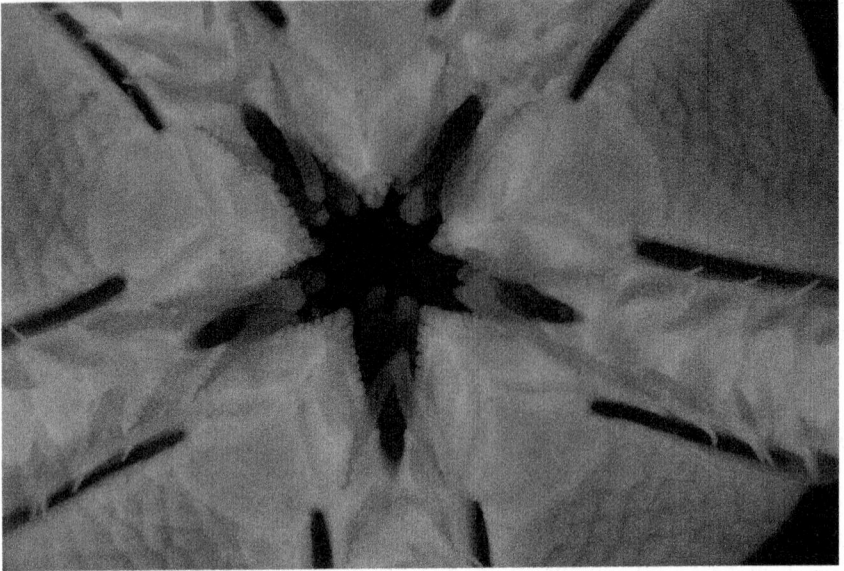

FIGURE 3.5 How to eat well: a close-up image of the mouth of a brittle star from APEI 7. Image courtesy of DeepCCZ Partners: University of Hawaii (US), Natural History Museum (UK) and University of Gothenburg (Sweden).

synecdoche of material predation, eating well, in the context of seabed mining, could entail not eating other beings whose communities are already extremely vulnerable due to overhunting or for whom too little is known of their life worlds and vulnerabilities. The point here is not to itemize potential solutions but rather to widen the conceptual aperture for imagining what being a good predator might entail, practically, ontologically, politically, and ethically in the context of the diminishing worlds of our prey.

Here I turn to philosopher Kelly Oliver's close analysis of Derrida's "limi-trophy" project in which Derrida proposes that the etymological associations of trophy are instructive to approaching how to eat well (both figuratively and literally):[93]

> In the semantics of trepho, trophe, or trophos, we should be able to find everything we need to speak about what we should be speaking about …: feeding, food, nursing, breeding, offspring, education, care and keeping of animals, training, upbringing, culture, living, and allowing to live by giving to live, be fed, grown.[94]

Picking up on the double meaning of trophe/trophy, Oliver examines its potential use in distinguishing "eating well or good eating from devouring the other in poor taste."[95] On a very literal level, the distinction can be understood as killing animals for food and nutrition (trophe) versus killing animals for sport and photo opportunities (trophy). Can the distinctions between (meta-phorically) eating for nutrition or trophy help to calibrate an ethics of material predation?

How might we distinguish between the violences necessarily inflicted in order to source the constituent metals needed to make prosthetic devices for communicating, knowledge making, and medical aids (trophe), versus doing the same thing but for economic accumulation (trophy)? Thinking with Derrida's conception of limitrophy as that which is "cultivated on the edges of a limit,"[96] what ethics emerge at the intersections of needing to protect the deep ocean and needing to enact violences upon it in order to source minerals? What life worlds are foreclosed at the edges of UNCLOS's interventions between mineral com-modities and more-than-human mineralized kinships? One way to approach the ethical nature of predation is by interrogating the possible motivations for extraction (eating others, metaphorically speaking) either as nourishment or tro-phy (noting that the distinction between these is not always clear).

Following Derrida, the invocation to *eat well* can be understood to mean that whatever and whichever one eats "must be nourishing."[97] Oliver provides a compelling explication of Derrida's intention that eating be understood as both the literal ingestion of food and the metonymical act of taking in or assimilat-ing others in all their forms and relations: family, ocean, friends, community, language, symbols, mountains, social codes, rivers.[98] In other words, eating the other entails taking in, at least partially, the worlds that constitute them. The ethical obligation that Derrida emphasizes is that by eating the other we ought to notice and extend hospitality to their world. Or to invoke Haraway again, "nothing comes without its world."[99] Seatruthing the unknowable and unseen "prey" of seabed mining as bio-ore, and flesh and kin waste, for example, brings these relations to the fore to remind us that the cobalt, magnesium, and other metals extracted for assimilation into prosthetics, or for the accumula-tions of private wealth, are more-than minerals—that come with multiple, mutually implicated others.

Derrida hinges the concept of nourishment to obligations of "infinite hospi-tality" to the other,[100] to which Oliver adds, "even those whom one ingests."[101] Accepting that, as material predators, we extinguish certain beings and rela-tions in the pursuit of necessary materials, how can we still ensure the con-ditions of possibility for life-world continuance? In the context of material predations of the seabed and deep ocean, some of the conditions that ought to be factored include how such activity is imbricated within cumulative anthro-pogenic impacts that are already changing the ocean; whether the physical extent of violence and its temporal continuance afford sufficient refuge and pause for lifeways to endure or recover; and whether enough is known of ocean realms to competently gauge the potential impacts of predation.[102] Seabed min-ing that doesn't take these conditions into account would amount to rapa-cious hunting that forgets its relations to the worlds of its prey— aligning with Derrida's notion of "trophy."

The seabed mining regime that is emerging within the legal framework of UNCLOS also aligns with "trophy"—with its emphasis on a model of profit-driven material predation in which responsibilities for environmental harms are monetized and capped by economic limits. The trophy validated within the CHM principle is the pursuit of seabed minerals in order to attain economic benefits principally for (corporate) humanity. If the concept of "humankind" envisaged by the CHM principle were to be re-imagined as embodied, materially vulnerable beings, who are interdependent with the ocean, then neither economic gain for corporations or the ecological harms resulting from mining, would guarantee them much benefit, or amount to nourishment or eating well.

Closing Notes

Obligations of hospitality call us to interrogate and modify the scale and motivations of our material predations if we are to co-occupy a transitioning futurity with the ocean. The predatory ontology sedimented within UNCLOS's legislative framework is already changing ocean worlds and the conditions of livability for our prey.[103] Its ecological force augurs still more by legitimating economically driven seabed mining on a planet already ravaged by extractive capital and in deep-ocean worlds thoroughly unprepared for corporate humanity. The material predations of seabed mining threaten to bring about what environmental philosopher Deborah Bird Rose saliently described as "double death," that is, an "amplification of death, so that the balance between life and death is overrun."[104]

This chapter has introduced elements of my approach to ocean justice in which material embodiment and vulnerability are situated at the seafloor. It has swung a spotlight beam outward to the midnight realm of the abyssal ocean and glimpsed the multiplicities of wondrous lives enfolded in multiple mineral and chemical kinships. Were seabed mining to extract too much from these relations it would risk what Rose describes as an "irreparable loss not only of the living but of the multiplicity of forms of life and of the capacity of evolutionary processes to regenerate life."[105] The chapter has also cast its beam inward to recognize, with equal compassion, the inalienable violence hinged to human material vulnerability and the realization that our status as exceptional predators comes with exceptional responsibilities of care.

Acknowledgments

This chapter was written on Yugambeh and Gadigal Country. It was made possible with support from *The Seed Box: A Mistra-Formas Environmental Humanities Collaboratory*. My thanks to the editor Irus Braverman and the authors in this collection for their collegiate interlocutions. Particular thanks to Vito De Lucia for his close reading and generous feedback. My thanks for the ocean's gentle inspiration.

Notes

1 A vast and biodiverse region up to 6,000 in depth that covers 83 percent of the total area of the ocean (https://en.wikipedia.org/wiki/Abyssal_zone) and 58 percent of the earth's surface. "The Second World Ocean Assessment: Volume II," United Nations, 2021, https://www.un.org/regularprocess/sites/www.un.org.regular-process/files/2011859-e-woa-ii-vol-ii.pdf, 455.

2 For more details about vent communities see Maria C. Baker et al., "Biogeography, Ecology, and Vulnerability of Chemosynthetic Ecosystems in the Deep Sea," in *Life in the World's Oceans*, ed. Alasdair D. McIntyre (Oxford: Blackwell Publishing Ltd., 2010), 161–182; seabed detrivores: Craig R. Smith and Amy R. Baco, "Ecology of Whale Falls at the Deep-Sea Floor," *Oceanography and Marine Biology: An Annual Review* 41 (2003): 311–354; Nicholas D. Higgs et al., "Fish Food in the Deep Sea: Revisiting the Role of Large Food-Falls," *PLOS ONE* 9, no. 5 (2014): e96016; rains of detritus: Peter Herring, *The Biology of the Deep Ocean* (Oxford: Oxford University Press, 2002); Thomas Kiørboe, "Formation and Fate of Marine Snow: Small-Scale Processes with Large-Scale Implications," *Scietia Marina* 65, supp. 2 (2001): 57–61.

3 "The Second World Ocean Assessment: Volume 1," United Nations, 2021, https://www.un.org/regularprocess/sites/www.un.org.regularprocess/files/2011859-e-woa-ii-vol-i.pdf., 463.

4 For more on human impacts: Eva Ramirez-Llodra et al., "Man and the Last Great Wilderness: Human Impact on the Deep Sea," *PLOS ONE* 6, no. 8 (2011): e22588; and for Anthropocene seas: Stacy Alaimo, "The Anthropocene at Sea," in *The Routledge Companion to the Environmental Humanities Routledge*, eds. Ursula K. Heise et al. (London: Routledge, 2017), 153–161.

5 United Nations, "The Second World Ocean Assessment: Volume 1," 464.

6 For more on anthropogenic impacts: Elliott A. Norse et al., *Marine Conservation Biology: The Science of Maintaining the Sea's Biodiversity* (Washington: Island Press, 2005), 4, 59; United Nations, "The Second World Ocean Assessment: Volume 1"; United Nations, "The Second World Ocean Assessment: Volume II," United Nations, 2021, https://www.un.org/regularprocess/sites/www.un.org.regular-process/files/2011859-e-woa-ii-vol-ii.pdf; "Special Report on the Ocean and Cryosphere in a Changing Climate," IPCC, accessed September 24, 2021, https://www.ipcc.ch/srocc/.

7 Todd Woody, "Seabed Mining: The 30 People Who Could Decide the Fate of the Deep Ocean," *New Humanitarian*, September 6, 2017, https://deeply.thenewhuman-itarian.org/oceans/articles/2017/09/06/seabed-mining-the-24-people-who-could-decide-the-fate-of-the-deep-ocean.

8 United Nations, "The Second World Ocean Assessment: Volume II," 279. For more about risk of species extinction: Lara Macheriotou et al., "Phylogenetic Clustering and Rarity Imply Risk of Local Species Extinction in Prospective Deep-Sea Mining Areas of the Clarion-Clipperton Fracture Zone," *Proceedings of the Royal Society B: Biological Sciences* 287 (2020): 20192666; slow recovery after human-made disturbance: Erik Simon-Lledó et al., "Biological Effects 26 Years after Simulated Deep-Sea Mining," *Scientific Reports* 9, no. 1 (2019): 8040.

9 For example, the impacts of mining in Bougainville: John C. Cannon, "Decades-Old Mine in Bougainville Exacts Devastating Human Toll: Report," *Mongabay Environmental News*, April 17, 2020, https://news.mongabay.com/2020/04/dec-ades-old-mine-in-bougainville-exacts-devastating-human-toll-report/; "One of World's Worst Mine Disasters Gets Worse – BHP Admits Massive Environmental Damage at Ok Tedi Mine in Papua New Guinea, Says Mine Should Never Have Opened," *Mining Watch Canada*, Nov. 8, 1999, https://miningwatch.ca/news/1999/8/11/one-worlds-worst-mine-disasters-gets-worse-bhp-admits-massive-environ-mental-damage-ok.

10 United Nations Convention on the Law of the Sea, December 10, 1982, 1833 UNTS 397 ("UNCLOS").

11 Wylie Spicer, "Legalising Seabed Access," Deep Sea Mining Campaign, October 18, 2013. http://www.deepseaminingoutofourdepth.org/legalising-seabed-access/.

12 See Kate Lyons, "Mining's New Frontier: Pacific Nations Caught in the Rush for Deep-Sea Riches," *Guardian*, June 23, 2021, http://www.theguardian.com/world/2021/jun/23/minings-new-frontier-pacific-nations-caught-in-the-rush-for-deep-sea-riches; Deep Sea Mining Campaign et al., "Why the Rush? Seabed Mining in the Pacific Ocean," 2019, http://www.deepseaminingoutofourdepth.org/wp-content/uploads/Why-the-Rush.pdf; Blue Ocean Law and The Pacific Network on Globalisation, "Resource Roulette: How Deep Sea Mining and Inadequate Regulatory Frameworks Imperil the Pacific and Its People," 2016, http://nabf219anw2q7dgn1rt14bu4.wpengine.netdnacdn.com/files/2016/06/Resource_Roulette-1.pdf.

13 See, for example, World Bank and United Nations, "The Potential of the Blue Economy: Increasing Long-Term Benefits of the Sustainable Use of Marine Resources for Small Island Developing States and Coastal Least Developed Countries," (Washington, D.C.: World Bank and United Nations, 2017); "Achieving Blue Growth Building Vibrant Fisheries and Aquaculture Communities," Food and Agriculture Organisation of the United Nations, 2018, http://www.fao.org/3/CA0268EN/ca0268en.pdf.

14 For related critiques concerning capitalism's blue turn: John Childs, "Performing 'Blue Degrowth': Critiquing Seabed Mining in Papua New Guinea through Creative Practice," *Sustainability Science* 15, no. 1 (2020): 117–129; "Surfacing the Agendas of the Blue Economy," *DAWN* (blog), March 19, 2019, https://dawnnet.org/2019/03/surfacing-the-agendas-of-the-blue-economy/; DAWNfeminist, "DAWN/PANG Blue Economy Panel Fiji," YouTube, March 22, 2019, video, 1:39:59 https://www.youtube.com/watch?v=6ICLPABe7JE.; Jean-Baptiste Jouffray et al., "The Blue Acceleration: The Trajectory of Human Expansion into the Ocean," *One Earth* 2, no. 1 (2020): 43–54; also Elizabeth DeLoughery's related critique of "blue capital" in this chapter.

15 The (ISA) is "an autonomous international organization" established under UNCLOS and the 1994 Agreement relating to the Implementation of Part XI of the United Nations Convention on the Law of the Sea, December 10, 1982 (July 28, 1994), 1836 UNTS 42 (1994 Implementation Agreement).

16 ISA Secretary General Michael Lodge speaking at a 2017 workshop jointly hosted by the ISA and the UN Department of Economic and Social Affairs in Tonga, https://www.isa.org.jm/vc/abyssal-initiative-blue-growth. For more on resistance to blue capitalism by Pacific Island nations and communities, see also DAWNfeminist, "DAWN/PANG Blue Economy Panel Fiji."

17 "Pacific Island Governments Cautioned About Seabed Mining Impacts," PIANGO, 2019, http://www.piango.org/our-news-events/latest-news/2019/pacific-island-governments-cautioned-about-seabed-mining-impacts/. Growing public resistance to seabed mining from multiple sectors can be seen, for example, in the campaigns of Pacific Blue Line: https://www.pacificblueline.org/.

18 David Heydon, CEO of mining corporation DeepGreen claims that seabed mining will "save the planet." Wil S. Hylton, "History's Largest Mining Operation is About to Begin," *Atlantic*, 2020, https://www.theatlantic.com/magazine/archive/2020/01/20000-feet-under-the-sea/603040/.

19 Sven Teske et al., "Renewable Energy and Deep-Sea Mining: Supply, Demand and Scenarios," ISF Report, 2016.

20 Ibid., 5, 37.

21 Henry Sanderson, "BMW, Volvo and Google Vow to Exclude Use of Ocean-Mined Metals," *Financial Times*, March 31, 2021, https://www.ft.com/content/e618a555-2d21-4f33-b6b5-46564197f834.

22 C.L. Van Dover et al., "Biodiversity Loss from Deep-Sea Mining," *Nature Geoscience* 10, no. 7 (2017): 464–465; C.L. Van Dover et al., "Research Is Needed to Inform Environmental Management of Hydrothermally Inactive and Extinct Polymetallic Sulfide (PMS) Deposits," *Marine Policy* 121 (2020): 104183.

23 Holly J. Niner et al., "Deep-Sea Mining with No Net Loss of Biodiversity—An Impossible Aim," *Frontiers in Marine Science* 5, no. 53 (2018), https://doi.org/10.3389 /fmars.2018.00053.

24 Michael Lodge, "The International Seabed Authority and Deep Seabed Mining," *UN Chronicle*, 2016, https://www.un.org/en/chronicle/article/international-sea- bed-authority-and-deep-seabed-mining.

25 Craig R. Smith et al., "The Near Future of the Deep-Sea Floor Ecosystems," in *Aquatic Ecosystems: Trends and Global Prospects*, ed. Nicholas V. C. Polunin (Cambridge: Cambridge University Press, 2008), 334–350; Niner, "Deep-Sea Mining."

26 Drazen cited in Hylton, "History's Largest Mining Operation."

27 Hylton, "History's Largest Mining Operation."

28 For a related discussion about the conversion of nonhuman natures as raw material for the expansion of capital: Jason W. Moore, *Capitalism in the Web of Life: Ecology and the Accumulation of Capital* (London: Verso, 2015).

29 UNCLOS; Part XI read together with the 1994 Implementation Agreement.

30 See, for example, Isabel Feichtner and Surabhi Ranganathan, "International Law and Economic Exploitation in the Global Commons: Introduction," *European Journal of International Law* 30, no. 2 (2019): 541–546; see also Surabhi Ranganathan, this volume; Anna Zalik, "Mining the Seabed, Enclosing the Area: Ocean Grabbing, Proprietary Knowledge and the Geopolitics of the Extractive Frontier beyond National Jurisdiction," *International Social Science Journal* 68, nos. 229–230 (2018): 343–359.

31 Lodge, "The International Seabed Authority."

32 For a related critique of the Enlightenment values underpinning international ocean law: Alice Ollino, "Feminism, Nature and the Post-Human: Toward a Critical Analysis of the International Law of the Sea Governing Marine Living Resources Management," *Gender and the Law of the Sea* 88 (2019): 204–228.

33 Ngaire Naffine, "Who Are Law's Persons? From Cheshire Cats to Responsible Subjects," *Modern Law Review* 66, no. 3 (2003): 346–367; A. Grear, "The Closures of Legal Subjectivity: Why Examining 'Law's Person' Is Critical to an Understanding of Injustice in an Age of Climate Crisis," in *Research Handbook on Human Rights and the Environment*, eds. Anna Grear and Louis J. Kotzé (Cheltenham: Edward Elgar Publishing, 2015), 79–101.

34 Elena Blanco and Anna Grear, "Personhood, Jurisdiction and Injustice: Law, Colonialities and the Global Order," *Journal of Human Rights and the Environment* 10, no. 1 (2019): 86–117.

35 "Regulatory capture" refers to when a regulatory agency created to further public interest, instead advances the commercial or political interests of the very entities it is mandated to regulate.

36 UNCLOS Article 157.

37 UNCLOS Article 150.

38 General obligations to protect and preserve the marine environment (Article 192); and in relation to the Area, the ISA's obligation under Article 145 and the 1994 Implementing Agreement.

39 The LTC is responsible for developing the seabed mining regulations, also known as the Mining Code: ISA, The Mining Code, regularly updated at https://www.isa .org.jm/mining-code.

40 Louisa Casson, "Deep Trouble: The Murky World of the Deep Sea Mining Industry," Greenpeace International, 2020, https://www.greenpeace.org/static/ planet4-international-stateless/c86ff110-pto-deep-trouble-report-final-1.pdf.

41 Casson, "Deep Trouble"; Woody, "Seabed Mining"; Blue Ocean Law and the Pacific Network on Globalisation, "Resource Roulette."

42 For a related critique see Deep Sea Mining Campaign et al., "Why the Rush?"

43 Aline Jaeckel, "An Environmental Management Strategy for the International Seabed Authority—The Legal Basis," *International Journal of Marine and Coastal Law* 30, no. 1 (2015): 93–119; Aline Jaeckel, "Strategic Environmental Planning for Deep Seabed Mining in the Area," *Marine Policy* 114 (2020): 103423.

44 Draft Regulations on Exploitation of Mineral Resources in the Area, ISA Doc. ISBA/25/C/WP.1 (March 25, 2019), 54 and 55(a) [hereinafter: Draft Exploitation Regulations].

45 Draft Exploitation Regulations, 55(e).

46 Draft Exploitation Regulations, Schedule: Use of Terms and Scope.

47 Aline Jaeckel et al., "Conserving the Common Heritage of Humankind—Options for the Deep-Seabed Mining Regime," *Marine Policy* 78 (2017): 150–157.

48 For example, references to "all peoples of the world," UNCLOS Preamble, First and Seventh Recitals; and "mankind as a whole," UNCLOS Preamble, First Recital. These references also gesture to the constitutional-based approach to ocean governance developed by one of UNCLOS's co-architects, Elisabeth Mann Borgese, and others. Betsy Baker, "Uncommon Heritage: Elisabeth Mann Borgese, Pacem in Maribus, the International Ocean Institute and Preparations for UNCLOS III," *Ocean Yearbook Online* 26 (2011): 11–34.

49 In relation to ocean governance/stewardship see: M. Jackson, "Indigenous Law and the Sea," in *Freedom for the Seas in the 21st Century: Ocean Governance and Environmental Harmony*, eds. Jon M. Van Dyke et al. (Washington, D.C.: Island Press, 1993), 41–48; Lara Taylor and Meg Parsons, "Why Indigenous Knowledge Should Be an Essential Part of How We Govern the World's Oceans," *The Conversation*, 2021, http://theconversation.com/why-indigenous-knowledge-should-be-an-essential-part-of-how-we-govern-the-worlds-oceans-161649; for discussion of the exclusion of Indigenous cultural, and ecological knowledge and ocean governance approaches: Marjo K. Vierros et al., "Considering Indigenous Peoples and Local Communities in Governance of the Global Ocean Commons," *Marine Policy* 119 (2020): 104039, https://doi.org/10.1016/j.marpol.2020.104039.

50 UNCLOS Articles 137, 2.

51 The Area comprises the ocean floor beneath the high seas, beyond national jurisdiction.

52 Amongst other rules, the CHM prevents a state or legal person (i.e., corporation) from appropriating any parts or resources of the Area. UNCLOS, Article 137(1).

53 UNCLOS Article 136; Preamble, para 6. Note that this chapter uses the gender neutral "humankind" from here on.

54 Tullio Treves, "Introductory Note, United Nations Convention on the Law of the Sea," United Nations, 2008, https://legal.un.org/avl/ha/uncls/uncls.html (emphasis added).

55 Acknowledging that UNCLOS was at a time when little was known about the deep ocean's rich biodiversity. See also Elliot Norse et al., *Marine Conservation Biology: The Science of Maintaining the Sea's Biodiversity* (Washington: Island Press, 2005): 415.

56 "Area" means the seabed and ocean floor and subsoil thereof, beyond the limits of national jurisdiction. UNCLOS Article 1(1).

57 See "About ISA," International Seabed Authority, 2022, https://www.isa.org.jm/about-isa.

58 See, for example, Andreas Philippopoulos-Mihalopoulos, "Towards a Critical Environmental Law," in *Law and Ecology, New Environmental Foundations*, ed. Andreas Philippopoulos-Mihalopoulos (New York: Routledge, 2011), 18–38.

59 The Metals Company, "DeepGreen: Metals for Our Future," Vimeo, Aug. 27, 2018, video, 3:49, https://vimeo.com/286936275 (in particular, from 2:24 until 2:43).

60 See also Philippopoulos-Mihalopoulos for a discussion of how law is situated within an "open ecology": Andreas Philippopoulos-Mihalopoulos, "Looking for the Space Between Law and Ecology," in *Law and Ecology, New Environmental Foundations*, ed. Andreas Philippopoulos-Mihalopoulos (New York: Routledge, 2011), 2.
61 Draft Exploitation Regulations, 15.3(b)(i).
62 UNCLOS Article 1(1), n. 35.
63 UNCLOS Article 133(a).
64 For related discussions in the context of the continental shelf regime and UNCLOS Article 77(4): Susan Reid, "Solwara 1 and the Sessile Ones," in *Blue Legalities*, eds. Irus Braverman and Elizabeth R. Johnson (Durham: Duke University Press, 2020), 25–44; hydrothermal vents, Ranganathan, this volume; and marine genetic resources, Braverman, this volume.
65 Draft Exploitation Regulations, Schedule: Use of Terms and Scope.
66 Here I draw on the work of feminist philosopher of epistemology Linda Alcoff where she discusses the need to look at "where the speech goes and what it does there." Linda Alcoff, "The Problem of Speaking for Others," *Cultural Critique* no. 20 (1991): 5–32, 26.
67 Melody Jue, *Wild Blue Media: Thinking through Seawater* (Durham: Duke University Press, 2020), 4–6.
68 Frida Maria Armas Pfirter, "The Management of Seabed Living Resources in 'The Area' Under UNCLOS," *Electronic Journal of International Studies*, no. 11 (2006): 21. https://www.academia.edu/16461528/The_Management_of_Seabed_Living_Resources_in_The_Area_under_UNCLOS.
69 Autun Purser et al., "Association of Deep-Sea Incirrate Octopods with Manganese Crusts and Nodule Fields in the Pacific Ocean," *Current Biology* 26, no. 24 (2016): R1268–R1269.
70 UNCLOS Article 133(a).
71 Draft Exploitation Regulations, Schedule: Use of Terms and Scope.
72 L.A. Levin and A. J. Gooday, "Possible Roles for Xenophyophores in Deep-Sea Carbon Cycling," in *Deep-Sea Food Chains and the Global Carbon Cycle*, eds. Gilbert T. Rowe and Vita Pariente (Dordrecht: Springer Netherlands, 1992), 93–104; A.J. Gooday et al., "Xenophyophores (Rhizaria, Foraminifera) from the Nazaré Canyon (Portuguese Margin, NE Atlantic)," *Deep-Sea Research: Part II, Topical Studies in Oceanography* 58, no. 23 (2011): 2401–2419.
73 "Foraminifera," Postgraduate Unit of Micropalaeontology, University College London, MIRACLE, 2002, https://www.ucl.ac.uk/GeolSci/micropal/foram.html.
74 Lisa A. Levin and Greg W. Rouse, "Giant Protists (Xenophyophores) Function as Fish Nurseries," *Ecology (Durham)* 101, no. 4 (2020): e02933-n/a, https://doi.org/10.1002/ecy.2933.
75 As declared in Resolution 2749 (XXV) in 1970 and adopted in UNCLOS's Preamble at para 6.
76 See, for example, Cecilia Jamasmie, "DeepGreen to Make Run for Battery Metals from Seafloor," *MINING [DOT] COM*, April 7, 2020, https://www.mining.com/deepgreen-makes-run-at-battery-metals-from-seafloor/.
77 For related critiques of ocean governance and the CHM regime see Vito De Lucia, "Ocean Commons, Law of the Sea and Rights For The Sea," *Canadian Journal of Law and Jurisprudence* 32, no. 1 (2019): 45–57; Surabhi Ranganathan, "The Common Heritage of Mankind: Annotations on a Battle," in *The Battle for International Law: South-North Perspectives on the Decolonization Era*, eds. Jochen von Bernstorff and Philipp Dann (Oxford: Oxford University Press, 2019); Julie Hunter et al., "Broadening Common Heritage: Addressing Gaps in the Deep Sea Mining Regulatory Regime," *Harvard Environmental Law Review*, April 16, 2018, https://harvardelr.com/2018/04/16/broadening-common-heritage/.
78 Sarah-Louise Ruder and Sophia Rose Sanniti, "Transcending the Learned Ignorance of Predatory Ontologies: A Research Agenda for an Ecofeminist-Informed

Ecological Economics," *Sustainability* (Basel, Switzerland) 11, no. 5 (2019): 1479. For a related and rich discussion of how Western imaginaries of mastery and extractivism externalize, or "background," environmental connections and harms, see also Val Plumwood, "Shadow Places and the Politics of Dwelling," *Australian Humanities Review*, no. 44 (2008): 139150, http://www.australianhumanitiesreview.org/archive/Issue-March-2008/plumwood.html.

79 Janine MacLeod, "Water and the Material Imagination: Reading the Sea of Memory against the Flows of Capital," in *Thinking with Water*, eds. Cecilia Chen et al. (Montreal: McGill-Queen's University Press, 2013), 40–60, 43.

80 Alexis Shotwell, *Against Purity: Living Ethically in Compromised Times* (Minneapolis: University of Minnesota Press, 2016).

81 For wider discussions of the concepts of predation, eating, and being eaten see: Val Plumwood, "Integrating Ethical Frameworks for Animals, Humans, and Nature: A Critical Feminist Eco-Socialist Analysis," *Ethics & Environment* 5, no. 2 (2000): 285–322, 292; Franklin Ginn et al., "Flourishing with Awkward Creatures: Togetherness, Vulnerability, Killing," *Environmental Humanities* 4, no. 1 (2014): 113–123.

82 See also Kathryn Yusoff, "Aesthetics of Loss: Biodiversity, Banal Violence and Biotic Subjects," *Transactions of the Institute of British Geographers* 37, no. 4 (2012): 578–592.

83 Lorraine Code, *Ecological Thinking; the Politics of Epistemic Location* (New York: Oxford University Press, 2006); Donna J. Haraway, *Staying with the Trouble: Making Kin in the Chthulucene* (Durham: Duke University Press, 2016).

84 Donna J. Haraway, "Situated Knowledges: The Science Question in Feminism and the Privilege of Partial Perspective," *Feminist Studies* 14, no. 3 (1988): 575–599.

85 Jacques Derrida, "The Animal That Therefore I Am (More to Follow)," trans. D. Wills, *Critical Inquiry* 28, no. 2 (2002): 369–418.

86 Alexis Shotwell, *Against Purity: Living Ethically in Compromised Times* (Minneapolis: University of Minnesota Press, 2016).

87 For more on relations of violence with more-than-human others see: Yusoff, "Aesthetics of Loss."

88 Donna J. Haraway, *When Species Meet* (Minneapolis: University of Minnesota Press, 2008), 4.

89 Donna J. Haraway, *Simians, Cyborgs, and Women: The Reinvention of Nature* (New York: Routledge, 1991), 178.

90 Astrida Neimanis, "Feminist Subjectivity, Watered," *Feminist Review*, no. 103 (2013): 23–41, 25.

91 Jacques Derrida, "'Eating Well,' or the Calculation of the Subject: An Interview with Jacques Derrida," in *Who Comes After the Subject?*, eds. Eduardo Cadava, Peter Connor, and Jean-Luc Nancy (New York: Routledge, 1991), 96–119, 114–115.

92 See also Kelly Oliver, "Derrida and Eating," in *Encyclopedia of Food and Agricultural Ethics*, eds. Paul B. Thompson and David M. Kaplan (Dordrecht: Springer Netherlands, 2014), 459–465, 462.

93 Oliver, "Derrida and Eating," 463.

94 Derrida, "The Animal," 397.

95 Oliver, "Derrida and Eating," 463.

96 Derrida, "The Animal," 397.

97 Derrida, "'Eating Well,'" 115.

98 Kelly Oliver, "Tropho Ethics: Derrida's Homeophatic Purity," *Harvard Review of Philosophy* 15, no. 1 (2007): 37–57, 43; Oliver, "Derrida and Eating," 460.

99 Donna J. Haraway, *Modest_Witness@Second_Millennium.FemaleMan©_Meets_OncoMouse™* (New York: Routledge, 1997), 137.

100 Derrida, "'Eating Well,'" 115.

101 Oliver, "Derrida and Eating," 460.

102 In June 2021, despite scientific and community concerns about insufficient knowledge of deep ocean ecologies, Nauru triggered a legal mechanism under the 1994 Implementation Agreement, by notifying the ISA of its intention to commence mining in two years through NORI, a subsidiary of a Canadian firm, The Metals Company (previously known as DeepGreen): Jonathan Watts, "Race to the Bottom: The Disastrous, Blindfolded Rush to Mine the Deep Sea," *The Guardian*, September 27, 2021, https://www.theguardian.com/environment/2021/sep/27/race -to-the-bottom-the-disastrous-blindfolded-rush-to-mine-the-deep-sea. See also "Nauru Requests the President of ISA Council to Complete the Adoption of Rules, Regulations and Procedures Necessary to Facilitate the Approval of Plans of Work for Exploitation in the Area," International Seabed Authority, June 29, 2021, https://isa .org.jm/news/nauru-requests-president-isa-council-complete-adoption-rules-regu- lations-and-procedures; and for more, see Ranganathan, this volume.

103 For example, the predatory fisheries regime enabled by UNCLOS's high seas provisions mandate the maximum sustainable yield measure to be used in fisheries management (Article 119(1)(a)), a measure exploited by Regional Fisheries Organisations that contributes to the decimation of fish populations and biodiversity as discussed by Telesca, for example, in relation to tuna fisheries: Jennifer E. Telesca, *Red Gold* (Minneapolis: Minnesota Press, 2020). For more on the diminishing fisheries see: Christensen Pauly et al., "Fishing Down Marine Food Webs," *Science* 279, no. 5352 (1998): 860–863; Daniel Pauly, *Vanishing Fish: Shifting Baselines and the Future of Global Fisheries* (Vancouver: Greystone Books, 2019); Daniel Pauly, "Aquacalypse Now," *New Republic*, Sept. 28, 2009, https://newrepublic.com/article /69712/aquacalypse-now; and relatedly: Elspeth Probyn, *Eating the Ocean* (Durham: Duke University Press, 2016); marine biodiversity: Boris Worm et al., "Impacts of Biodiversity Loss on Ocean Ecosystem Services," *Science* 314, no. 5800 (2006): 787–790; and United Nations, "The Second World Ocean Assessment: Volume I"; United Nations, "The Second World Ocean Assessment: Volume II."

104 Deborah Bird Rose, "What If the Angel of History Were a Dog? [Paper in: Art and Ecology]," *Cultural Studies Review* 12, no. 1 (2006): 67–78, 75.

105 Deborah Bird Rose, "Multispecies Knots of Ethical Time," *Environmental Philosophy* 9, no. 1 (2012): 127–140, 128.

4

GENETIC FREEDOM OF THE SEAS IN THE AGE OF EXTRACTIVISM

Marine Genetic Resources in Areas Beyond National Jurisdiction

Irus Braverman

Characterized by scientists and mainstream media alike for being "utterly alien," newly discovered undersea life-forms are no longer gigantic, but microbial.

—Mariana Silva, "Mining the Deep Sea"[1]

Introduction

Nearly two-thirds of commercial pharmaceutical medicine originate from so-called natural products.[2] While terrestrial organisms have been used in medicine for millennia, the use of marine biological "resources" for purposes other than food, otherwise referred to as "marine bioprospecting," is more recent—and booming.[3] Marine ecosystems are particularly suited for bioprospecting: they are about twice as likely to yield at least one gene in a patent than their terrestrial counterparts. In fact, "the success rate in finding previously undescribed active chemicals in marine organisms is 500 times higher than that for terrestrial species."[4]

The extraction of marine genetic resources (referred to as MGRs in the expert jargon—but I will try to keep acronyms to a minimum in this chapter for legibility purposes) is growing rapidly, with over 38,506 natural products and 4,900 patents associated with genes of marine organisms, the latter increasing at a rate of 12 percent per year.[5] Scientists found, along these lines, that the "appropriation of MGRs is progressing much faster than the already impressive

FIGURE 4.1 The ROPOS manipulator arm holds a sample of an inactive chimney in which fossilized tubeworms are embedded, an extremely rare find. Ring of Fire 2002 Expedition. Image courtesy of NOAA.

DOI: 10.4324/9781003205173-5

rate of domestication for aquaculture."[6] Marine genetic resources are, in other words, "a growing source of biotechnological and business opportunities"[7]—the new, and perhaps final, frontier[8] of what has been lauded, and also criticized, as the blue economy.

The discovery of marine organisms containing molecules and genes of commercial interest has proceeded alongside the scientific explorations of marine biodiversity. The term bioprospecting is often used in this context to refer to the search for living organisms as a source of commercially exploitable products, such as medicinal drugs. However, there is a considerable divergence of opinion within the international community as to the precise meaning of this term and whether it includes non-commercial products. At least in the context of marine genetic resources, the term is typically defined as including the entire research and development process from sample extraction by public scientific and academic research institutions (which are generally, but not exclusively, funded by governments) all the way to full-scale commercialization and marketing by biotechnology and other companies.[9] The focus of research on marine genetic resources is geographically broad as well, encompassing deep-sea genetic resources alongside genetic resources from other areas of the sea.[10]

Despite their growing significance as the new ocean frontier, there is currently no internationally agreed upon legal or scientific definition of marine genetic resources. The meaning of this term in the marine context has been inferred from definitions of genetic resources (not specifically devised in the marine context) in the 1992 Convention on Biological Diversity (also "Convention" herein) and the 2010 Nagoya Protocol on Access to Genetic Resources and the Fair and Equitable Sharing of the Benefits Arising from their Utilisation (herein, the "Nagoya Protocol").[11] Because they do not contain DNA, the Convention's definition of marine genetic resources leaves out derivatives (natural products, proteins, toxins, et cetera), which can be highly valuable for commercial ends. The utilization of derivatives is regulated by the complementary Nagoya Protocol.

Both the Convention on Biological Diversity and the Nagoya Protocol apply only to genetic resources sourced from *within* national jurisdictions and do not apply to Areas Beyond National Jurisdiction (herein in lowercase).[12] Such areas, which encompass over 95 percent of the oceans' volume, are defined by the United Nations as the open ocean waters that lie beyond the economic zones and jurisdiction of any one country.[13] Over the past few decades, fishing and mineral exploitation expanded into these areas. Meanwhile, the International Seabed Authority has recently granted licenses to 29 mining contractors for exploitation activities there. Deep-sea scientists point out that there is currently no legal regime that protects biodiversity in areas beyond national jurisdiction, cautioning that "over a 15-year period, a single mining operation could damage marine systems over an area of 50,000 square kilometers."[14]

Similar to fishing and mineral exploitation in areas beyond national jurisdiction, instances of marine genetic resources sourced from these areas are also becoming more frequent, and likewise lacking regulation. Still, as of 2019, only one commercial product on the market was derived from marine genetic resources in areas beyond national jurisdiction.[15] To address the legal lacuna pertaining to the management of the multiple "resources" in areas beyond national jurisdiction, the United Nations decided in Resolution 72/249 of December 24, 2017 to convene an intergovernmental conference and has, since then, been negotiating a legally binding agreement, entitled the International Legally Binding Instrument Under the United Nations Convention on the Law of the Sea on the Conservation and Sustainable Use of Marine Biological Diversity of Areas Beyond National Jurisdiction (herein, the "BBNJ"). The negotiations aim to address several topics, including marine protected areas and marine biological diversity.[16] Many have characterized marine genetic resources as the most challenging topic being discussed under the BBNJ treaty.

This chapter draws on interviews conducted with 20 deep-sea scientist and legal experts, as well as on observations of their work at one session of the BBNJ treaty negotiations that took place in the United Nations headquarters in New York City in 2017, to highlight the uneasy symbiotic relationship between scientists, legal experts, and policy makers. The chapter explores this relationship through the debates regarding the scope of marine genetic resources, which have involved questions about the significance of their place of origin, the standards regarding their documentation, and whether or not they should encompass digital sequence information (also "DSI"). The discrepancy between law's "terracentric"[17] need to fix bodies in place to better govern them and the more fluid materiality of the ocean is on display here, providing an opportunity to reflect on the underlying tensions between law and science as embodied and expressed by lawyers and scientists in the BBNJ context.

Alongside the visible contestations between scientists and legal experts regarding the definition of marine genetic resources, there are the less visibly shared assumptions that underlie this definition. The most obvious assumption is the very use of the term resources in this context. Arguably, defining life forms in this anthropocentric and utilitarian way already lends itself to extractivist regimes, which draw on colonial paradigm, worldview, and technologies to "reduce, constrain, and convert life into commodities."[18] It is therefore not surprising that the question for most of the scientists and legal practitioners engaged in this work has been *how* to utilize marine genetic resources and not *if* to do so. This, despite the shared understanding that wild harvests of marine organisms are undesirable from a conservation standpoint "because it is not always possible to predict their impact accurately."[19] That said, some marine scientists contend that "most marine bioprospecting does not harvest large amounts of materials. Once the original gene [or] derivative is identified, it can be reproduced in the lab without having to obtain more of it from the sea."[20]

The term *mare geneticum* was recently coined[21] to celebrate the newly found freedoms of the sea, more than 400 years after the coinage by Hugo Grotius of *mare liberum*. Whereas the original "freedom" referred to journeys of (certain) ships across the ocean's surface, the journey undertaken here follows live matter as it travels from source, into data, and finally into information. The abstraction and extraction of marine life is enabled through its decontextualization as part of this "data travel."[22] Here, life is suspended from its bodily matter and ecological context and reconfigured as genetic sequences that can thus become mobile commodities for exploitation.

The chapter ends with an urgent call by marine experts, both legal and scientific, to seize the precious opportunity of crafting a new treaty for areas beyond national jurisdiction so as to challenge the pervasive extractivist logic that currently underlies ocean governance. Instead of abstracting, fragmenting, and decontextualizing ocean lifeworlds, an alternative way of relating to these more-than-human lives is called for.[23] A scientist I interviewed for this project reflected on how we might address the fragmented state of ocean law, which he blamed on lawyers.[24] From his perspective, a new treaty that encompasses both land and sea regimes would be an important step in the right direction.

Marine Genetic Resources *Within* National Jurisdiction

Article 2 of the Convention on Biological Diversity defines genetic resources as "material from plants, algae, animals, and microbial or other organisms, and parts thereof containing functional *units of heredity of actual or potential value*."[25] Such "actual or potential" value can be considered in environmental, economic, societal, and scientific terms, and is based on "the many ways in which biological materials (also referred to as biomolecules) function and how organisms interact to transform chemicals and change their environments."[26]

One of the key innovations of the Convention on Biological Diversity is the way it mapped out key principles regarding the access and benefit-sharing (also "ABS") of genetic resources. For example, Article 15 of the Convention states that each "contracting party" shall endeavor to create conditions to facilitate access to genetic resources for environmentally sound uses by other contracting parties. Moreover, each contracting party is required to take legislative, administrative, and policy measures with the aim of sharing in a fair and equitable way the results of research and development and the benefits arising from the commercial and other utilization of genetic resources. The equitability factor is one of the most contentious aspects of the Convention because of the limited experience of countries in dealing with access and benefit-sharing and the rather uneven administrations that have developed as a result.[27]

The Nagoya Protocol added the concept "derivatives" to the Convention on Biological Diversity's definition of genetic resources. According to Article 2 of the Nagoya Protocol, such derivatives are defined as any "*naturally occurring* biochemical compound resulting from the genetic expression or metabolism of biological or genetic resources, even if it does not contain functional units of heredity, therefore also encompassing secondary metabolites, enzymes, and natural products."[28] This clause has elicited major debates, mostly focusing on whether derivatives are themselves a genetic resource, or whether they should only be considered when discussing how a genetic resource is utilized and, in turn, which access and benefit-sharing regime would be relevant to it.

If derivatives were not confusing enough, the term "digital sequence information" was introduced to the two treaty regimes in decisions CBD XIII/16 and NP-2/14,[29] and is currently negotiated under the auspices of the Convention on Biological Diversity. Although different from derivatives, the scope of digital sequence information is no less contested. Some definitions include only DNA, RNA, and protein sequences, while others encompass additional elements that are further removed from the "original" genetic matter.[30] Expanding the definition of a marine genetic resource to include the broadest scope of digital sequence information would bring under the treaty a wide range of sample types—from entire organisms, through environmental samples of water, ice, or sediment, all the way to samples derived from any of these, such as extracted DNA or tissue preparations preserved to enable utilization.[31]

Genetic resource "samples" and "data" are intrinsically connected and so deploying a rigid legal distinction between a resource, its derivative, and digital information is challenging on a scientific level. The exclusion of digital sequence information from the definition of marine genetic resources would also lead to "biotechnology companies profiting from use of the 'global commons' without redistribution to those states with a reduced capacity to undertake such work themselves."[32] At the same time, embracing a definition of marine genetic resources that includes a broad definition of digital sequence information might result in restrictions on access to data that is currently openly available, which could in turn hamper scientific research.[33] For this reason, the scientific community has often not been too keen about using the term digital sequence information to expand the scope and usages of genetic resources.[34]

Take, for example, a sponge that produces a toxin, and that toxin (but not the sponge) is used in pharmaceutical research. Under the Convention on Biological Diversity, this toxin is not classified as a (marine) genetic resource because it does not contain DNA, and thus no legal conditions or requirements are placed over its extraction from areas within national jurisdiction. However, under the Nagoya Protocol, and possibly also under the BBNJ, the sponge can be categorized as a derivative and at least its utilization would thus be covered and regulated.[35] Marcel Jaspars, the co-leader of one working group in the Deep-Ocean Stewardship Initiative, explained why an expansive approach toward the definition of genetic resources is most desirable:

> In my mind, a tree sap belongs to the tree. It came from the tree's biosynthetic process. Therefore, although it is not alive in itself, it's something that derived from that, it's a derivative and should be covered. But apparently, it's not necessarily covered [because] it's not a genetic resource. ... Beer is an example. It's made by bacteria but there is no bacteria in it. It has no DNA in it, yet it's a very valuable product ... it's a multi-billion-dollar

market. Same with other things, like cosmetics that are made from plants. They won't contain the DNA in the plant oils, for instance, but they contain the oil. Their value is not [necessarily derived from] using the genes, it's actually about having the physical plant that you can grow and get the oil from. That's the value.[36]

Jeffrey Marlow, Assistant Professor of Biology at Boston University, similarly advocates for a broad definition of marine genetic resources that includes derivatives:

> If you don't include the derivatives, then things like antibiotics or proteins would not be included [in the definition of marine genetic resources]. Which means that you could harvest them, reuse them, or sell them without going through this legal framework. A more expansive definition would encompass all of that.[37]

According to Marlow, marine genetic resources should therefore be defined broadly in the BBNJ so that they encompass "all information associated with or extracted from a physical MGR sample, specifically including any genetic sequence information, in both raw and processed form."[38] From this point of view, the main objective of the new treaty is not to prevent turning matter into resource, but rather to more tightly regulate this process.

Marine Genetic Resources in Areas *Beyond* National Jurisdiction

Areas beyond national jurisdiction are the largest environment on the planet, encompassing 64 percent of the world oceans and 47 percent of the earth's surface.[39] The definitional scope of marine genetic resources in areas beyond national jurisdiction thus carries significant legal, scientific, and economic implications. Under the BBNJ negotiations, some states have insisted that marine genetic resources on the seabed in areas beyond national jurisdiction (otherwise known as the "Area") are comparable to mineral resources as defined by the United Nations Convention on the Law of the Sea ("UNCLOS"), and should therefore be encompassed by the common heritage of mankind principle[40] and subject to benefit-sharing regimes.[41] Other states have interpreted corresponding articles in UNCLOS as excluding biological resources and thus advocated to apply the freedom of the high seas principle to them, implying that no legal obligation exists to share the benefits arising from their exploitation.[42] To bridge these approaches, marine policy advisors have suggested that the BBNJ adopt a novel *sui generis* regime that would provide for unique access and benefit-sharing mechanisms.[43]

A central question deliberated by the BBNJ policy makers and scientists is whether benefits associated with exploitation of marine genetic resources should be shared by the entire international community, or whether they should only

FIGURE 4.2 Gold coral on pillow lava in over 1,000 feet depth off Hawai'i, 1988. Pillow lava is commonly cited as the most abundant geological landform on earth's surface. Credit: OAR/National Undersea Research Program (NURP).

be shared by the wealthy developed states that have the technological capacity to exploit such resources, which are typically so difficult to access and require considerable investment. There are currently two broad approaches and mechanisms in response to this question: bilateral and multilateral. Both the Convention on Biological Diversity and the Nagoya Protocol have adopted a bilateral access and benefit-sharing approach. Under this approach, access and benefit-sharing transactions are defined as existing between the state where the marine genetic resource is found (one provider) and an individual or entity that requests access to this resource to use it for research and development (one user). The provider is obliged to facilitate access to the genetic resources found within its national jurisdiction, but maintains the sovereign right to make such access subject to the granting of prior informed consent (usually a permit) and mutually agreed upon terms (the conditions identified in an access and benefit-sharing contract).[44] The user must share benefits with the provider in an equitable and fair way, based on the terms established between the two parties.

Ocean experts have been debating which governance model to apply to marine genetic resources in areas beyond national jurisdiction. The bilateral approach is not readily applicable to marine genetic resources from areas beyond national jurisdiction under the existing UNCLOS regime, as these resources neither fall under the jurisdiction of a particular state nor under the authority of a global entity that could grant its consent and negotiate an access and benefit-sharing agreement with an interested user.[45] A multilateral access and benefit-sharing system would create a common pool, or a "global commons,"[46] and then establish access rules.[47] But the multilateral approach is much less commonly used

FIGURE 4.3   ROV Deep Discoverer observes a cliff that marks the edge of a coral platform in American Samoa. Image courtesy of the NOAA Office of Ocean Exploration and Research, Mountains in the Deep: Exploring the Central Pacific Basin.

and has been put to practice only with regard to a limited number of genetic resources.[48] While the appropriate BBNJ regime is being debated,[49] a small group of countries and transnational companies are already disproportionately influencing production volumes and revenues from the bioprospecting of marine genetic resource.[50]

Another issue that has not been adequately considered when negotiating the governance model for marine genetic resources in areas beyond national jurisdiction is the desirable patent regime. One way of securing exclusive access to a marine genetic resource is to patent it.[51] Patenting is especially significant in areas beyond national jurisdiction because the granting of a patent always occurs within a national jurisdiction and is thus determined by the domestic law of that state, regardless of where the marine genetic resource was sourced.[52] Consequently, rights in relation to patents (as opposed to *access* rights) are not affected by the absence of a specific regulatory regime in areas beyond national jurisdiction.[53] By 2018, 862 marine species were associated with patents, the majority of which pertain to microbial species, and 221 companies registered 84 percent of all patents pertaining to marine species (universities accounted for 12 patents).[54] One single corporation, the German multinational chemical manufacturer BASF, registered 47 percent of these patent sequences. Sophie Arnaud-Haond, a researcher at a French marine research institute, explained the problematic implications of the current patent regime on the documentation of marine genetic resources:

> If we want a fair sharing of benefits, it's not the sampling step [that] we have to [regulate]—it's the patenting step. And we don't. We pretend to ignore

all that. The World Trade Office [WTO] has refused, in the last ten or fifteen years, to mandate [the recording of] the origin of the samples from which a particular MGR derived. ... The solution is not in our hands, it's in the hands of the WTO, and they should do something about it.[55]

Despite its obvious relevance and importance to it, the existing patent regime is not currently considered under the BBNJ. The question, then, is whether the new legal regime negotiated under the BBNJ is "even capable of fostering greater equity and ocean stewardship, or is it too deeply seeped in a broader mode of extractive governmentality."[56] According to scholar Macarena Gómez-Barris, the answer is clear. Law, for her, is "embedded within a global political economic and interstate system that does not serve as a steward to the natural world but sells it to the highest commodity market."[57] The fact that patents are off the table in the BBNJ negotiations seems to support this criticism. Expressing her related critique toward international law's significant role in the creation of the neoliberal global economy while recognizing its powerful potential to steer the future in a positive direction, international law scholar Janne Nijman wrote: "To save the legitimate popular grievances from exploitation by extreme Right nationalist politicians, justice in all its dimensions needs to underpin the international rule of law."[58]

Marine Genetic Resources Under the Future BBNJ Regime

As of 2019, the draft of the BBNJ treaty defined marine genetic resources as "any material of marine plant, animal, microbial or other origin, [found in or] originating from areas beyond national jurisdiction and containing functional units of heredity with actual or potential value of their genetic and biochemical properties."[59] Although he had originally advocated for a broader definition, Marcel Jaspars thought that this definition of marine genetic resources under the draft of the BBNJ was *too* broad. Muriel Rabone is the Data and Sample Collector at London's Natural History Museum, and works with the Museum's Deep-Sea Systematics and Ecology Research Group. Similar to Jaspars, Rabone was concerned about the unintended consequences of the BBNJ possibly expanding the definition of marine genetic resources too broadly. She explained in our interview that "genetic databases are the lifeblood of biological science." If genetic resources were to be defined broadly and the BBNJ mandated subscription to access the database, for example, this "would not be acceptable to the scientific community [as it would] hamper that open data [archive]." She explained that "this is where the disquiet and the concern among scientists toward the BBNJ is coming from."[60]

Based on lessons learned from negotiations of prior international treaties, and the Nagoya Protocol in particular, the conveners of the BBNJ have set out to ensure that the consultation process included input from the scientific community.[61] But the scientific involvement in the legal process has often been challenging, as the two expert groups differ on even the most basic issues of terminology and classification. For example, scientists generally use the term "deep-sea" to refer to any part of the ocean, pelagic or benthic, deeper than 200 meters, without reference to the national boundaries within this space. In legal terms, however, this range is called the "High Seas," and is separated from the regime on the sea floor, which is referred to as "the Area."[62] Working in the field, marine scientists will often sample organisms across jurisdictional boundaries, even during one single expedition. Margaret Spring is a deep-sea scientist from the Monterey Bay Aquarium and former Chief of Staff at the United States National Oceanic and Atmospheric Administration (NOAA). For her, the central question is: "how do you deal with the fact that [marine] genetic resources move across boundaries when you're doing science?"[63] In our interview, Arnaud-Haond explained that with most marine animals,

> you have absolutely no control over the fact [that] they live within an [exclusive economic zone or EEZ], or [that] their genetic material can be found at one stage or the other of their lifespan in areas beyond national jurisdiction. It's like birds. All oceans are connected, whereas different continents are not.[64]

Since their organisms typically don't recognize legal borders and jurisdictional lines, the marine scientists were consistently amazed at the legal experts' insistence on asserting linear, telluric geographies. Here from Jaspars:

> At the very beginning, [one lawyer] asked me, in all seriousness, "Can you tell if an animal originates from the EEZ or ABNJ?" And I said, "No. It can swim from one to the other." That was a revelation to him. [L]awyers are very smart people and they ask lots of questions … [But] a lot of them are based on absolutely no science whatsoever. [In] my interaction with lawyers [I will] often write the same message again and again and again, but maybe in different ways, until they get it.[65]

The underlying tensions between lawyers and scientists in the BBNJ context reveal the discrepancy between law's terracentric need to fix governable bodies in place and the more fluid realities of oceans and their lifeworlds as reflected in scientific knowledge.

To make sense of the negative attitude toward lawyers expressed in some of these quotes by the marine scientists, it might help to reflect on the earlier dynamics that accompanied the Convention on Biological Diversity and the Nagoya Protocol. Evidently, these treaty processes excluded the scientists from important decision-making processes that were in turn handled exclusively by lawyers and policy makers. This, despite the fact that the importance of an interdisciplinary decision-making process was acknowledged by some of the central figures already during the negotiations of UNCLOS. As Elisabeth Mann Borgese put it in 1993:

> If the issues under consideration are interdisciplinary, the decision-making process must be interdisciplinary. It cannot be implemented by just one discipline, for example, the lawyers and the politicians (who generally are lawyers). It must involve scientists, economists, industrial managers, and all others whose disciplines are involved.[66]

While the treaty-making processes that ensued were less inclusive than was called for, it is also the case that when the two expert communities have worked together in such contexts, this work would often result in deep frustrations. Related to the incompatibility between law and science as pertaining to place-making practices, the legal and scientific communities have also been incongruent in their approach toward time. "The scientific landscape is developing much more rapidly than associated legislation," one of the scientists told me.[67] Including scientists in the process would, in this view, provide the additional benefit of accounting for the rapid pace of scientific development, in synthetic biology in particular.[68]

This approach, which is shared widely among scientists and lawyers alike, represents an important assumption about the relationship between science and law—namely, that science leads the way, while law follows.[69] Sheila Jasanoff problematized this assumption when she wrote that "the law today not only interprets the social impacts of science and technology but also constructs the very environment in which science and technology come to have meaning, utility, and force."[70] And so whereas law is often depicted as separate from and as constantly racing to catch up with science, the two are in fact deeply interdependent and even coproduced—as is strongly evident in the context of marine genetic resources.

Further differences—and commonalities—between law and science have emerged during the BBNJ negotiations over the definition of marine genetic resources. One of the main concerns in this context, which is linked to the characterization of law as lagging behind science, is that the regulation of marine genetic resources would stifle scientific advancement. As one of my interlocutors put it: "Regulation must ensure to not stifle innovation or impede research progress in any way. Input from the scientific community is ... crucial to the BBNJ discussions."[71]

Negotiating the BBNJ: Lawyers, Scientists, and Policy Makers

The BBNJ process has revealed some of the differences in the legal and scientific modes of knowing the world. Jaspars offered in this context that:

> The lawyers are often very keen on having a box to put around something in which they can say, "Everything that belongs inside the box belongs inside the box. Everything outside is somewhere else." And they start testing the boundaries' conditions, asking how you would get from inside the box to outside the box. That takes creative thinking. It's very nice.

They ask lots of really good questions. I just don't know how to answer most of the questions they're asking.[72]

Further reflecting on the difference between legal and scientific thinking, Jaspars told me:

I was shown four or five definitions by [one of the BBNJ] delegations. ... The lawyers saw a big difference between them. For me they were just very similar definitions that meant roughly the same thing. It was about jurisdictional boundaries. So, [the debate was] whether you use the word ocean or marine or whether you use habitat or environment, things like that. To me, that was interesting. But I'm a very visual thinker. I like to draw diagrams and create pictures. Lawyers, on the other hand, like to use lots of words.[73]

From Jaspars's perspective, then, the difference between law and science is not only reflected in their disparate definitions, but also goes deep into the way that these groups are trained to see the world. Fran Humphries is a legal researcher of marine biodiversity situated in Brisbane. Her interview shed additional light on the relationship between the ways that legal and scientific experts construct and convey knowledge. In her words: "Sometimes the information from scientists can be so technical that the policy makers can't see what is relevant for their particular policy. They're looking for a more broad-brushed kind of thing."[74] Jaspars explained, along these lines, that: "lawyers don't want something that is too heavily defined because you get in trouble with a clear scientific definition. You don't want something that can be misinterpreted."[75]

Alongside scientists and lawyers, policy makers have played an important, albeit often under-explored, role in negotiating international treaties, and the BBNJ treaty is no exception. Humphries explained that there are three actors at work in the BBNJ, which breaks up what others have often perceived as one indistinct group of lawyers into two very different groups. In her words:

There are the lawyers who want to get into the nitty gritty, watertight definitions that can capture things and be held up. ... Then you have the diplomats who focus on the diplomatic language, making it broad enough. [Thirdly,] from the scientists' perspective, this text [becomes] so broad that it's unworkable because it can incorporate many activities. ... So it's actually the policy makers who broaden it out to make it so big that it's frightening for both scientists and lawyers because they can't exclude things. ... The draft text of the BBNJ treaty is a simple form [of] diplomatic language. It's not even a legal text, it's what we call a "legal framework." It's not like national laws [that are] able to capture those sorts of nuances.[76]

The trick when negotiating a treaty, according to Humphries, is to find language that is specific enough to mean something and broad enough to withstand changes in time and differences in interpretation among nation-states. Jaspars clarified that: "A lot of treaty law, in order to be passed, needs to be very general, so that it's acceptable to everyone and everyone can implement it in such a way that their national legislation can recognize."[77]

According to Humphries, then, the problem of the cross-disciplinary BBNJ negotiations is not, as so many others depicted, the tense relationship between law and science, which she sees as actually being deeply compatible, but instead lies within the diplomatic or political realm. Furthermore, Humphries divides diplomats into two subgroups: policy advisors from national governments, and "career negotiators," who "don't necessarily have any grasp on the topic, but ... their skills are essentially to make deals with other countries behind closed doors." For Humphries, the aspect of diplomacy foregrounds the uniqueness of international law, which requires specific expertise in the art of negotiation itself. The generalized international law negotiator who is not an expert in substance but in form is a perfect example for the power of abstraction in contemporary law of the sea—its decontextualizing away from matter and into a legal realm where it can then be detached from lively situated properties into a two-dimensional platform for exploitation and extraction.

Humphries's interpretation resonates with discussions within international law about the rule of law and its legitimacy vis-à-vis politics. If Hugo Grotius argued in the 17th century that "recourse to law would take international actors away from the divisive and dangerous field of 'politics' and into the world of abstract and neutral rules to be applied by impartial courts and expert arbitrators,"[78] then contemporary Finnish legal scholar Martti Koskenniemi claimed that law is fundamentally political. In his words: "even as law did offer a specialist vocabulary and a set of institutions that would enable the translation of raw interests into the language of rules, the way those rules then operated remained still dependent on contestable (and often contested) assumptions about the world."[79] The idea that international law is separate from politics was also criticized by Janne Nijman. In her words:

> to produce authoritative interpretations and eventually substantive resolutions of conflicts, international law ultimately depends on political choices made in the daily practices of international law. The international rule of law as a—profoundly liberal—flight from politics was revealed to be an illusion of sorts.[80]

International law's politics are quite specific: it adopts a Western liberal political theory that presents itself as universal, rather than utilizing a regional, cultural, and historically-specific language. This also explains the yet underrecognized importance of local groups in international law. Humphries emphasized in this context that the BBNJ has actually included a group that focuses on Indigenous knowledge. In her words:

> People are genuinely trying to grapple with incorporating traditional knowledge into the BBNJ, which is fabulous. In this forum, policy makers [are] really trying to understand what traditional knowledge means for this

treaty and how we would use the knowledge and protect it, not just [for] marine genetic resources, but also for the other elements.[81]

The question one might pose in this context is whether Indigenous knowledge can be meaningfully incorporated into an existing political regime that defines living organisms as resources and that speaks in the utilitarian language of benefit-sharing. At the same time, for local and Indigenous groups to not be part of the BBNJ negotiations would mean not having a voice at the regulatory table. This is a common dilemma for Indigenous experts as they navigate the existing governance structures that they so fundamentally oppose.

Digital Sequence Information: A Legal-Scientific Journey

The debates over the adequate infrastructure, procedures, and standards pertaining to the production and management of marine genetic resources can also be referred to as debates over "data regimes." Here, the foundational distinction of Western science between nature and culture maps onto the corresponding distinction between raw sample and data to introduce a further distinction, this time between data and information.[82] According to this distinction, data describes the inherent properties of material artifacts as distinguished from "research outputs or other value-adding steps."[83] Operating within this logic, data is presumed to be neutral and objective until a human or an algorithm transforms it into information. The distinction between data and information thus relies on two interrelated assumptions: first, that DNA is the original and most natural and neutral matter, and second, that human intervention interrupts this naturalness and "contaminates" it.[84]

Should digital sequence information be confined to "representational data" (such as the DNA sequence GTACCTGA)? And, if not, to what extent should it include processing activities performed with that data by data producers, curators, and users to generate information? According to experts from the Deep-Ocean Stewardship Initiative, "different amounts of work are needed to convert different types of data into information."[85] Under this approach, to decide whether something falls into the definition of a digital sequence information, one would need to consider its proximity to the "original" genetic resource as well as the degree of "biological processing" it has undergone. In other words, one should ascertain "how far along the flow from genetic resource onwards to DNA, RNA, protein sequences and metabolites DSI can be considered to extend."[86]

Additionally, human labor is distinguished based on its complexity. Data that is intensely worked upon and thus severed from the material source becomes information. The availability of easy tools for DNA or protein conversion defines the resulting knowledge as data, whereas the difficulty of 3-D protein folding models defines it as information. By this logic, DNA sequences that are considered data today would have been defined as information 50 years ago, when the act of examining and documenting a sample was a process of human-guided

FIGURE 4.4 The top portion of a tubeworm from the Brine Pool, photographed in the deep sea of the northern Gulf of Mexico with white light by Operation Deep Scope Expedition, 2004. Credit: NOAA/OAR/OE.

interpretation. If a given tool of interpretation is sufficiently ubiquitous and its meaning standardized within the scientific community then it is no longer seen as generating information, only data. In other words, the distinctions between data and information—and the nature-culture assumptions that underpin it—change over time. Information today will become data tomorrow (but not the other way around). And so while data seems uninteresting, in fact: "the real source of innovation in current biology is the attention paid to data handling and dissemination practices and the ways in which such practices mirror economic and political modes of interaction and decision making, rather than the emergence of big data and associated methods per se."[87]

Enfolded into seemingly technical deliberations about data, the major scientific debates of this digital era illuminate deeper debates about the definition and scope of scientific knowledge and its relationship with other social and legal practices. In the words of philosopher Sabina Leonelli:

> Data are at once technical and social objects, local products and global commodities, common goods to be freely shared and strategic investments to be defended, potential evidence to be explored and meaningless clutter to be eliminated—and the tension between these conflicting and yet perfectly adequate interpretations is what keeps debates around data and their role in science so lively and indicative of the multifaceted nature of scientific and technological expertise.[88]

"Data journeys" entail processes of decontextualization ("to make sure that data are extracted from their original birthplace") and recontextualization ("to make it possible for researchers unfamiliar with these data to assess their evidential value and use them for their own research purposes").[89] Specifically, the legal abstraction of marine life occurs through a threefold legal-scientific process: first, organisms are configured as marine genetic resources, then as data, and finally, as information. This legal abstraction is what enables the decontextualization of marine life as an object of juridical and scientific knowledge and, thus, the exploitation of this life through capitalist extraction. The debates over the scope of marine genetic resources and their relation to digital sequence information illustrate that the scientific definition of data and its regulation are tied together and even coproduced. The next section moves to consider these data debates as they apply to the specific context of the BBNJ.

Archiving Marine Genetic Resources: The Devil Is in the Data

Although the BBNJ negotiations are still underway, access to marine genetic resources is already happening on the ground and, along with it, the production of massive marine databases. Deep-sea records currently available from areas beyond national jurisdiction include 371,890 records of 10,437 species, observed between 1866 and 2018.[90] However, the records are not consistent. Generally, there is a shortage in deep-sea taxonomists and in funding for taxonomic research. Specifically, the existing data reflects taxonomic priorities and geographic biases, such as the extensive sampling in the North East Atlantic.[91] Consequently, there are taxonomic data gaps in certain parts of the world's oceans.[92] Along these lines, Jaspars described large ocean patches, especially in the Pacific, from which data was never collected and sequenced. A recent expedition to one such site involved the discovery of microbes that are estimated to be one-hundred million years old.[93]

Alongside the concerns about the equitable uses of data archives in the context of the deep sea, two fundamental questions remain in this context: first, what should be classified as data? and, second, which standards should govern the production of this data? While a consensus is emerging among deep-sea scientists about the importance of data openness and transparency, practices vary with regard to sharing information on marine scientific research activities, and no central global cruise registry currently exists to facilitate such information sharing.[94] As part of this tendency, a growing number of sequences are deposited without reference to their formal scientific names, what scientists call "operational taxonomic units." This practice has resulted in an explosion of "dark taxa"—species in databases such as GenBank that lack useful scientific reference data[95]—and the generation of excess "taxonomic entities" with limited scientific meaning.

A related issue that has come up in the negotiations for the BBNJ treaty is the "lack of site and other core associated data connected to genetic data"[96]— what is also referred to as "contextual data." While the database guidelines often recommend that contextual data be uploaded alongside the sequences, there often is no obligation to provide any more data about the specimen than a mandatory specimen ID number. This has resulted in a proliferation of sequences deposited at genetic data repositories without sample collection information. The disconnect between sequence and contextual data is perceived by many deep-sea scientists as a significant problem.[97] From the other end, an abundance of data requirements will often translate into terabytes of data for just one single study—which introduces yet another set of problems.[98] The practical implications of imposing standards for contextual data across the board are also questionable, as commercial products are rarely, if ever, traced back to their source in areas beyond national jurisdiction.[99]

In sum, the data journey that defines certain forms of life as marine genetic resources or as digital sequence information involves decontextualization from their material entity and their recontextualization as *res juridicus*. The price of this journey is the alienation of this source, which renders it irrelevant for protection as a form of ocean life. This potential loss of protection is especially relevant in the context of environmental or e-DNA—that is, the DNA extracted from environmental samples (such as soil, water or air) without requiring a sample from an identified organism source.[100] If DNA matters in databases only when it can be recontextualized "back" to the living organism from which it was sourced,[101] then in the case of e-DNA asserting such a contextual link becomes even more challenging.

Such expert discussions have become especially important due to the technological advancements that allow not only a journey away from the organic source of the DNA but also a journey back to the source through advanced synthetic biology. Indeed, synthetic biology could potentially combine useful gene sequences from different organisms and insert them into a host organism. The gene sequences in turn "become very difficult to trace back and it is important to find a workable solution to link all of the parts together."[102] Law holds an affirmative biopolitical power to (re)make life from information by regulating the obligatory passage points in these various data journeys.

Mare Geneticum as the "Final Frontier": Final Thoughts

This chapter has explored the concept of marine genetic resources, or MGRs, using this exploration as a lens into the complex interrelations between science and law and among scientists and legal experts—especially as this has manifested in their work on drafting the new treaty for areas beyond national jurisdiction, referred to as the BBNJ. Up to this point, much of the debate on the legal status of marine genetic resources obtained from areas beyond national jurisdiction has been concerned with the monetary benefits that could arise from their commercial utilization.[103] Yet some have come to see marine genetic resources as the "last frontier," highlighting that their central benefits lie far beyond the financial sphere.

The concept *mare geneticum* has been applied to all forms of the frontier, monetary and otherwise. *Mare geneticum* is a modernized version of Hugo Grotius's freedom of the seas, or *mare liberum*. This updated version of Grotius's concept contends that freedom includes not only the right to freely travel upon the ocean's surface but also to "shar[e] its natural resources in an organized and regulated fashion, in particular commodities like fish and minerals."[104] Beyond the conventional commodities associated with the sea, this chapter has highlighted another form of extraction: marine genetic resources. As the marine experts who have coined the term put it: "The first benefit to be shared under *mare geneticum* is enabling and facilitating access to marine genetic resources and associated data, thus empowering humankind to make the best of the *last frontier that is the ABNJ* [areas beyond national jurisdiction]."[105] In their words: "the *Mare Liberum* of the 17th century finds its echo in the *Mare Geneticum* of the 21st century."[106]

Upon reading this chapter, one might question, or at least be more suspicious of, the celebratory mode of *mare geneticum* proponents in the BBNJ context. Rather than paving a novel path forward that departs from anthropocentric ways of viewing the ocean as a resource for humans to grab and extract from, the BBNJ seems to be mirroring and duplicating prior principles and treaties, especially the freedom of the seas principle, the Convention on Biological Diversity, the Nagoya Protocol, and, of course, UNCLOS.

Drawing on my interviews with marine experts and recording their concerns, I have argued here that such approaches toward governing the ocean are largely anachronistic. Faced with this opportunity for change, the current trajectory of the BBNJ seems to be falling into the same potholes of the prior regimes. Jeffrey Marlow commented along these lines that: "The treaty is a relatively conservative document in that it is borrowing a lot from what's been done before, without questioning the first principles of how the science has evolved and how the environment has changed since a lot of these things were initially written."[107]

Existing international treaty law usually inflicts fragmentation and violence by tearing ocean life and matter out of spatial and temporal contexts. This is performed in the traditional way, by physically extracting fish and minerals; but it is also done by extracting living DNA samples from the ocean, transforming them into digital sequences, and then recontextualizing them into commodities. Operating under the auspices of the BBNJ, certain deep-sea scientists have recently cautioned about the dangers of such an extractivist approach.

Indeed, several years into the BBNJ negotiations, some deep-sea scientists are now considerably less thrilled about genetic freedoms and more concerned about whether the new treaty will resolve the problems of the prior legal treaties,[108] and the Nagoya Protocol in particular. The criticism of the deep-sea scientists toward the Nagoya Protocol is multifold and includes its overwhelming economic focus,[109] its stifling bureaucratic processes,[110] and its disregard of scientists and their concerns.[111] Arnaud-Haond put it in the bluntest terms:

Nagoya is a complete disaster. ... What Nagoya did that was extremely deleterious was that it imposed the very same amount of bureaucracy on companies and on the conservationists and researchers. [This, although we] do not have the means of the pharmaceutical companies: we don't have lawyers; we don't have all this administrative bunch of people doing the administrative work for us. We do it ourselves.[112]

More broadly, the deep-sea scientists I spoke with pointed to the challenges faced by the access and benefit-sharing model of the Convention on Biological Diversity and the Nagoya Protocol. For them, the still nascent consideration of digital sequence information under the BBNJ might provide an opportunity to sever benefit-sharing from access so as to mitigate the biotechnological inequalities of access.[113] This "would secure benefits while maintaining open science and generating funds from taxes, levies, or tiered approaches that feed a multilateral fund."[114]

But if the BBNJ treaty indeed moves away from the Nagoya Protocol and its related mechanisms, this would introduce incongruencies between terrestrial and marine regimes, which would in turn result in what Jaspars described as "loopholes that would allow us to do something with marine species that we couldn't do with terrestrial species." Such loopholes, he explained, might end up facilitating problematic jurisdiction shopping between land and sea.[115]

In light of these myriad problems and after negotiating multiple international biodiversity protection treaties, Jaspars has come to altogether question the fundamental binary between land and ocean both reflected and reinforced by the existing treaty regimes. Since these treaties impact each other in deep ways, it is time, in his opinion, to consolidate the legal protection of life—both on land and at sea—into one organic law. In the introduction to this volume, I referred to such an approach that moves from the binaries of land/sea to more fluid land-sea regimes as "amphibious legal geographies."[116] To conclude, I circle back to Jaspars' vision for the future of the BBNJ: "I would honestly start again with a new treaty from the bottom up. I would make it multilateral; I would make it about sharing; and I would make it all-inclusive."[117] How to do so without repeating the mistakes of UNCLOS is the challenge of ocean governance at this precarious time.

Acknowledgements

This chapter was mostly written on the territory of the Seneca Nation, a member of the Haudenosaunee/Six Nations Confederacy. I am grateful to the many marine experts who so readily engaged with me on this project, and especially to Kristina Gjerde, Sophie Arnaud-Haond, Muriel Rabone, and Marcel Jaspars for their contagious passion. My gratitude also extends to the participants of the Laws of the Sea workshops, and to Susan Reid, Henry Jones, Vito De Lucia, and Katrina Brown in particular for their helpful comments on the chapter's draft. Thanks also to Matthew Booker of the National Humanities Center for his insights and to the Center's fellows and administrators for making the final task of writing this chapter possible and even pleasurable. Margaret Drzewiecki and Daniel Piersa provided invaluable research assistance. Finally, I am grateful to the Research Council of Norway for granting the funding for the fieldwork executed for this chapter, as part of the research project *Funding Future Welfare: Bioeconomy as the "New Oil" and the Sharing of Benefits from Natural Resources* ("Bioshare," grant no. 294867), led by Ruralis.

Notes

1 Mariana Silva, "Mining the Deep Sea," *e-flux Journal* 109 (2020): 1–9, 3. http://worker01.e-flux.com/pdf/article_331369.pdf.
2 Atanas G. Atanasov et al., "Discovery and Resupply of Pharmacologically Active Plant-Derived Natural Products: A Review," *Biotechnology Advances* 33, no. 8 (2015): 1582–1614, 1584.
3 Jesús M. Arrieta et al., "What Lies Underneath: Conserving the Oceans' Genetic Resources," *PNAS* 107, no. 43 (2010): 18318–18324.
4 Ibid.
5 Ibid. The first figure is updated regularly on MarinLit, accessed April 8, 2022. https://marinlit.rsc.org/.
6 Ibid., 18139.
7 Ibid.
8 Andrew Merrie et al., "An Ocean of Surprises—Trends in Human Use, Unexpected Dynamics and Governance Challenges in Areas Beyond National Jurisdiction," *Global Environmental Change* 27 (2014): 19–31.
9 Balakrishna Pisupati et al., "Access and Benefit Sharing: Issues Related to Marine Genetic Resources," *Asian Biotechnology and Development Review* 10, no. 3 (2008): 49–68; Joanna Mossop, "Marine Bioprospecting," in *The Oxford Handbook of the Law of the Sea*, eds. Donald Rothwell et al. (Oxford: Oxford University Press, 2015), 825–842.
10 G. Bala et al., "Impact of Geoengineering Schemes on the Global Hydrological Cycle," *PNAS* 105, no. 22 (2008): 7664–7669.
11 Muriel Rabone et al., "Access to Marine Genetic Resources (MGR): Raising Awareness of Best-Practice Through a New Agreement for Biodiversity Beyond National Jurisdiction (BBNJ)," *Frontiers in Marine Science* 6, no. 520 (2019): 1–22, 3. See also Marjo Vierros et al., "Who Owns the Ocean? Policy Issues Surrounding Marine Genetic Resources," *Oceanography and Limnology Bulletin* 25, no. 2 (2016): 29–35; Harriet Harden-Davies, "Deep-Sea Genetic Resources: New Frontiers for Science and Stewardship in Areas Beyond National Jurisdiction," *Deep Sea Research Part II: Topical Studies in Oceanography* 137 (2017): 504–513.
12 Obviously, it is often unclear where the marine genetic resource was sourced. Harriet Harden-Davies (Post-Doctoral Research Fellow, Nereus Program at the Australian National Centre for Ocean Resources and Security, University of Wollongong & delegate of Deep-Ocean Stewardship Initiative), telephone interview by author, November 24, 2019. See also "Advancing Science-Based Policy," Deep-Ocean Stewardship Initiative, accessed October 2, 2020, https://www.dosi-project.org/.
13 Sophie Arnaud-Haond, "Mind the Gap Between Biological Samples and Marine Genetic Resources in Areas Beyond National Jurisdiction: Lessons from Land," in *New Knowledge and Changing Circumstances in the Law of the Sea*, ed. Tomas Heidar

(Leiden: Brill Nijhoff, 2020), 29–39. See also the UN Convention on the Law of the Sea, Articles 1, 87–90, December 10, 1982, 1833 UNTS 397.

14 "Ocean Genomics Horizon Scan: Executive Summary," Revive & Restore, July 2019, https://ocean.reviverestore.org/, at 20. In the context of marine protected areas, see Kristina M. Gjerde and Anna Rulska-Domino, "Marine Protected Areas beyond National Jurisdiction: Some Practical Perspectives for Moving Ahead," *International Journal of Marine and Coastal Law* 27 (2012): 351–373.

15 Rabone et al., "Access to Marine Genetic Resources," 3.

16 From the resolution: "the conservation and sustainable use of marine biological diversity of areas beyond national jurisdiction, in particular, together and as a whole, marine genetic resources, including questions on the sharing of benefits, measures such as area-based management tools, including marine protected areas environmental impact assessments and capacity-building and the transfer of marine technology." GA Res 72/249, ¶ 2 (December 24, 2017); see also "Ocean Genomics," 20.

17 Marcus Rediker, *Outlaws of the Atlantic: Sailors, Pirates, and Motley Crews in the Age of Sail* (Boston: Beacon Press, 2014), 2.

18 Macarena Gómez-Barris, *The Extractive Zone* (Durham: Duke University Press, 2017), xix.

19 Arrieta et al., "What Lies Underneath," 18321.

20 Marcel Jaspars (Co-Lead, Deep Sea Genetic Resources Working Group of the Deep-Ocean Stewardship Initiative), e-mail communication with author, April 7, 2022.

21 See Arianna Broggiato et al., "*Mare Geneticum*: Balancing Governance of Marine Genetic Resources in International Waters," *International Journal of Marine and Coastal Law* 33, no. 1 (2018): 3–33.

22 Sabina Leonelli, *Data-Centric Biology: A Philosophical Study* (Chicago: University of Chicago Press, 2016).

23 See, along these lines, Ranganathan and Reid's discussions of deep-sea mining as well as Henry Jones' work on extractivism, all in this volume. See also Surabhi Ranganathan, "Ocean Floor Grab: International Law and the Making of an Extractive Imaginary," *European Journal of International Law* 30, no. 2 (2019): 573–600.

24 Jaspars, Zoom interview by author, September 15, 2020.

25 The Convention on Biological Diversity, June 5, 1992, 1760 UNTS 69.

26 Jeffrey Marlow et al., "The Full Value of Marine Genetic Resources," Deep-Ocean Stewardship Initiative, March 2019, http://dosi-project.org/wp-content/uploads/2018/05/Full-value-mgr-March2019.pdf, 1.

27 Gurdial Singh Nijar, "The Nagoya Protocol on Access and Benefit Sharing of Genetic Resources: Analysis and Implementation Options for Developing Countries," South Centre, 2011, https://www.southcentre.int/wp-content/uploads/2013/08/Ev_130201_GNjar1.pdf, 8; "New Elements of the International Regime on Access and Benefit-Sharing of Genetic Resources: The Role of Certificates of Origin," Federal Agency for Nature Conservation, 2005, https://www.ecosystemmarketplace.com/wp-content/uploads/archive/documents/Doc_413.pdf, 48; Jane Eva Collins et al., "Stakeholder Perspectives on Access and Benefit-Sharing for Areas Beyond National Jurisdiction," *Frontiers in Marine Science* 7 (2020), https://doi.org/10.3389/fmars.2020.00265.

28 United Nations Nagoya Protocol on Access to Genetic Resources and the Fair and Equitable Sharing of Benefits Arising from their Utilization to the Convention on Biological Diversity, October 29, 2010, 3008 UNTS 1 (herein: "Nagoya Protocol"). See also Rabone et al., "Access to Marine Genetic Resources," 3 (emphasis added).

29 *Decision Adopted by the Conference of the Parties to the Convention on Biological Diversity,* GA Dec XIII/16, UN Doc CBD/COP/DEC/XIII/16 (December 16, 2006); Nagoya Protocol.

30 "Digital Sequence Information—Clarifying Concepts," Deep-Ocean Stewardship Initiative, March 2020, https://www.dosi-project.org/wp-content/uploads/070 -DSI-Policy-brief-V4-WEB.pdf, 2.
31 Rabone et al., "Access to Marine Genetic Resources," 3.
32 Ibid., 7–8. For this reason, the Consortium of European Taxonomic Facilities has recently proposed to replace the term "DSI" with "NSD"—Nucleotide Sequence Data—which, they argue, is a more precise term as it specifically refers to *raw* sequences. "Digital Sequence Information on Genetic Resources – Concept and Benefit-Sharing," Consortium of European Taxonomic Facilities, 2019, https:// www.cbd.int/abs/DSI-views/2019/CETAF-DSI.pdf.
33 "2017, GGBN Letter to CBD on DSI," Global Genome Biodiversity Network, May 6, 2020, https://library.ggbn.org/share/s/RWfhVdVKSriNmj8MPvG8qw, 1; "Digital Sequence Information," International Chamber of Commerce Commission on Intellectual Property, May 30, 2017, https://iccwbo.org/con-tent/uploads/sites/3/2017/05/ICC-IP-position-paper-on-digital-sequence -information.pdf, 1. The scientific community has pointed to current barriers to taxonomic related biodiversity research arising from ABS regulations within national jurisdictions. See Rohan Pethiyagoda, "Biodiversity Law Has Had Some Unintended Effects," *Nature* 429, no. 6988 (2004): 129; Meegan-Kumar, "The Nagoya Protocol," 31–35; Dirk Neumann et al., "Global Biodiversity Research Tied Up by Juridical Interpretations of Access and Benefit Sharing," *Organisms Diversity & Evolution* 18 (2018): 1–12; Prathapan et al., "When the Cure Kills," 1405–1406. Similar concerns have been raised by several states that any future reg-ulation of access to marine genetic resources in areas beyond national jurisdiction could hinder marine scientific research. See International Institute for Sustainable Development, "BBNJ IGC-1 Highlights," *Earth Negotiations Bulletin* 25, no. 176 (2018), https://enb.iisd.org/vol25/enb25176e.html.
34 Sarah A. Laird and Rachel Wynberg, "Fact-Finding and Scoping Study on Digital Sequence Information on Genetic Resources in the Context of the Convention on Biological Diversity and the Nagoya Protocol," *Convention on Biological Diversity*, January 12, 2018, https://www.cbd.int/doc/c/e95a/4ddd/4baea2ec772be28 edcd10358/dsi-ahteg-2018-01-03-en.pdf. See also Jaspars, interview. In the BBNJ context, different delegations opted to replace the term DSI with the term "*in silico*," to indicate a more spatially specific and thus more restrictive definition. This approach was adopted in the Access to Marine Genetic Resources Report, which stated that: "The accessibility of MGR from sources in situ (on site), ex situ (samples in collec-tions) and in silico (information in databases) is key to the functioning of the deep-sea research community." Rabone et al., "Access to Marine Genetic Resources," 7.
35 Harden-Davies, interview.
36 Jaspars, interview.
37 Jeffrey Marlow (Assistant Professor of Biology, Boston University), telephone inter-view by author, October 17, 2019.
38 Rabone et al., "Access to Marine Genetic Resources," 4–7.
39 "Areas Beyond National Jurisdiction," Global Environmental Facility, 2021, https:// www.thegef.org/topics/areas-beyond-national-jurisdiction; see also Alex G. Oude Elferink, "Exploring the Future of the Institutional Landscape of the Oceans Beyond National Jurisdiction," *Special Issue: New Frontiers in Ocean Environmental Governance* 28, no. 3 (2019): 236–243; Alex G. Oude Elferink, "Coastal States and MPAs in ABNJ: Ensuring Consistency with the LOSC," *International Journal of Marine and Coastal Law* 33 (2018): 437–466; Vito De Lucia, "The Question of the Common Heritage of Mankind and the Negotiations towards a Global Treaty on Marine Biodiversity in Areas Beyond National Jurisdiction: No End in Sight?" *McGill International Journal of Sustainable Development Law and Policy* (2020), http:// dx.doi.org/10.2139/ssrn.3542384.

40 See Marie Bourel et al., "The Common of Heritage of Mankind as a Means to Assess and Advance Equity in Deep Sea Mining," *Marine Policy* 95 (2018): 311–316. See also Prue Taylor, "The Common Heritage of Mankind: Expanding the Oceanic Circle," in *The Future of Ocean Governance and Capacity Development*, ed. International Ocean Institute—Canada (Nijhoff: Brill, 2018), 142–150.

41 Joanna Mossop, "The Relationship Between the Continental Shelf Regime and a New International Instrument for Protecting Marine Biodiversity in Areas Beyond National Jurisdiction," *ICES Journal of Marine Science* 75, no. 1 (2017): 444–450. See also Leary, *International Law and the Genetic Resources*; T. Treves, "Principles and Objectives of the Legal Regime Governing Areas Beyond National Jurisdiction," in *The International Legal Regime of Areas Beyond National Jurisdiction: Current and Future Developments*, eds. Erik J. Molenaar and Alex G. Oude Elferink (Leiden: Brill, 2010), 5–25.

42 See Taylor, "The Common Heritage of Mankind."

43 Elizabeth Druel and Kristina Gjerde, "Sustaining Marine Life Beyond Boundaries: Options for an Implementing Agreement for Marine Biodiversity Beyond National Jurisdiction Under the United Nations Convention on the Law of the Sea," *Marine Policy* 49 (2014): 90–97.

44 "Introduction to Access and Benefit-Sharing," Secretariat of the Convention on Biological Diversity, 2011, https://www.cbd.int/abs/infokit/revised/web/all-files-en.pdf.

45 Thomas Greiber, "An International Instrument on Conservation and Sustainable Use of Biodiversity in Marine Areas beyond National Jurisdiction," IUCN, accessed October 6, 2021, https://www.iucn.org/sites/dev/files/import/downloads/paper_iii___options_and_approaches_for_access_and_benefit_sharing.pdf, 3.

46 From the United Nations, the global commons "have been traditionally defined as those parts of the planet that fall outside national jurisdictions and to which all nations have access." In "Global Governance and Governance of the Global Commons in the Global Partnership for Development Beyond 2015," United Nations, 2013, https://www.un.org/en/development/desa/policy/untaskteam_undf/thinkpieces/24_thinkpiece_global_governance.pdf.

47 Under this approach, ocean experts have been advocating for a differentiated view in relation to access to *in situ* resources, *ex situ* resources, *in silico* analysis, and relevant technology. See Greiber, "An International Instrument," 4.

48 Namely, plant genetic resources for food and agriculture listed in Annex I of the ITPGRFA and the influenza viruses covered by the PIPF. See "International Treaty on Plant Genetic Resources for Food and Agriculture," United Nations, 2009, http://www.fao.org/3/i0510e/i0510e.pdf.

49 Because of high investment costs, uneven distribution of technologies and expertise, and legal ambiguity. See David Kenneth Leary, *International Law and the Genetic Resources of the Deep Sea* (Leiden: Martinus Nijhoff, 2007), 170. See also Morten Walløe Tvedt, "Chapter 11 Marine Genetic Resources: A Practical Legal Approach to Stimulate Research, Conservation and Benefit Sharing," in *The Law of the Seabed*, ed. Catherine Banet (Nijhoff: Brill, 2020), 238–254; and Robert Blasiak et al., "Corporate Control and Global Governance of Marine Genetic Resources," *Science Advances* 4, no. 6 (2018): eaar5237.

50 Blasiak et al., "Corporate Control"; Merrie, "An Ocean of Surprises," 23. This situation is highly characteristic of an extractivist mentality, "where corporate entities and states are indistinguishable in their economic interests and activities." Gómez-Barris, *The Extractive Zone*, xviii.

51 Blasiak et al., "Corporate Control."

52 Leary, *International Law and the Genetic Resources of the Deep Sea*, 170.

53 Ibid.

54 Ibid.

55 Arnaud-Haond, interview. See also Arnaud-Haond, "Mind the Gap," 29–39.

56 Gómez-Barris, *The Extractive Zone*, 29.

57 Ibid. Although she is talking about South America, her analysis is not less poignant in the deep sea.

58 Janne E. Nijman, "Grotius' 'Rule of Law' and the Human Sense of Justice: An Afterword to Martti Koskenniemi's Foreword," *European Journal of International Law* 30, no. 4 (2019): 1105–1114, 1114.

59 UN General Assembly, Revised Draft Text of an Agreement Under the United Nations Convention on the Law of the Sea on the Conservation and Sustainable Use of Marine Biological Diversity of Areas Beyond National Jurisdiction, ¶ 1, November 18, 2019, A/CONF.232/2020/3.

60 Muriel Rabone (Data and Sample Collector for schistosomiasis collection, London's Natural History Museum), Zoom interview by author, September 28, 2020.

61 Kristina Gjerde et al., "Protect the Neglected Half of Our Blue Planet," *Nature* 554 (2018): 163–165.

62 United Nations Convention on the Law of the Sea art. 87, December 10, 1982, 1833 UNTS 432. See also Rabone et al., "Access to Marine Genetic Resources," 8.

63 Margaret Spring (Chief Conservation and Science Officer, Monterey Bay Aquarium & former chief of staff, National Oceanic and Atmospheric Administration under the first Obama Administration), telephone interview by author, October 15, 2019.

64 Arnaud-Haond, interview.

65 Jaspars, interview.

66 Elisabeth Mann Borgese, "The Process of Creating an International Ocean Regime to Protect the Ocean's Resources," in *Freedom for the Seas in the 21st Century: Ocean Governance and Environmental Harmony*, eds. Jon M. Van Dyke et al. (Washington, D.C.: Island Press, 1993), 23–37, 29.

67 Meegan-Kumar, "The Nagoya Protocol," 33.

68 Ibid. According to Meegan-Kumar, the over legalistic approach that left the scientists out was one of the pitfalls currently facing the Nagoya Protocol. See also Rabone et al., "Access to Marine Genetic Resources," 15.

69 As one scientist writes: "The scientific landscape is developing much more rapidly than associated legislation." Meegan-Kumar, "The Nagoya Protocol," 33. See also Rachel Weinberg and Sarah Laird, "Fast Science and Sluggish Policy: The Herculean Task of Regulating Biodiscovery," *Trends in Biotechnology* 36, no. 1 (2018): 1–3.

70 Sheila Jasanoff, *Science at the Bar: Law, Science, and Technology in America* (Cambridge: Harvard University Press, 1996), 11–12. See also Sheila Jasanoff, "The Idiom of Co-Production," in *States of Knowledge: The Co-Production of Science and the Social Order*, ed. Sheila Jasanoff (London: Routledge, 2014), 1–12. For a further discussion of this idea see Irus Braverman, *Gene Editing, Law, and the Environment: Life Beyond the Human* (Oxford: Routledge, 2017), 13–14.

71 Rabone et al., "Access to Marine Genetic Resources," 15. See Vierros et al., "Who Owns the Ocean?" See also Harriet Harden-Davies, "The Next Wave of Science Diplomacy: Marine Biodiversity Beyond National Jurisdiction," *ICES Journal of Marine Science* 75, no. 1 (2018): 426–434.

72 Jaspars, interview.

73 Ibid.

74 Fran Humphries (Senior Research Fellow and Program Leader of the Law and Nature Research Program, Griffith University), Zoom interview by author, October 6, 2020.

75 Jaspers, interview.

76 Humphries, interview.

77 Jaspars, interview.

78 Paraphrased in Martti Koskenniemi, "EJIL Foreword: Imagining the Rule of Law: Rereading the Grotian 'Tradition,'" *European Journal of International Law* 30, no. 1 (2019): 17–52, 17.

79 Ibid.

80 Nijman, "Grotius' 'Rule of Law,'" 1107.

81 Humphries, interview.

82 DOSI is a "global network of experts which seeks to integrate science, technology, policy, law and economics to advise on ecosystem-based management of resource use in the deep ocean." Deep-Ocean Stewardship Initiative, "Advancing Science-Based Policy."

83 DOSI, "Digital Sequence Information," 2.

84 The United States Supreme Court case *Chakrabarti* decided on similar grounds that what is "patentable" is based on whether or not human intervention into natural life occurred. *Diamond v. Chakrabarty*, 447 U.S. 303 (1980). There, the Court held that a live, human-made microorganism is patentable subject matter under statute providing for issuance of a patent to a person who invents or discovers any new or useful manufacture or composition of matter. The material in question was distinguished from Newton's discovery of gravity, Einstein's discovery of the formula $E = mc^2$, and any other discovery of the "handiwork of nature," since it did not exist in nature before being produced by Chakrabarty.

85 Deep-Ocean Stewardship Initiative, "Advancing Science-Based Policy."

86 DOSI, "Digital Sequence Information," 3.

87 Leonelli, *Data-Centric Biology*, 2.

88 Ibid., 195.

89 Ibid., 194.

90 Rabone et al., "Access to Marine Genetic Resources," 11. Data was extracted from areas beyond national jurisdiction at depths of 500 meters and greater from the Ocean Biogeographic Information System (OBIS).

91 Rabone et al., "Access to Marine Genetic Resources," 11.

92 European Commission, "Report of the Workshop," 17.

93 Jaspars further noted that although access to such a place is "phenomenally expensive, the sedimentation rates are so low in the Pacific, [that] what they found is that one meter or two meters [down] is already one million years." But these extraordinarily ancient microbes are situated relatively close to the seafloor. "They went to this water in the middle of nowhere, they drilled a hole a few hundred meters down, and they went down hundreds of millions of years [finding] a 100-million-year-old bacteria. It was a big news splash a few weeks ago." Jaspars, interview. News reports about this event include Elizabeth Pennisi, "Scientists Pull Living Microbes, Possibly 100 Million Years Old, From Beneath the Sea," *Science*, July 28, 2020, https://www.sciencemag.org/news/2020/07/scientists-pull -living-microbes-100-million-years-beneath-sea.

94 Muriel Rabone et al., "Assessing and Sharing Benefits from Marine Genetic Resources from Areas Beyond National Jurisdiction: Building on Best Practices in the Scientific Community," Deep-Ocean Stewardship Initiative, March 2019, https://www.researchgate.net/publication/332152611, 2.

95 Rabone et al., "Access to Marine Genetic Resources," 10. See Roderic D.M. Page, "DNA Barcoding and Taxonomy: Dark Taxa and Dark Texts," *Philosophical Transactions of the Royal Society B* 371, no. 1702 (2016): 20150334.

96 Rabone et al., "Access to Marine Genetic Resources," 10. See also Lisa C. Pope et al., "Not the Time or the Place: The Missing Spatio-Temporal Link in Publicly Available Genetic Data," *Molecular Energy* 24, no. 15 (2015): 3802–3809.

97 Rabone et al., "Access to Marine Genetic Resources," 10. See also Pope et al., "Not the Time or the Place," 3802–3809; Paolo Gratton et al., "A World of Sequences: Can We Use Georeferenced Nucleotide Databases for a Robust Automated Phylogeography?" *Journal of Biogeoraphy* 44, no. 2 (2016): 475–486; John Deck et al., "The Genomic Observatories Metadatabase (GeOMe): A New Repository for Field and Sampling Event Metadata Associated with Genetic Samples," *PLoS Biology* 15, no. 8 (2017): e2002925.

98 Rabone et al., "Access to Marine Genetic Resources," 10.

99 According to Jaspars, they were only traced back once, in the case of a face cream. Jaspars, interview, see also "Kiehl's Abyssine Cream +," Birchbox, accessed October 21, 2020, https://www.birchbox.com/product/1793/kiehls-abyssine-cream.

100 Pierre Taberlet et al., "Environmental DNA," *Molecular Ecology* 21, no. 8 (2012): 1789–1793.

101 Harden-Davies, interview; see also European Commission, "Report of the Workshop," 10.

102 European Commission, "Report of the Workshop," 16.

103 Yet some marine scientists have pointed out that this is unfounded. As one put it: "these monetary benefits have gained a prominence out of scale to their likelihood and without any evidence of their importance to date." Rabone et al., "Access to Marine Genetic Resources," 7.

104 Broggiato et al., "*Mare Geneticum*," 6.

105 Ibid., 22–23.

106 Ibid., 6.

107 Marlow, interview.

108 Jaspars, interview.

109 Humphries, interview; Kristina Gjerde (High Seas Policy Advisor, the International Union for Conservation of Nature Global Marine and Polar Program), in-person interview by author, Cambridge, MA, September 25, 2019.

110 Arnaud-Haond, interview.

111 Rabone et al., "Access to Marine Genetic Resources," 15.

112 Arnaud-Haond, interview. See also Prathapan et al., "When the Cure Kills," 1405–1406. The authors contend that "the NP [Nagoya Protocol] and resulting national ambitions on Access and Benefit Sharing (ABS) of genetic resources have generated several national regulatory regimes fraught with unintended consequences. Anticipated benefits from the commercial use of genetic resources, especially those that might flow to local or indigenous communities because of regulated access to those resources, have largely been exaggerated and not yet realized. Instead, national regulations created in anticipation of commercial benefits, particularly in many countries that are rich in biodiversity, have curtailed biodiversity research by in-country scientists as well as international collaboration."

113 Sarah Laird et al., "Rethink the Expansion of Access and Benefit Sharing," *Science* 367, no. 6483 (2020): 1200–1202.

114 Ibid., 1202. See Charles Lawson et al., "The Future of Information Under the CBD, Nagoya Protocol, Plant Treaty, and PIP Framework," *Journal of World Intellectual Property* 22, nos. 3–4 (2019): 103–119.

115 Jaspars, interview.

116 For a detailed exploration of the land-sea binary and its impacts see the Introduction to this volume. See also Moana Jackson, "Indigenous Law and the Sea," in *Freedom for the Seas in the 21st Century: Ocean Governance and Environmental Harmony*, eds. Jon M. Van Dyke et al. (Washington, D.C.: Island Press, 1993), 41–48; Stephen Allen, Nigel Bankes, and Øyvind Ravna, *The Rights of Indigenous Peoples in Marine Areas* (London: Bloomsbury Publishing, 2019).

117 Jaspars, interview.

5

OCEANIC HETEROLEGALITIES?

Ocean Commons and the Heterotopias of Sovereign Legality

Vito De Lucia

Introduction

Sovereignty is waning. Sovereignty is waxing. It is ebbing, and it is flowing, caught in a double movement, an undecidable oscillation, a tidal dynamic that finds no resolution. While eroded, deconstructed, unbundled, and challenged in multiple ways, sovereignty also remains, perhaps stubbornly, the fundamental idea of the international legal order. Even where it is absent, where it doesn't fully reach, where it is only a tentative presence, it looms, demands, constrains.

While sovereignty has projected its legality over land with ease, through occupation, appropriation, conquest, and exchange, at sea sovereignty has *always* struggled. It has been washed out by the waves, it has disappeared in them; it has tried to grasp oceanic waters, yet it has been repeatedly eluded by them. Its primary operation, the material inscription and thus the legal "characterization"[1] of land could not be reproduced at sea. Already in the 17th century, Dutch legal scholar Hugo Grotius understood this: a "ship sailing over the sea no more leaves behind itself a legal right than it leaves a permanent track."[2]

Yet sovereignty has constantly returned as a telluric[3] gaze projecting its will to dominate, know, and impose material marks over the fluidity and mobility of the sea. Timidly at first, solidly anchored on land,[4] then increasingly bold until it now reaches out far, far from land,[5] albeit in exchange for increasing dilutions of its intensity the farther from land sovereignty has ventured.[6] To support its expansion at sea, "impervious" to the territorial expressions of sovereign legality,[7] sovereignty has had to latch onto legal abstractions and geometric delineations. Only on the ocean floor has it been able to capture oceanic space in a firm

FIGURE 5.1 The curly-cue shape is a characteristic of this *chrysogorgid* octocoral, called *Iridogorgia*. Image courtesy of the NOAA Office of Ocean Exploration and Research, Discovering the Deep: Exploring Remote Pacific MPAs.

DOI: 10.4324/9781003205173-6

territorial embrace.[8] It has thus resorted to "shapeshifting,"[9] to "liquid"[10] and "terraqueous"[11] modalities, trying to adapt to the peculiar materialities of the ocean. It has unbundled itself across the spatio-legal delineations carved though the maritime zoning of the oceans enshrined in the United Nations Convention on the Law of the Sea (UNCLOS),[12] manifesting itself in various configurations and combinations of the three classical elements: *imperium* (public authority), *dominium* (ownership), and *jurisdictio* (jurisdiction).

Sovereign legality has, however, produced a particular "lawscaping"[13] of the sea, trying to normalize, mathematize, calculate, and delimit it in unique ways. The sea has "pushed back," in part thanks to the fundamental misalignment between the wet, fluid, slippery, mobile materiality of the oceans—well-illustrated in the work of political geographers such as Philip Steinberg and Kimberley Peters[14]—and the firm, telluric enfolding of sovereign legality.[15]

All of these contrasting tendencies are visible in the ongoing negotiations on a new global treaty on marine biodiversity in areas beyond national jurisdiction (BBNJ), which is to further project sovereign legality on ocean commons that—particularly the high seas, that is those sea areas beyond the limits of national jurisdiction—have for long escaped the firm grip of sovereignty. While formally *beyond* the reach of any and all individual sovereign claims, the ocean commons are in fact still arguably approached through a *logic* of sovereignty that tries to impose on the oceans the territorial ordering codified in the UNCLOS.[16] Such sovereign *legality* aims to order and orient the water, the ocean floor, marine life, and marine ecosystems according to a double logic of abstract geometrization and cartographic delineation.

Yet oceanic spaces remain materially and legally recalcitrant (something somewhat already recognized in UNCLOS), continuing to present both material and legal "monsters"[17] that the sovereign legality underpinning UNCLOS cannot displace or govern, except through a process of de-intensification and self-problematization. Accordingly, UNCLOS contains "gradients of sovereignty"[18] in the different maritime zones it establishes, from the full territorial sovereignty of the territorial sea to the mobile flag state jurisdiction on the high seas. UNCLOS, then, already contains a "heterotopic" spatiality (to anticipate an important theme of this chapter) linked to its mobility, perhaps its slipperiness, one at any rate contrasted with the firm, telluric, singular, and ultimately hegemonic sovereign spatiality.

Against this background, this chapter will suggest that oceanic spaces, particularly those beyond national jurisdiction now being subsumed under novel regulatory regimes in the context of the BBNJ negotiations, can be usefully approached through the notion of heterotopias. The concept of heterotopia, proposed by philosopher Michel Foucault as a way to address the problematic of space, refers to any site in which space is "other," including, importantly, those counter-sites that resist the dissolution of social space produced by the calculating and abstracting spatiality of modernity that now represents normal space.[19] And as space and law are entangled, and mutually constitute each other, heterotopias may reflect also other legalities that may emerge and resist the embrace of sovereign legality and its normalizing spatiality. The future BBNJ treaty attempts to regulate, govern, and regularize the last open oceanic spaces by utilizing marine environmental protection—biodiversity conservation, more precisely—in ways that resonate with how coastal states have legitimized their encroachment at sea with regard to the continental shelf[20] and the exclusive economic zone,[21] as both claims found their original, unilateral justification in the need to protect marine resources.

The concept of heterotopia will be tested to understand whether it may help read oceanic spaces as subversive sites or counter-sites vis-à-vis hegemonic legality, as spaces resisting against and problematizing the sovereign construction of space at sea, as spaces of *heterolegalities*—that is, alternate forms of the legal—that in the context of the crises of the Anthropocene—epistemological, aesthetic, and legal[22]—are a crucial necessity in order to be able to navigate precisely such a novel and unprecedented geopolitical epoch; or whether they may remain entangled with such hegemonic legality and even facilitate its reproduction, and thus perhaps best be understood as spaces of biopolitical "encaring," which, as I will show, is a novel horizon of sovereign legality in the Anthropocene.

The chapter will thus first outline the notion of heterotopias and then explore an oceanic inflection of the notion. Next, the chapter will discuss sovereign legality in terms of its territorializing thrust and its extension and transformation from territorialization to biopolitical encaring. Finally, I will bring to bear the preceding discussions on the ongoing BBNJ negotiations, with the view of testing and exploring in some details the productive potential of heterotopia in that context, particularly for the articulation of "heterolegalities." Toward the end, the chapter will outline some tentative conclusions.

Heterotopias

The concept of "heterotopia," which originally appeared in the realm of medicine,[23] was appropriated by Michel Foucault during his turn to what he called the "epoch of space."[24] In a lecture entitled, "Of other spaces: Utopias and Heterotopias," Foucault offered a set of remarks aimed at articulating theoretical principles for thinking about space. The lecture, held in 1967 in French, was only published in 1984 in French and in English translation in 1986.[25] While Foucault himself would not address further the question of space in such an explicit manner,[26] nor would he give much consideration to the essay,[27] the conceptual frame developed in that lecture has proved apt at being further articulated both in relation to the principles outlined in the essay—which would constitute the initial basis for what Foucault called "heterotopology," that is the "study, analysis, description, and 'reading' of ... these other places"[28]—and in novel directions.[29] Foucault's starting point was the emergence of the modern concept of space as a geometric extension, and the corresponding re-organization of places as points in a grid of abstract coordinates. This "desantification"[30] of space, as he calls it, which indicates the dissolution of the medieval concept of social space (that is a hierarchical ensemble of places, a "space of emplacement")[31] determines the de-locatization of things and the re-articulation of places as sets of points on a grid of infinitely open space. This transformation allows the appropriation, delimitation, formalization, and management of space through technical means.[32] However, this dissolution is not complete.[33] Modern space remains traversed by a significant number of places still localized and operating along a counter trajectory. These sites have the peculiar characteristic of "being linked to all other sites," but in a way that "neutralize[s]" or "contradict[s]" them.[34] These are the heterotopias. Heterotopias, for Foucault, share a set of characters with utopias, in that they are both "counter-sites."[35] Yet heterotopias are, in some ways and to varying degrees, "effectively enacted utopias in which the real sites, all the other real sites that can be found within the culture, are simultaneously represented, contested, and inverted."[36]

Foucault identifies six principles that characterize heterotopias. Heterotopias are a constant for every human group. Each heterotopia has a specific function; heterotopias can juxtapose in a single real place a multiplicity of other places (theater, cinemas, et cetera). Heterotopias are linked to particular "slices in time," ruptures of the regular temporality[37] (and as such they have been described as "spatial *and* temporal disruptions").[38] Heterotopias, additionally, "always presuppose a system of opening and closing that both isolates them and makes them penetrable," either by way of compulsion (e.g., prison) or through purifications.[39] Heterotopias, finally, are either places of illusion (that expose as illusory every other real place) or of compensation, that is places that are other and real, yet better arranged (from a multiplicity of perspectives) than the hegemonic and normalized space.

Foucault's thinking about other spaces "is not a neat theoretical package ready to be applied."[40] Indeed the concept of heterotopia (and "Foucault's geography"[41] more broadly) has been described as "nebulous and nomadic,"[42] "confusing," "briefly sketched," "provisional,"[43] and ultimately "inadequate ... for analysing spatial difference."[44] Yet, Foucault's spatial thinking remains "inspirational for theoretical reflection"[45] and a useful voice to "stir into"[46] the debate about space (and law, in our context) precisely for its unaligned, unorthodox, and theoretically productive perspective on other spaces. It is probably precisely the "nomadism and fuzziness" of his spatiality that has made Foucault's thinking about space so susceptible of further elaboration, not the least as it insightfully captures the power-knowledge intersections at each of the multiple spatial scales the notion of heterotopia may latch onto,[47] thus prompting commentators to emphasize Foucault's "hugely productive encounter with geography."[48]

Additionally, Foucault's first utilization of the concept of heterotopia, in the book *The Order of Things*, also suggests other ways to further elaborate the concept that are productive for the purpose of this chapter, which is to explore possibilities of "other" legalities for ocean commons. In Foucault's book, heterotopias are described as "disruptive classifications, or instances where referential language breaks down."[49] Foucault utilized Borges's famous Chinese taxonomy,[50] a taxonomic monstrosity, to illustrate a discursive space driven by a "deviant ... logic,"[51] a discursive and taxonomic non-space that disrupts and subverts rational classifications. It is thus a *discursive* space which is introduced in such manner, one that challenges hegemonic modes of organizing and classifying the world. Later, Foucault would deploy the concept of heterotopias and their disruptive capacity in a more squarely spatial dimension. He would however retain heterotopias' subversive counter-logic, a counter-logic that then traverses both physical space and the discursive space of language, taxonomies, and classifications. And as space is re-read through the filter of heterotopias, it is also important to consider that the discursive heterotopias from which Foucault started his reflections on topological disruptions, remain crucial to understand precisely the counter-logic that ocean spaces offer up against sovereign legality, as I will try to illustrate.

In this respect, in this chapter I will be primarily concerned with the general idea that arguably traverses all inflections of the idea of heterotopias, namely the idea of heterotopias as "counter-sites," and will try to elaborate on this potential with respect to a distinct oceanic perspective. The character of counter-sites descends from the fact that heterotopias can be understood as examples "of resistance against the homogenization and normalization of space."[52] Heterotopias, qua "other places," are a key resource for imagining "other possibilities for the present and the future,"[53] and more specifically for our purposes may create an opening for disarticulating the stability of hegemonic legality in its construction of a singular space. As critical legal scholar Andreas Kotsakis has suggested when discussing the role of heterotopias for a critical environmental law, through their "peculiar spatial function, heterotopias expose and oppose, invert and divert, disassemble and upset legal and political"[54] hegemonic closures and "counteract and subvert the dominant classifications"[55] of what I am calling here sovereign legality. This last point offers a clear if implicit reference to the discursive dimension of heterotopias, and thus allows us to combine the *spatial* alterity of oceanic spaces vis-à-vis sovereign spatiality with the disruptive counter-logic of oceanic materialities vis-à-vis sovereign classificatory logic.

After this general outline, I will discuss whether and how heterotopias may have a uniquely *oceanic* dimension or, at any rate, whether and how it is possible to outline specifically *oceanic* heterotopias.

Oceanic Heterotopias

Foucault ends his essay on heterotopias by presenting the "ship" as a heterotopia "par excellence."[56] This example, obviously linked to oceanic imaginaries, has been picked up and expanded on, as it combines, at once, all principles of heterotopias identified by Foucault.[57] Other authors have reinforced this view of the privileged heterotopic character of the ship, described as the "heterotopia of heterotopias,"[58] for example in light of the ship being "a summary of the world as it is not."[59] Foucault himself had made this sort of connection between the ship and heterotopias when discussing the "ship of fools" as a spatial and discursive other in his book *Madness and Civilization*.[60] Indeed, the very book has been described as "very much a spatialized account of the institutionalized medical confinement of madness."[61] The notion of heterotopia has further inspired accounts of ocean liners,[62] and other alternate spatialities linked to seafaring vessels of different types.[63]

The notion of heterotopia has also, however, and more importantly for our purposes, prompted theoretical reflections on the broader idea of oceanic space: the sea as "an 'other place' with special transformative power,"[64] partly linked to its double character of "element and *locus*."[65] In this latter respect, it is the high seas that dominate the heterotopic imaginary of scholarly literature since it exists (juridically) outside of national territorial boundaries, and as such it has been often considered "a space of multiplicity, outside the law of any one sovereign power."[66]

The "heterotopic dimension of the open sea"[67] has also inspired practices such as the Exterritory Project, an experiment of cultural and political production to be localized in the "heterotopic" space of the high seas, with the view of finding inspiration in those oceanic areas outside the reach of sovereign legality for new political discourses and imaginaries—or for new classifications and rationalities. Yet these endeavors—these ideas and these practices—did not engage with the notion of heterotopias as a critical legal theoretical register, and either focused on ships as discrete objects that navigate the seas, or generally outlined how the high seas, as such (rather than as an ensemble of multiplicities and particularities itself), may function as a heterotopia.

FIGURE 5.2 The Southern Sea as a "heterotopic dimension of the open sea." Photo by Christopher Michel, licensed under the Creative Commons Attribution 3.0.

Philip Steinberg also has discussed marine heterotopias in his seminal book *The Social Construction of the Ocean.*[68] Inspired perhaps more by Hetherington than by Foucault, Steinberg also approaches the open space of the sea as a heterotopia, though cautioning against the unidimensional consideration of heterotopias as "places of resistance," and instead underlying, again with Hetherington, how such alternate spatial articulations operate just as easily as sites that can facilitate the reproduction of capitalist power relations. Rightly so, the prison, one of the paradigmatic heterotopias for Foucault (and for subsequent commentators), opens alternative (or better, specialized) spatial and legal configurations—as well as alternative, or better, specialized *legal* configurations—that are *instrumental* to the exercise of hegemonic sovereign legality. And this ambiguity between resistance and hegemony is precisely a crucial element that will enable a nuanced exploration, from the perspective of heterotopias, of marine protected areas (MPAs) later in this chapter.

Steinberg also attempts to problematize the notion of alternative spatialities in relation to relevant legal constructions such as the principle of common heritage. However, and besides the question of whether the common heritage qualifies as a heterotopia (both in discursive and spatial terms), Steinberg did not fully explore the productive potential of the notion of heterotopias from the perspective of critical legal imaginaries. Such perspective may wish to articulate a series of problematizations in relation to the *production* of the high seas as a heterotopia, the genealogy of this production[69] (something which Ranganathan has done, to an extent, with regard to the ocean floor), the implication of its being a product of the delineations of sovereign legality (and power) through maritime zoning, and its qualification as ocean commons and the implication of the purported juxtaposition of ocean commons to sovereign legality.

In light of these questions, it appears that the full productive potential of the notion of heterotopias has yet to be tapped into, particularly from a legal perspective. There are multiple degrees of intensity and of spatial and temporal resolution that can guide the exploration of *oceanic* heterotopias, and this chapter can only offer some initial illustrations, without pretending to be exhaustive. On the contrary, this initial attempt at outlining a tentative map of heterotopias seeks to only outline *some* of the potential the concept contains for the purposes of imagining what I am calling "heterolegalities"[70]—that is, transposing onto the legal plain the otherness of space identified by Foucault as potentially subversive and counter-hegemonic. Such outline is intended to be supplemented, complemented, adjusted, corrected, and added to. Then, by way of fuller, multi-perspectival approaches to the oceanic potential of the notion of heterotopias—also expanding on the initial Foucauldian catalogue—it may become clearer whether and how far heterotopias can stimulate theoretically rich engagement with forms of "blue" legalities[71] that remain effectively outside the reach of, or divert, invert, oppose, and resist hegemonic, sovereign legality.

In this respect, the history of sovereignty at sea can in some ways be thought of as the history of the (attempted) territorialization (and desanctification!) of the oceans,[72] and in parallel as the history of the continuous normalization and colonization of heterotopias—yet heterotopias only exist in a relational sense, as the obverse of hegemonic spaces, as the contrasting opposite or as the other of the normalizing sovereign legality. In fact, we could even say that law, qua sovereign legality, constructs heterotopias by determining borders and determining spatialities that run against, or intersect, traverse, and conflict with the chaotic, fluid, deep spatiality of the oceans.[73] Thus we can paradoxically say that the existence, resolution, and intensity of heterotopias are a manifestation or a reflection, at any given juncture, of the degree of territorialization of the seas. Simultaneously, however, the existence, resolution, and intensity of heterotopias express a degree of recalcitrance of ocean spaces and oceanic life to the sovereign command, particularly as the sovereign command re-articulates itself as biopolitical "encaring."[74] Sovereign legality, we can venture, produces heterotopias as its waste, as that excess that it cannot subsume, consume, and digest. Heterotopias thus remind us that hegemony is always tentative and incomplete.

In the next section, I will draw on a longer argument that I developed elsewhere[75] to outline how sovereign legality is bound with histories of territorialization of the sea, despite the latter's recalcitrance and fluid disarticulation of the attempted sovereign grasp. Subsequently, I will briefly show how sovereign legality has itself stretched to operate as biopolitical "encaring." This is a crucial move that is particularly apt at capturing the heterotopic byproduct of the regulation of ocean commons that is being sought through the ongoing negotiations toward a global treaty on marine biodiversity in areas beyond national jurisdiction. The latter will be the focus of the last section of the chapter.

Sovereign Legality: From Territorialization to Biopolitical *Encaring*

The history of sovereignty at sea can be considered as the history of the (attempted) territorialization of the sea.[76] The telluric projection from land to sea had been a constant desire of sovereignty and remains fundamental today,[77] despite the political (and economic) expediencies associated with the notion of the freedoms of the seas. In the modern law of the sea, this idea is captured by the principle that land dominates the sea, according to which "the land is the legal source of the power which a State may exercise over territorial extensions to seaward."[78] This principle is arguably the basis of every "encroachment on maritime expanses"[79] and fuels and shapes the process of territorialization of the sea that has accompanied, with ebbs and flows, the entire history of the law of the sea, but that exploded in earnest in the 1940s and is proceeding apace even today.[80]

Drawing on the work of political geographer Stuart Elden,[81] territorialization is understood here as a process of transforming places into space and then into territory. So, in terms of process, there are two transformations involved. One from places, understood as a multiplicity of living and lived habitats, to a

singular, regularized, geometric space, and entails the transformation of the living into the measurable. The second transforms space into territory, and entails
the transformation of the measurable into the controllable. For the purposes
of this chapter, territory is understood, with Elden, as a "political technology"
aimed at "measuring land and controlling terrain."[82] Territorialization requires
the combination of the technical (for measuring) and the legal (for controlling),
or, in other words, knowledge and power. Territorializing the sea means the
transformation of living spaces, —what Massey would call "simultaneit[ies] of
stories"[83] and *"spatio-temporal event[s]"*[84] —into measurable space, where the purpose of measurement is control. The sea is thus "disciplined," made "predictable
and comprehensible."[85]

Without delving into the historical trajectories of territorialization, something which I did in some details elsewhere,[86] it will be sufficient to recall
briefly the spatio-legal architecture ultimately enshrined in UNCLOS.[87] Under
UNCLOS, territorial waters may extend up to 12 nautical miles from the baseline.[88] However, UNCLOS has further fragmented oceanic space through a grid
of maritime zones enabled by the ability to measure, an ability which in turn
has enabled the control of maritime space and its telluric marking. The area that
extends from the end of the territorial sea and up to 200 nautical miles has been
regulated as the exclusive economic zone (EEZ) of coastal states,[89] which enjoy
sovereign rights (though balanced by some duties) over marine living resources,
while certain aspects of the high seas freedoms remain intact (e.g., navigation).
The seabed and its resources have been placed either under state sovereignty,
through the notion of continental shelf[90]—a sub-marine continuation of a land's
geophysical base—or under a common property regime regulating access to the
resources of the seabed and its subsoil beyond the limits of national jurisdiction,
through the regime of the Common Heritage of Mankind.[91]

It seems, in brief, that a demarcation of boundaries has been established on the
seas, despite the Grotian argument that ships leave no absolute track from which
occupancy may be asserted and territories delineated. Technology has certainly
had a crucial role in this legal demarcation of the oceans. But the underlying
framework, its crucial legal underpinning, remains arguably the compulsion for
territorialization, for the sovereign ordering of space through law.

In the context of the Anthropocene, meanwhile, territorialization is also a step
that preludes and is an antecedent to a more subtle, fluid, and abstracting form of
sovereign legality: biopolitical "encaring." The latter is less contingent on a strict
territorial control (although it has been used to justify sovereign encroachments
on the high seas),[92] which at sea has always faced material challenges and could
only be accomplished in the rarefied plane of abstract measurements and calculations, with the consequence that sovereign legality has had inevitably to reduce
its intensity the farther from shores it tried to reach.[93]

Biopolitical encaring, in brief, refers to the positive inflection of the exercise
of power that finds its legitimation and its goal in caring for nature, and, in our
case, for the oceans and their health.[94] Once nature is attracted within the sphere

of knowledge and control of sovereign legality, a further need, turned into goal, presented itself to postmodernity in the context of the Anthropocene: the world suffers the pervasiveness of humankind's impact on the earth, whose "body," including its liquid oceans, has been physically marked and reshaped. The imaginary of the earth has been however equally reshaped: from a hostile realm to be subjugated and exploited through sovereign commands, the earth—and its oceanic space —is reframed as a vulnerable domain in need of protection. In this context, "[p]ower governs no longer through sovereign command, but through technical norms and scientific regimes of knowledge. Power, in its biopolitical inflection, becomes equivalent with *Earth-care*."[95] However, as I have argued elsewhere, in order to care for earth, "biopolitics must engulf nature in its entirety under a conceptual and regulatory framework where care and subjugation, vulnerability and productivity, life and death are constantly entangled in a reciprocal and inevitable relation."[96] In this novel context, sovereignty while apparently displaced, "returns inevitably as a key modality of biopolitical intervention"[97] every time nature is unbiddable, unpredictable, or dangerous. Biopolitics remains thus caught in an undecidable dilemma, where care and subjugation are inevitably entangled, and at the same time determines the continuous oscillation between the "displacement of sovereignty and its eternal return."[98]

This dilemma, this continuous oscillation, is precisely what I have called encaring, a term that draws inspiration from Heidegger's concept of enframing[99] but is applied to the biopolitical focus of power as an exercise of life-affirming interventions in the world, that is, power *as* care. This, I will explain in the next section, is a relevant way to frame the discussion of the heterotopias that are being produced in the context of the BBNJ negotiations (and especially marine protected areas). In that context, in fact, heterotopias are produced and reproduced at the crossroad between sovereignty, biopolitics, and resistance, and it is in the center of that crossroad that it may be possible to read ocean commons differently—as heterolegalities.

The Production of Legal Heterotopias in the BBNJ Negotiations

Against the background outlined so far, this section will explore the ways in which the BBNJ negotiations, as the latest expression of the (biopolitical) territorialization of the oceans, may be usefully approached through the concept of heterotopia. The existence of important legal and governance gaps related to BBNJ has been the focus of a UN process for almost two decades. In 2003, the UN Open-ended Informal Consultative Process on Oceans and the Law of the Sea (UNICPOLOS) underlined the urgency of developing norms and mechanisms aimed at protecting vulnerable marine ecosystems, especially in areas beyond national jurisdiction.[100] In 2004, the General Assembly of the United Nations (UNGA) established an Ad Hoc Open-ended Informal Working Group to study issues relating to the conservation and sustainable use of marine biological diversity of areas beyond national jurisdiction (BBNJ WG).[101] In its 2011 report, the BBNJ WG recommended that a "process be initiated" by UNGA that could include, among other options, the development of a multilateral agreement under UNCLOS on marine biodiversity in areas beyond national jurisdiction.[102]

The report identified four substantive areas to address urgently, "together and as a whole":[103] marine genetic resources (MGRs), including questions on the sharing of benefits; measures such as area-based management tools (ABMTs), including marine protected areas (MPAs); environmental impact assessments (EIAs); and capacity-building and the transfer of marine technology.[104] On the basis of the final report of the BBNJ WG, submitted in 2015,[105] UNGA decided to "develop an international legally binding instrument under the United Nations Convention on the Law of the Sea on the conservation and sustainable use of marine biological diversity of areas beyond national jurisdiction" (ILBI).[106] UNGA launched thus a preparatory committee (PrepCom) to "make substantive recommendations" to UNGA on elements of an ILBI.[107] The PrepCom submitted its report in July 2017.[108]

Finally, on the basis of the recommendation of the PrepCom,[109] UNGA launched an intergovernmental conference (IGC) on December 24, 2017.[110] At the time of writing, the IGC has held three of the four substantive sessions scheduled in the resolution. A fourth session was schedule for March/April 2020, but was postponed due to the Covid-19 pandemic. Subsequently, an intersessional meeting was held in the Fall of 2020, to maintain momentum, while the last IGC has now been tentatively rescheduled to take place in the "earliest possible available date in 2022."[111] The geographical scope of the ILBI is likely to include the high seas water column and the seabed beyond the limits of national jurisdiction (i.e., the Area)[112]—that is, those areas that are usually referred to as global ocean commons.[113]

As anticipated, the BBNJ negotiations offer a useful illustration of how sovereignty and its attendant sovereign legality manages to loom, demand, constrain even where it doesn't fully reach, where it is only a tentative presence. Indeed, the goal of the BBNJ negotiations is the adoption of a governance and regulatory framework for those areas that, while formally *beyond* the reach of sovereign claims—the ocean commons, the high seas, the open space of the oceanic expanse that much literature singles out as a heterotopia, as mentioned above—are still approached through a *logic* of sovereignty, of a territorial ordering that, codified in UNCLOS, shapes the oceans in a very telluric sense, that orders and orients the water, the ocean floor, marine life, and marine ecosystems.

The BBNJ negotiations, in other words, are approaching the ocean commons through the perspective of sovereign legality, through the will to territorialize, in the sense of rendering calculable and controllable the high seas (and to a less extent the Area, already in part territorialized).[114] Thus, with regard to the high seas as a space historically "impervious to human law,"[115] the BBNJ process operates precisely as an instrument of sovereign legality, as its starting point and its ultimate goal is indeed to fill existing legal and governance gaps, and thus to normalize unruly space and bring into the fold of sovereign legality the high seas:

not entirely, completely, or without excess, but with more intensity than is the case today, with the particular aim of allowing partial territorializations that are in turn functional to the biopolitical encaring of oceanic spaces and oceanic life that is the embedded goal of biodiversity conservation.

In other words, the BBNJ process is an important institutional and political context where sovereign legality is simultaneously intensifying its territorializing reach, reformulating itself as biopolitical encaring *and* producing heterotopias. Such heterotopias may conform to one or more of the principles outlined by Foucault and discussed earlier in this chapter, or may more broadly be a manifestation of a heterotopic counter-logic that sits at the intersection of space and discourse and problematizes the classificatory logic of sovereign legality through taxonomic monstrosities.

A first example related to how the new treaty will lay out the rules to govern hydrothermal vents, which are liminal sites between land (ocean floor) and sea (water column). Hydrothermal vents traverse multiple materialities and inhabit multiple worlds: water, gas, rock, soil, living, and nonliving; fluid and solid. Vents occupy thus multiple spatialities and are constituted by a multiplicity of stories across multiple temporalities. Their complexity eludes sovereign legality, obsessively bent on sorting out to which maritime zone hydrothermal vents belong, to which domain (the living or the nonliving) and thus, consequently, which legal regime is applicable.[116]

The transversal spatiality of hydrothermal vents, their inhabiting multiple worlds, locates them in a spatial horror that can't be subsumed neatly within any of the spatial delineations of sovereign legality, which it transgresses. Additionally, the same transversal localization activates precisely that taxonomic counter-logic

FIGURE 5.3 Looking through the shimmering water of a hydrothermal vent in the Caribbean Sea, August 2011. Credit: NOAA Okeanos Explorer Program, Mid-Cayman Rise Expedition 2011.

that challenges sovereign legality and upsets its normative closures, thus opening space for another reading of oceanic materialities that in turn may offer an
opportunity to read other legalities, that emerge from the particular materialities
of the ocean itself, not reducible to sovereign legality.

This spatial alienation, this heterotopic dimension that resists simultaneously the normalization of space and is deviant of the attendant classificatory
logic, is precisely *determined* by sovereign legality as it tries to normalize and
categorize space. It is indeed law that produces such other spaces as it cannot simply integrate them within its existing spatial architecture characterized by "singularity, stability and closure,"[117] as evidenced also by doctrinal
interventions.[118]

MGRs offer another useful example of a heterotopia that is at once spatial, material, and discursive. MGRs remain in fact difficult to integrate in the
framework of sovereign legality due to their multiple ontological natures, spatial origins, transversal locations, and regulatory classifications. MGRs present
themselves to the regulatory gaze through practices that approach them simultaneously as embodied in living organisms or as abstracted digital information.
They may be harvested *in situ*, that is in the place they inhabit, delocalized in a
collection (*ex situ*), or accessed in a spaceless digitalized form (*in silico*).[119] MGRs
may be additionally present around hydrothermal vents, making their attribution to one or another maritime zones problematic, if not impossible. Finally,
the same materiality may be subsumed by different regulatory regimes depending on whether the material substrate (a living organism) is considered from
the perspective of fisheries and thus as commodity or from the perspective of
bioprospecting, and thus as genetic material.

Here, again, we encounter heterotopias that are primarily spatial and those that
fuse the spatial and the discursive. Again, we encounter incompatible spatialities
but also incompatible taxonomies. In both these instances the notion of heterotopia denotes additionally a *discursive* space, which challenges hegemonic modes
of organizing and classifying the world, in ways similar to Borges's taxonomic
monstrosity. The distinctions between living organisms and rock, or between
water and gas, or between material organism and its digitalized information, are
precisely the type of hegemonic taxonomies that are upset and resisted by heterotopias in their capacity of counter-sites. Additionally, however, these heterotopias also produce heterolegalities, inasmuch as they offer alternative spatialities
and deviant classificatory logics that contain a subversive normative dimension.
If life and rock, or water and gas cannot be segregated analytically, materially,
or legally, the entire distinction underpinning the law of the sea between living
and nonliving resources, with its attendant separate legal regimes, falls apart.
The very language of law, confronted with such taxonomic monstrosities, may
break down in a manner similar to the linguistic and logic breakdown caused by
Borges's Chinese taxonomy.

A third example of how sovereign legality produces heterotopias in the context of the BBNJ negotiations is MPAs. With the caveat that such productivity is

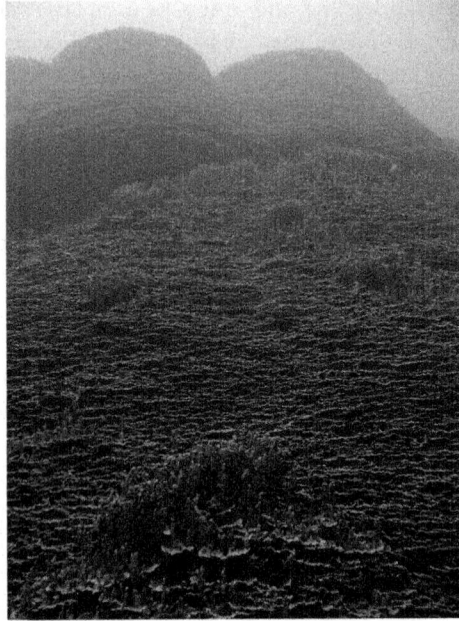

FIGURE 5.4 Scleractinian coral structures in the Commonwealth of Northern Mariana Islands in the Pacific Ocean. Photo by Kevin Lino, 2009. Courtesy of NOAA/NMFS/PIFSC/ESD.

not exclusive to the high seas, given that (marine) protected areas are a consolidated conservation tool especially for areas *within* national jurisdiction,[120] what is peculiar with regard to the BBNJ context is that the topic of MPAs is simultaneously conceived as a mechanisms for the protection of global ocean commons—itself a heterotopia produced by sovereign legality and one that normalizes it through what Kathryn Milun has aptly called the "uncommons"[121]—and integrated within the continuous albeit graded process of territorialization of the high seas.

MPAs reflect the dynamic oscillations between heterotopias as sites of resistance and subversion and, by contrast, as sites of reproduction of hegemony. The envisioned regulatory framework for the designation of MPAs—and indeed the very concept of protected area[122]—will enable the further fragmentation of space through the ambiguous bracketing of ocean areas for their particular ecological and biological sensitivity. This spatial delineation will in turn operate as a "force or line designed to keep chaos … at arm's length,"[123] separating "the sacred [from] the abject."[124] In this light, UNCLOS's general rule on the protection and the preservation of marine environment, as further implemented through the future ILBI, will rely upon acts of partition and classification in which "islands of wildness … are conceivable only on the basis of an ongoing and generalized ecological violence."[125] This, suggests critical criminologist Mark Hasley, is a dialectic between monstrous and majestic that traverses environmental law.[126]

From this perspective, it may thus seem that MPAs are sites that facilitate the functioning of sovereign legality *elsewhere*. Thus MPAs could conform to Foucault's second principle of heterotopias, that is, that of the alignment with a "precise and determined function."[127] This function can be read either as the formally intended function of protecting and preserving the marine environment in the oceanic spaces beyond national jurisdiction, or, from a rather more critical angle, as the heterotopic inversion of capitalist practices, an inversion whose role is precisely to enable widespread ecological violence in "normal" space. Additionally, MPAs are arguably also subject to the fifth principle of heterotopias, in that they are underpinned by a "system of opening and closing that both isolates them and makes them penetrable."[128] This is reflected in the MPAs management rules, the seasonal closing, the particular regulatory details that may be attached to the particular site, and in relation to particular features, temporalities, technologies, resources, activities that are allowed, disallowed, restricted within particular spatio-temporal horizons, et cetera.

But perhaps for imagining MPAs as heterotopias it is Foucault's last principle that is most immediately relevant: heterotopias "have a function in relation to all the space that remains."[129] As heterotopias of encaring, MPAs embody in fact the tragic dilemma of biopolitical sovereignty while also operating as "spatial *and* temporal disruptions" that *simultaneously* reflect and legitimize normal space. And it is precisely in that respect—the legitimation of normal space and its subjection to sovereign legality—that MPAs have a particular function in relation to all other spaces.

Indeed, MPAs create illusions of environmental stewardship, producing "ecological fixes," as Ramesh and Ray call it,[130] but ultimately remain enmeshed in the biopolitical encaring of sovereign legality as it offers compensation for precisely the widespread distribution of ecological violence in all other places and spaces. As has been observed already by the literature, MPAs are sites of sovereign enclosure for the purposes of conservation.[131] Yet this is a more common phenomenon in coastal waters, where MPAs effectively exclude subsistence access to resources and displace local communities.[132] The same logic may become operational for areas beyond national jurisdiction in the context of the BBNJ negotiations.[133]

At the same time, MPAs are a crucial site where subjugation and care intersect, where nature, and oceanic nature in particular, is protected and mastered, nurtured and controlled, restored and rendered productive. This spatial and discursive ambiguity is the particular distinctive element that makes it possible to understand MPAs as heterotopias of encaring, where the two dimensions of subjugation and care are caught in a continuous and unsolvable oscillation, transforming MPAs in a sort of infinity mirror room.[134]

The question then, as in other contexts, is whether or not it is possible to displace and disarticulate this tragic dilemma, which seems to enfold the whole field of international environmental law,[135] or whether MPAs as heterotopias of encaring—a characterization that also subsumes the preceding consideration of how MPAs may be described through Foucault's catalogue of heterotopias—are tragically destined to ultimately align to the new biopolitical articulation of sovereign legality that is arguably becoming prevalent in the Anthropocene.[136]

Conclusions

Heterotopias are ambiguous sites, they represent openings, ruptures in the fabric of sovereign legality and its spatial ordering, yet also simultaneously may facilitate the reproduction of such hegemonic legality, they simultaneously "reflect and contest."[137] They defy classificatory closures and yet remain enmeshed in regulatory taxonomies though which only ocean commons can be protected and marine biodiversity conserved. In this chapter, I have endeavored to test the concept of heterotopias in the context of the processes of territorialization of the oceans that, enshrined in UNCLOS maritime zoning, are being arguably further imposed on areas beyond national jurisdiction—the ocean commons—through the BBNJ negotiations.

The chapter has focused particularly on three examples—material, regulatory, or blended—that are central to the ongoing BBNJ negotiations. Thinking these three examples, hydrothermal vents, MGRs, and MPAs, through the concept of heterotopia has prompted different reflections as to the counter-spatiality potential each of these three examples possesses. If hydrothermal vents and MGRs more readily seem to upset or defy sovereign spatial delineations and regulatory classifications, thus expressing a clear subversive potential, MPAs, I have argued, reproduce the logic of territorial sovereign legality as they delineate oceanic spaces and regulate them on the basis of fixed boundaries. Moreover, MPAs operate as a crucial site where subjugation and care intersect, where oceanic materialities are protected and mastered, nurtured and controlled, restored and rendered productive. This spatial and discursive ambiguity makes it possible to understand MPAs as heterotopias of encaring, a novel spatio-discursive entanglement where the two dimensions of subjugation and care are caught in a continuous and unsolvable oscillation, transforming MPAs, as mentioned earlier, in a sort of infinity mirror room.

Each example of heterotopias discussed, however, arguably contains in different degrees seeds of subversion and productive potential for imagining oceanic heterolegalities. As Youatt has insightfully shown,[138] nature is recalcitrant. This recalcitrance, at once spatial and discursive, resists both the territorial normalization of space of sovereign legality and its Anthropocenic biopolitical inflection, producing hybridizations that may activate productive political (and legal) potential that is localized in, in our case, heterotopias that are simultaneously sites of facilitation of sovereign legality and counter-sites where such legality is resisted from what Negri would call a position of "*dentro e contro*."[139] The expression—which literally means "inside and against" —is borrowed from the political language of Italian operaism, and in particular Mario Tronti[140] and Antonio Negri[141] (albeit already Pashukanis moved his critique of law from a perspective of "*dentro e contro*").[142] The precise reconstruction of the historical emergence of the expression exceeds the scope and interest of this chapter. What is important to note, however, is the present significance (and expediency) of the expression as a *methodological* approach. In the context of this chapter, this expression refers to the double discursive and spatial localization of each heterotopia, as both *inside* and *against* sovereign legality, as both *inside* and, potentially, *against* the logic of biopolitical encaring. They reflect the territorial enclosure of space, indeed the transformation of inhabited, living places into bounded territories, while simultaneously resisting such enclosing gesture, as neither life nor oceanic mobilities can ever be fully apprehended within the sovereign embrace.

And it is precisely these counter-spatialities that offer opportunities for making legible alternative legalities—heterolegalities—emerging from the very subversive materialities of oceanic heterotopias. This potential, emerging from the inherent recalcitrance of nature, and expressed through its ungovernable excesses—be such excesses a reflection of oceanic mobilities or of the resistance of life to be contained by sovereign legality—allows us to read oceanic heterotopias as crucial points of intersections between the oscillating poles of the biopolitical encaring, and it is precisely *inside* this tension that it may be possible to decipher heterolegalities, alternate forms of the legal that are novel, plural, and emerging from the suspension of the conflict between subjugation and care, or perhaps from the acknowledgement that it is in the very point of intersection between the two that it is possible to articulate new legalities that are both *inside* and *against* sovereign legality and that may contain the seeds for its disarticulation.

Acknowledgments

I am grateful to all of the contributors to this volume for the enriching conversations throughout the project and for the comments on the initial drafts of this chapter. I am especially indebted to Surabhi Ranganathan for her insightful comments on an earlier draft and particularly grateful to Irus Braverman for her care in leading the project and in reviewing and editing this chapter. All errors and shortcomings remain mine.

Notes

1 Referring to the susceptibility of being materially scratched. The sea lacked this "character," observed Schmitt. Carl Schmitt, *The Nomos of the Earth in the International Law of the Jus Public European* (New York: Telos Press Publishing, 2006), 42–43.
2 Hugo Grotius, *Commentary on the Law of Prize and Booty* (Indianapolis: Liberty Fund, 2006), 334.
3 The adjective telluric derives from *Tellus*, which is the Latin name of the goddess of the earth: hence telluric indicates something earthly; see also "Telluric," Merriam-Webster, accessed October 9, 2021, www.merriam-webster.com/dictionary/telluric.
4 I am referring to the cannon shot rule.
5 Reflecting the historical development of the law of the sea, as territorial claims first extended up to 12 nm, through the codification of the territorial sea in UNCLOS,

then further up to 200 nm from the baseline, through the so-called Exclusive Economic Zone (EEZ).

6 In the EEZ, for example, coastal states *only* have sovereign rights (as well as duties) with respect to the utilization of resources, as opposed to full sovereignty (which however, is also subject to the right of innocent passage of third states' vessels).

7 Schmitt, *Nomos*, 181.

8 Surabhi Ranganathan, "Ocean Floor Grab: International Law and the Making of an Extractive Imaginary," *European Journal of International Law* 30, no. 2 (2019): 573–600.

9 See e.g., "Sovereignty: A Shape-Shifting Concept in Ocean Governance?" Expert Workshop, June 13, 2019, https://www.uu.nl/sites/default/files/rebo-sovereignty _expert_workshop_programme.pdf.

10 Costas Constantinou and Maria Hadjimichael, "Liquid Entitlement: Sea, Terra, Law, Commons," *Global Society* 35, no. 3 (2021): 351–372.

11 Liam Campling and Alejandro Colás, "Capitalism and the Sea: Sovereignty, Territory and Appropriation in the Global Ocean," *Environment and Planning D: Society and* Space 36, no. 4 (2018): 776–794.

12 United Nations Convention on the Law of the Sea, December 10, 1982, 1833 UNTS 397.

13 On the lawscape, see e.g., Nicole Graham, *Lawscape: Property, Environment, Law* (Abingdon: Routledge, 2011); Andreas Philippopoulos-Mihalopoulos, *Spatial Justice Body, Lawscape, Atmosphere* (Abingdon: Routledge, 2015).

14 See e.g., Philip Steinberg and Kimberley Peters, "Wet Ontologies, Fluid Spaces: Giving Depth to Volume through Oceanic Thinking," *Environment and Planning D: Society and Space* 33, no. 2 (2019). 247 264; Kimberley Peters and Philip Steinberg, "The Ocean in Excess: Towards a More-than-Wet Ontology," *Dialogues in Human Geography* 9, no. 3 (2019): 293–307.

15 This particular theme is discussed in more details in Vito De Lucia, "Ocean Commons and an 'Ethological' Nomos of the Sea," in *International Law and Areas Beyond National Jurisdiction: Reflections on Power, Knowledge, Space and Justice*, eds. Vito De Lucia et al. (Leiden: Brill, in press): 15–44.

16 See De Lucia, "Ocean Commons," 20ff.

17 Chet Van Duzer, "Hic Sunt Dracones: The Geography and Cartography of Monsters," in *The Ashgate Research Companion to Monsters and the Monstrous*, eds. Asa Simon Mittman and Peter Dendle (Farnham: Ashgate, 2013).

18 De Lucia, "Ocean Commons," 33.

19 Michel Foucault, "Of Other Spaces: Utopias and Heterotopias," *Architecture/ Mouvement/ Continuité*, trans. Jay Miskowiec (1984).

20 Proclamation No. 2667, "Policy of the United States with Respect to the Natural Resources of the Subsoil and Sea Bed of the Continental Shelf," September 28, 1945, https://www.trumanlibrary.gov/library/proclamations/2667/policy-united -states-respect-natural-resources-subsoil-and-sea-bed.

21 See Proclamation No. 2668, "Policy of the United States with Respect to Coastal Fisheries in Certain Areas of the High Seas," September 28, 1945, https://www.tru- manlibrary.gov/library/proclamations/2668/policy-united-states-respect-coastal -fisheries-certain-areas-high-seas-1.

22 For some references see e.g., Davor Vidas, "The Anthropocene and the International Law of the Sea," *Philosophical Transactions of the Royal Society A* 369, no. 1938 (2011): 909–925, 925; Daniel Matthews, "Law and Aesthetics in the Anthropocene: From the Rights of Nature to the Aesthesis of Obligations," *Law, Culture and the Humanities* (2019), https://doi.org/10.1177/1743872119871830; Vito De Lucia, "Rethinking the Encounter Between Law and Nature in the Anthropocene: From Biopolitical Sovereignty to Wonder," *Law and Critique* 31 (2020): 329–349.

23 In medicine, a heterotopia is defined as "the presence of normal cells in an improper location." Errol Francis, "Michel Foucault's Concept of Heterotopia and

Postcolonial Artistic Responses to Museum Spaces" (doctoral thesis, Slade School of Fine Art UCL, 2018), 16.

24 Foucault, "Of Other Spaces," 1.

25 For a detailed historical and conceptual reconstruction of the emergence of the notion of heterotopia, see Francis, *Michel Foucault's Concept*, especially Chapter 2.

26 Though Foucault would think spatially in much of his work in a more implicit manner, see e.g., Christopher Philo, "Foucault's Geography," *Environment and Planning D: Society and Space* 10 (1992): 137–161.

27 The essay was only added to the official corpus of his work "in extremis." Francis, *Michel Foucault's Concept*, 29. See also Andreas Kotsakis, "Heterotopias of the Environment: Law's Forgotten Spaces," in *Law and Ecology*, ed. Andreas Philippopoulos-Mihalopoulos (Abingdon: Routledge, 2011): 193–213, 196.

28 Foucault, "Of Other Spaces," 4.

29 See e.g., Kotsakis, "Heterotopias of the Environment," but also Matt Craven, "'Other Spaces': Constructing the Legal Architecture of a Cold War Commons and the Scientific-Technical Imaginary of Outer Space," *European Journal of International Law* 30, no. 2 (2019): 547–572. In the context of geography, Foucault's work on space has been widely discussed, criticized, and elaborated. See, among many, Philo, "Foucault's Geography"; Peter Johnson, "Unraveling Foucault's 'Different Spaces,'" *History of Human Sciences*, 19 no. 4 (2006): 75–90; Peter Johnson, "Foucault's Spatial Combat," *Environment and Planning D: Society and Space* 26, no. 4 (2008): 611–626; Arun Saldhana, "Heterotopia and Structuralism," *Environment and Planning A: Economy and Space* 40, no. 9 (2008): 2080–2096; Francis, *Michel Foucault's Concept*, especially the bibliography.

30 Foucault, "Of Other Spaces," 2.

31 Ibid., 1.

32 See also Stuart Elden, *The Birth of Territory* (Chicago: University of Chicago Press, 2013); Doreen Massey, *For Space* (London: SAGE, 2005).

33 Foucault speaks of a "theoretical" desanctification and of a "practical" desanctification of space, the latter having not yet fully occurred, though perhaps the practical desanctification has progressed apace since 1984. Foucault, "Of Other Spaces," 2.

34 Ibid., 2.

35 Ibid., 3.

36 Ibid.

37 Ibid., 6.

38 Johnson, "Unraveling," 75 (emphasis added).

39 Foucault, "Of Other Spaces," 7.

40 Francis, *Michel Foucault's Concept*, 18.

41 Philo, "Foucault's Geography."

42 Saldhana, "Heterotopia and Structuralism," 2080, citing a 1976 interview to Foucault.

43 Johnson, "Unraveling," 81.

44 Saldhana, "Heterotopia and Structuralism," 2081.

45 Christopher Philo, "Foucault's Geography," *Thinking Space*, eds. Mike Crang and Nigel Thrift (Abingdon: Routledge, 2000), 205–238, 206.

46 Ibid.

47 Foucault did locate the production of space with this power-knowledge complex, as he observed how "we live inside a set of relations that delineates sites which are irreducible to one another." Foucault, "Of Other Spaces," 3.

48 Saldhana, "Heterotopia and Structuralism," 2081.

49 Francis, *Michel Foucault's Concept*, 26; Michel Foucault, *The Order of Things: An Archeology of the Human Sciences* (Abingdon: Routledge, 2002), xvi–xxvi.

50 Where animals are divided into those who belong to the Emperor, those that are embalmed, those who are sirens, fabulous, frenzied, innumerable or that from afar look like flies, to mention a few of the taxa. Foucault, *Order of Things*, xvii.

51 Francis, *Michel Foucault's Concept*, 27.
52 Eric Darier, *Discourse on the Environment* (Hoboken: John Wiley & Sons, 1998), 29.
53 Ibid.
54 Kotsakis, "Heterotopias of the Environment," 196.
55 Ibid.
56 Foucault, "Of Other Spaces," 9.
57 Bernhardt Siegert, *Cultural Techniques: Grids, Filters, Doors, and Other Articulations of the Real* (New York: Fordham University Press, 2015), 69.
58 Cesare Casarino, *Modernity at Sea: Melville, Marx, Conrad in Crisis* (Minneapolis: University of Minnesota Press, 2002), 26.
59 Siegert, *Cultural Techniques*, 68.
60 Michel Foucault, *Madness and Civilization: A History of Insanity in the Age of Reason* (New York: Vintage, 1988).
61 Francis, *Michel Foucault's Concept*, 30. Francis argues additionally that "if we leave aside the word heterotopia, then all of Foucault's major projects were spatialised in terms of the way that he conceived of institutional spaces and ower/knowledge systems." Ibid., 31.
62 Jonathan Rankin and Francis Collins, "Enclosing Difference and Disruption: Assemblage, Heterotopia and the Cruise Ship," *Social and Cultural Geography* 18, no. 2 (2017): 224–244.
63 See e.g., Fella Benabed, "Marine Heterotopia and Odyssean Nomadism in Malika Mokeddem's N'zid," *Journal of North African Studies* 25, no. 1 (2020): 100–115 (where the ship is imagined as "a heterotopia of emancipation from patriarchal society and dogmatic sedentariness"); Deanna Grant-Smith and Robyn Mayes, "Freedom, Part-time Pirates, and Poo Police: Regulating the Heterotopic Space of the Recreational Boat," *Environment and Planning A: Economy and Space* 49, no. 6 (2017): 1379–1395 (about recreational boats being imagined as "heterotopias of compensation and/or illusion"); or, more generally, Casarino, *Modernity at Sea*.
64 Natalie Katsou, "Theatre Heterotopias: Sea on Stage," *Body, Space & Technology* 19, no. 1, (2020): 1–16, 2.
65 Ibid., 1.
66 Noah Simblist, "Heterotopia on the High Seas: Battles of Representation Around the Gaza Flotilla," *Art Papers* (2011): 24–29, 25.
67 Ibid., 26.
68 Philip Steinberg, *The Social Construction of the Oceans* (Cambridge: Cambridge University Press, 2001), 192ff.
69 And here it is perhaps particularly important to heed Kotsakis's invitation to consider the notion of heterotopias in the context of the rest of Foucault's conceptual and theoretical work.
70 I first deployed the term heterolegality, playing obviously on Foucault's heterotopia, in a paper presented at the Critical Legal Conference 2016 held at Kent University on September 1–3, 2016, entitled "Of Commons, Heterolegalities and Alter-Temporalities."
71 Irus Braverman and Elizabeth Johnson, eds., *Blue Legalities: The Life and Laws of the Sea* (Durham: Duke University Press, 2020).
72 For a more detailed discussion of the territorialization of the oceans see De Lucia, "Ocean Commons."
73 On the fluidity and depth of oceanic spaces, see e.g., Steinberg and Peters, "Wet Ontologies."
74 Encaring is a neologism that I coined in De Lucia, "Rethinking the Encounter," and that aims at describing the biopolitical inflection of sovereignty, that at once aims at subjugating nature and at caring for it. Indeed, biopolitical sovereignty paradoxically must subjugate nature in order to care for it. The term plays on the Heideggerian notion of enframing.

75 See, generally, De Lucia, "Ocean Commons."

76 For a fuller account, see ibid.

77 For more details, see De Lucia, "Ocean Commons," esp. 20ff.

78 *North Sea Continental Shelf*, Judgment, I.C.J. Reports 1969, p. 3, ¶ 96.

79 Ibid. 77

80 For details, again, see De Lucia, "Ocean Commons."

81 Stuart Elden, *The Birth of Territory*, (Chicago, University of Chicago Press, 2013) and Stuart Elden, "Land, Terrain, Territory," *Progress in Human Geography* 34, no. 6 (2010): 799–817.

82 Elden, "Land, Terrain, Territory," 799.

83 Massey, *For Space*, 130.

84 Ibid., emphasis in the original.

85 Henry Jones, "Lines in the Ocean: Thinking with the Sea about Territory and International Law," *London Review of International Law* 4, no. 2 (2016): 307–343.

86 De Lucia, "Ocean Commons."

87 But resulting from centuries' legal development, ibid.

88 UNCLOS Article 3.

89 UNCLOS Part V.

90 UNCLOS Part VI.

91 UNCLOS Part XI, and esp. Article 136.

92 The EEZ, for example, was born of unilateral states practice, such as the Truman Declarations or the Santiago Declaration, aimed at ensuring the protection of marine resources and the marine environment. De Lucia, "Ocean Commons."

93 This incomplete unfolding of sovereign legality has been by some commentators discussed in terms of "liquid entitlements" or "terraqueous territory." See respectively Constantinou and Hadjimichael, "Liquid Entitlement"; and Campling and Colás, "Capitalism and the Sea"; see also De Lucia, "Ocean Commons."

94 For a longer articulation of the relation between biopolitics and sovereignty, see esp. Vito De Lucia, "Bare Nature: The Biopolitical Logic of the International Regulation of Invasive Alien Species," *Journal of Environmental Law* 31, no. 1 (2018): 109–134; and for a more detailed outline of the idea of encaring see De Lucia, "Rethinking the Encounter."

95 De Lucia, "Rethinking the Encounter," 336 (footnotes omitted).

96 Ibid., 337.

97 Ibid.

98 Ibid.

99 Which indicates "making things technologically accessible," which in turns leads to the arraying of nature as a standing reserve of "resources to be exploited as means to ends." Preston Salisbury, "Ways of Being and Ways of Knowing: Heidegger's *The Question Concerning Technology* and Knowledge Organization," *NASKO* 7 (2019): 7–15, 7, 8.

100 Report of the Open-Ended Informal Consultative Process on Oceans and the Law of the Sea, June 26, 2003, UN Doc. A/58/95, ¶ 98ss.

101 GA Res "Oceans and the Law of the Sea," UN Doc. A/RES/59/24 at ¶ 73 (November 17, 2004).

102 Letter from the Co-Chairs of the Ad Hoc Open-Ended Informal Working Group to the President of the General Assembly, Annex, Section I "Recommendations," UN Doc A/66/119, ¶ 1(a) (hereinafter: "BBNJ WG 2011 Report").

103 This expression indicates the goal of pursuing the negotiating agenda as a package deal, that is, either there is agreement on all the elements or no agreement at all.

104 BBNJ WG 2011 Report, at ¶ 1(b).

105 Letter from the Co-Chairs of the Ad Hoc Open-Ended Informal Working Group to the President of the General Assembly, Annex, Section I "Recommendations," UN Doc A/69/780, at ¶ 1(e) (Feb. 13, 2015) (hereinafter: "BBNJ WG Recommendations").

142   Vito De Lucia

106 GA Res 69/292, "Development of an International Legally Binding Instrument Under the United Nations Convention on the Law of the Sea on the Conservation and Sustainable Use of Marine Biological Diversity of Areas Beyond National Jurisdiction," UN Doc A/69/292 (June 19, 2015).

107 Ibid.

108 Report of the Preparatory Committee established by GA Res. 69/292 (hereinafter: PrepCom Report).

109 PrepCom Report, Section III.

110 GA Res 72/429, "International Legally Binding Instrument Under the United Nations Convention on the Law of the Sea on the Conservation and Sustainable Use of Marine Biological Diversity of Areas Beyond National Jurisdiction," UN Doc A/RES/72/249 (December 24, 2017).

111 GA Res 75/570, UN Doc A/75/L.96 (March 9, 2020).

112 But the question is not yet settled, ibid.

113 For a problematization of the idea and legal meaning of ocean commons see Vito De Lucia, "The Concept of Commons and Marine Genetic Resources in Areas beyond National Jurisdiction," *Maritime Safety and Security Journal,* 5 (2018/2019): 1.

114 See in this respect also Daniel Lambach, "The Functional Territorialization of the High Seas" *Marine Policy* 130 (2021): 104579.

115 Schmitt, *Nomos of the Earth,* 181.

116 See also De Lucia, "Ocean Commons."

117 Massey, *For Space,* 109.

118 See, e.g., Chuanliang Wang, "On the Legal Status of Marine Genetic Resources in Areas Beyond National Jurisdiction," *Sustainability* 13, no. 14 (2021): 1–13.

119 For a detailed discussion of such "data journeys" of MGRs see Braverman, this volume.

120 E.g., under the Convention on Biological Diversity, Article 8.

121 Kathryn Milun, *The Political Uncommons: The Cross-Cultural Logic of the Global Commons* (Abington: Routledge, 2011).

122 Which, while not formally defined in international law, usually refers to a "clearly defined geographical space" subject to varying degrees of environmental regulation. Nigel Dudley, *Guidelines for Applying Protected Area Management Categories Including IUCN WCPA Best Practice Guidance on Recognising Protected Areas and Assigning Management Categories and Governance Types,* IUCN, 2013, 8.

123 Mark Hasley, "Majesty and Monstrosity: Deleuze and the Defence of Nature," in *Law and Ecology,* ed. Andreas Philippopoulos-Mihalopoulos (Abingdon: Routledge, 2011): 214–236, 218–219.

124 Ibid., 219. In a similar fashion, Lee Godden emphasizes how modernity constructs nature as other, and in doing so allows only one of two alternative views: either as an object of control—through property rights—or as "wilderness to be preserved apart from human society." Lee Godden, *Nature as Other: The Legal Ordering of the Natural World* (doctoral thesis, Griffith University, 2000), 2. See also Susan Chaplin, "Fictions of Origin: Law, Abjection, Difference," *Law and Critique* 16, no. 2 (2005): 161–180, 165–166, which describes law as a "dividing line that serves to exclude filth," that is, to separate the filth of industrial modernity from the purity that protected areas are supposed to guarantee and protect, through law.

125 Hasley, "Majesty and Monstrosity," 219.

126 Ibid.

127 Foucault, "Of Other Spaces," 5.

128 Ibid., 7.

129 Ibid., 8.

130 The expression "ecological fix" indicates "a token spatial solution that removes environmental barriers to the accumulation of capital." Madhuri Ramesh and Nitin Rai, "Trading on Conservation: A Marine Protected Area as an Ecological Fix," *Marine Policy* 82 (2017): 25–31. See also Surabhi Ranganathan, this volume.

131 See, e.g., Michelle Marvier et al., "Conservation in the Anthropocene Beyond Solitude and Fragility," *The Breakthrough Journal* (2012), https://thebreakthrough .org/journal/issue-2/conservation-in-the-anthropocene.

132 Daniel Brockington and James Igoe, "Eviction for Conservation: A Global Overview," *Conservation and Society* 4, no. 3 (2006): 424–470.

133 See, e.g., Fletcher Chmara-Huff, "Marine Protected Areas: Territorializing Objects and Subjectivities," *EchoGéo* 29 (2014), https://journals.openedition.org/echogeo /14040.

134 "Infinity Mirrored Room—Filled with the Brilliance of Life," TATE, accessed Sept. 26, 2021, https://www.tate.org.uk/art/artworks/kusama-infinity-mirrored -room-filled-with-the-brilliance-of-life-t15206.

135 See, e.g., Vito De Lucia, "Beyond Anthropocentrism and Ecocentrism: A Biopolitical Reading of Environmental Law," *Journal of Human Rights and the Environment* 8, no. 2 (2017): 181–202; and Vito De Lucia, *The Ecosystem Approach in International Environmental Law: Genealogy and Biopolitics* (Abington: Routledge, 2019).

136 As I argue in De Lucia, "Rethinking the Encounter."

137 Johnson, "Unraveling Foucault."

138 Rafi Youatt, "Counting Species: Biopower and the Global Biodiversity Census," *Environmental Values* 17, no. 3 (2008): 393–417.

139 Antonio Negri, *Dentro/contro il Diritto Sovrano. Dallo Stato dei Partiti ai Movimenti della Governance* (Verona: Ombre Corte, 2009).

140 Mario Tronti, *Operai e Capitale* (Roma: DeriveApprodi, 2006).

141 Negri, *Dentro/contro il Diritto Sovrano.*

142 Bruno Cava, "Pashukanis e Negri: Do Antidireito ao Direito do Comum," *Revista Direito e Praxis,* 4 (2013): 2–30.

6

MINING THE SEAS

Speculative Fictions and Futures

Elizabeth DeLoughrey

> The most valuable thing we extract from our oceans is our existence.
> —Sylvia Earle[1]

This chapter examines the recent oceanic turn in the humanities, particularly what French theorist Gaston Bachelard once termed the "depth imagination," and argues that it has been reconstituted by a new era of extraction, in both material and imaginary terms.[2] In the epigraph to this chapter, marine biologist Sylvia Earle reminds us of the true value of extraction as the possibility of species being. Extraction is also about futurity, narrative, technology, and speculation. Here I stage an interdisciplinary conversation between recent scholarship about the speculative practices of deep-sea mining ("DSM") and speculative fiction ("sf") that imagine techno-utopian futures of human life under the sea. In doing so, I raise questions about the ways in which particular kinds of literary genres and reading practices produce an extractive imaginary, and examine the uncomfortable overlap between the concept of innovation as a driver of the blue economy as well as the blue humanities.

I'll begin with an overview of what I've seen in the development of what is being called critical ocean studies or the blue humanities (which are different strands of scholarship) from the perspective of my training in postcolonial literary studies. I provide this critical background in order to make two provocative claims—first, that the turn to what is being called the "blue humanities," while certainly driven by our environmental crisis and the ecological/multispecies turn in scholarship, is also the product of the neoliberalization of academia and the rebranding of humanities work in an era of intellectual and economic downsizing.[3]

FIGURE 6.1 Signs posted in Huntington Beach, California after amplify energy leaked 144,000 gallons of crude oil into the ocean in 2021. Credit: Kat Schuster/Patch.com.

DOI: 10.4324/9781003205173-7

Second, that while there is currently a scramble for mineral rights and access to the seabed by transnational mining conglomerates purportedly due to the global shift toward "green" technologies, the oceanic turn in capitalism and scholarship seems to fulfill a desire for a material and intellectual (blue) "spatial fix." Consequently, this spatial fix is a critical current in the development of a contemporary depth imagination, a vision derived from both creative and extractive capital.

Critical ocean studies is an interdisciplinary method of thinking with, engaging, and submerging into the ontological, material, political, and cultural body of the largest part of our biosphere. Its ontological concerns might be illuminated by Gaston Bachelard's claim that "space, vast space, is the friend of being."[4] In recent years, the field has challenged the surface-based readings of oceanic representation, has dived deeply into complex multispecies entanglements, and has focused more pointedly on the logic of capital and its flows as well as its concordant militarization, from nuclear testing to the ways in which US naval forces "secure the volume" for the transit of oil.[5] Cold War politics have been critical to oceanic thinking; the oceanic turn in humanities scholarship was largely a response to the enclosure of the oceans through the Truman Proclamation of 1945 and the United Nations Convention on the Law of the Sea (UNCLOS), a global conversation and debate about 99 percent of the planet's biosphere.[6]

An oceanic current emerged in the 1990s at the peak of the new fields of globalization and diaspora studies. The general approach from anglophone scholars in History, Anthropology, and Cultural Studies such as James Clifford, Marcus Rediker and Peter Linebaugh, and Paul Gilroy was to think in terms of the fluidity and flow of migrants, refugees, pirates, sailors, and cosmopolitans as a vital counter-narrative to the fixity of the ethnic absolutisms that are entrenched in the structural racism of the nation-state.[7] While they focused on the concept of the ship as a chronotope and the flow of (heterosexual male) bodies across social and material borders, the metaphors of fluidity were not all together new. I argued in a book—which was very much a child of these discourses—that the ocean was a vital and ubiquitous trope of the flows and torrents of British expansion and trade in the 18th and 19th century, evident in British poetry as well as travel narratives.[8] There is a critical link between transoceanic empire, the rise of capitalism, and the imaginative grammar of fluidity and flow. In an important book on H20, Ivan Illich argued that the concept of the *circulation* of social fluids was imagined through images of blood, water, and commodities in 18th-century Europe. By 1750, the social came "to be imagined as a system of conduits," where the "liquidity" of bodies, labor, ideas, raw materials, capital, and products arose as a "dominant metaphor."[9] In sum, transoceanic empire helped constitute a fluid grammar for what Edward LiPuma and Benjamin Lee call "circulatory capitalism."[10] This liquidity was also constitutive of the discourse of globalization, postmodernity, and what sociologist Zygmunt Bauman famously termed "liquid modernity."[11]

This fluid turn was not necessarily engaged in the ontology of "wet matter," to borrow from geographers Philip Steinberg and Kimberley Peters.[12] In other words, the oceans were spaces to be traversed by (heteronormative) male agents, not necessarily to immerse or submerge in a dynamic relation to nonhuman matter (water) and more-than-human species as we see in more scholarship today.[13] The oceanic turn in scholarship after the 1990s was not just driven by the changing mobilities of human activity but also by the largest remapping of the planet since the Truman Proclamation which declared the length of a coastal "cannon-shot" (200 miles) as sovereign national territory. This created what Maltese Ambassador Arvind Pardo famously labeled a global "scramble for the seas" which was based on the expectation of new technologies for extracting strategic seabed minerals like manganese. This eventually led to the United Nations Convention on the Law of the Sea which, when ratified by over 165 states in 1994, enclosed the oceanic global commons.[14] Because this remapping includes the sea, subsea, and airspace, this is the largest juridical and cartographic change to the globe in human history. Just as 18th-century discourse adopted a lexicon reflecting the fluidity of empire, our 21st-century discourse is entangled with the aquatic flows of neoliberal extraction and "circulatory capitalism." Critical ocean studies is attentive to how the enclosure of the seas has discursive effects. In other words, aquatic space shapes our language just as we are shaped by the ocean, materially and ontologically. Consequently, in this shift to the "blue economy" it's not surprising that global regime changes are reflected in the maritime grammars we use to communicate about everything from fluidity to the "blue humanities."[15]

There is precedent to this argument that geopolitical and juridical changes impact academic disciplines as well as discourse. For example, the late capitalist era of globalization that characterized the 1990s "Asia Pacific pivot" led literature scholars such as Christopher Connery to theorize the utopian discourse of the Pacific Rim in relation to the increased visibility of transnational capital. Connery located the emergence of Pacific Rim studies as an academic reflection of US imperialism as it continues to fulfill its "manifest destiny" by expanding across the Pacific toward Asian capital, reflecting a similar teleology to that which led the US to overthrow the sovereign territory of Hawai`i in 1893.[16] Building on the work of David Harvey, Connery argued that:

> The concept of region, arising as it does within a binary logic of difference, is a semiotic utopia, a "spatial fix" for those faced with analyzing the always differentiating but always concealing logic of capital. The region, less encumbered by the various ideological or mythical mystifications that pervade the state, will be where history and analysis takes place.[17]

The ocean and its disciplinary reframings and investments also reflect a similar fixing of desire. This is a conceptual, spatial, and neoliberal fix, as I will explain. To Harvey, the problems of capital's excess are resolved (temporarily) through space: "the absorption of excess capital and labor (is achieved through) geographical expansion. This spatial fix ... entails the production of new spaces within which capitalist production can proceed."[18] Here we see capitalism's use of the oceanic body as an accumulation strategy or fix. Having exhausted terrestrial markets, capitalism co-creates and adapts technologies to turn to outer space and the so-called inner space of the oceanic realm. Yet this new era of unfixed capital, derivatives, and speculative futures raises new formal and conceptual questions about the oceanic turn.

In their work on *Financial Derivatives and the Globalization of Risk*, LiPuma and Lee claim that we are now in an era of "circulatory capitalism":

> speculative capital, circulated through risk driven derivatives, is currently restructuring the relationship between production and circulation by accelerating and expanding the spatial reach of the reproduction of capital ... We are witnessing the rise of a transformed form or new phase of capitalism in which production is (and remains) a crucial, indispensable, but now encompassed moment of a globalizing system that is striving toward a different type of totality.[19]

Their metaphors of fluidity about "cultures of circulation" and "streams of capital" point toward the ways in which technologies help produce the overaccumulation of capital and thus by extension, will need a spatial fix.[20] That spatial fix, increasingly, has become the world ocean.

In their article "The Blue Fix," the authors Zoe Brent, Mads Barbesgaard, and Carsten Pedersen provide a compelling argument about the ways in which the UNCLOS enclosure created a spatial fix for capitalism, a new frontier for raw materials and consumption. The neoliberal discourse of "blue growth, blue economy, blue revolution" as well as blue investments and blue mining seek to entice state and corporate investment in ocean technologies and extractive industries without addressing the social and technological propensity for devastating ecological loss.[21] This spatial fix is comprised of a "conservation fix" (ecosystems management), a "protein fix" (industrial fisheries), and an "energy/extractive fix" (offshore and deep-sea mining). Their particular concern is the commodification of the ocean and its resources and the ways in which states and corporate actors are working together to create a neoliberal blue economy. Turning briefly to an infographic from the World Bank, we see that the blue economy is defined as a "sustainable use of ocean resources for economic growth, improved livelihoods and jobs, and ocean ecosystem health" based on the management of renewable energy, fisheries, transportation, waste, and tourism.

The authors of "The Blue Fix" point out the International Seabed Authority (ISA) which manages what is called the "Area" beyond the exclusive economic zone ("EEZ")—nearly half the surface of the earth not to mention volume—distributes corporate mining rights through its Legal and Technical Commission (LTC) without transparency, even to its own member states.[22] This question of imagining the "Area"—a space far beyond terrestrial vision—is precisely the

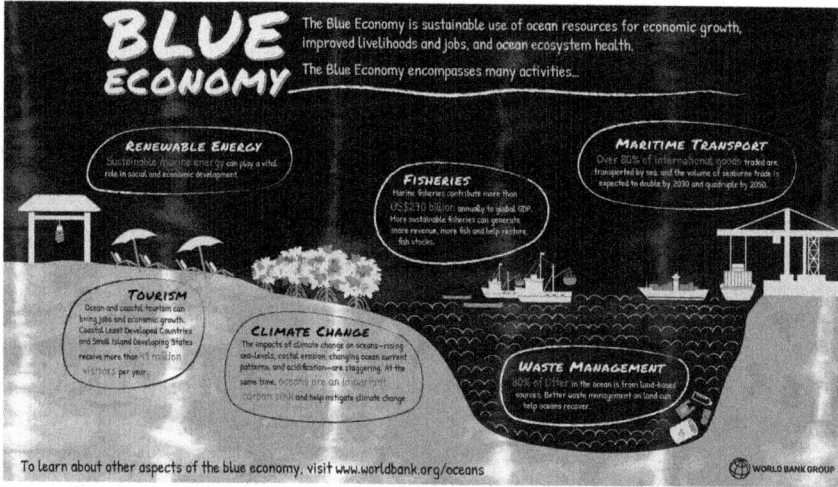

FIGURE 6.2 The Blue Economy, The World Bank. Used with permission.

concern of this chapter and apparently of the ISA itself, which hosted a visual arts competition for World Oceans Day in 2021 to represent the abyss.[23] This is a timely moment for exploring, imagining, and representing all things oceanic, which is increasingly branded as "blue." Brent, Barbesgaard, and Pedersen note the dizzying accumulation of new "blue" concepts from finance to revolution. While they do not mention the "blue humanities," certainly one must question the rebranding of disciplines during this unprecedented scramble for the minerals of the seabed and neoliberal downsizing of arts and humanities divisions and departments. Generally speaking, critical ocean studies foregrounds methodologies that examine the hydropower of militarism, empire, slavery, and extractivism to a greater extent than the scholarship engaged in the blue humanities, which has more literary and Eurocentric origins.[24] Here I will dive in a bit deeper into the extractive imaginary to engage its speculative futures.

There is critical new scholarship being produced about the enclosure of the ocean, "speculative capitalist futures,"[25] and the oceanic "techno-frontier" which is "always open and expanding."[26] Under the guise of neoliberal extractive regimes, the ocean has become a new space of the blue economy, a new commodity frontier in the scramble for rare earth elements and so-called green energy supplies, leading to new vocabularies and practices of deep-sea oil exploration, subsea carbon dioxide capture (CCS), and ocean carbon sequestration.[27] This has produced a new body of interdisciplinary scholarship with a critical eye on DSM. Of particular concern is the public-private alliance of transnational extractive industries with nation-states (that is, for drilling within the EEZ) or with the International Seabed Authority (ISA), when mining takes place in the "Area."[28] DSM is understood to be a range of practices in the seabed, water column, as well as processing on land and thus the scholarship presses against the industry's claims that this supposedly remote drilling will cause no social or ecological effects. The creation of this "blue frontier" and commodification of the ocean's minerals is speculative because the technologies are, as yet, untested outside of one project off the coast of Okinawa.[29] Nevertheless, an unprecedented number of exploratory permits have been granted by the ISA, and in the next year mining will commence in the Clarion Clipperton Zone, an abyssal plain of the Pacific that is 1.7 million square miles large, which is the width of the continental United States.[30]

The Canadian-based mining conglomerate Nautilus attempted to mine in the territorial waters of Papua New Guinea, naming themselves after the ship in the Jules Vernes 1870 adventure novel, *20,000 Leagues Under the Sea*, presumably as a way to frame extraction as adventure. In the scholarship on the rhetoric of extractive industries, scholars have pointed out that DSM poses challenges to what is called the "social license to operate" from the local community, because the mining itself takes place far offshore.[31] Thus the usual corporate social technologies that manufacture consent in extraction zones are challenged in an effort to create a "deep-sea community." As Carver, Childs, and Steinberg et al. argue, these companies trade in the discourse of nautical adventure and the blue economy through "blue growth discourse that (re)opens the ocean to imaginations of adventure, wherein new opportunities can be harnessed, and potential capital accumulated."[32] This new extractive imaginary[33] of a blue frontier is not only produced through the industries themselves. It is also evident in a 2019 sf (science or speculative fiction) ocean anthology commissioned by XPRIZE, an organization funded by Royal Dutch Shell PLC (Shell), entitled *Current Futures: A Sci-Fi Ocean Anthology*, which demonstrates the suturing of the extractive imaginary to the genre of sf itself.[34] Taken as a whole, the online short story collection and accompanying art gives us an altogether neoliberal vision of the "depth imagination" as it merges petrocapitalist extraction narratives with speculative fiction.[35] My claim is that because DSM and other forms of oceanic extraction take place outside of coastal vision, XPRIZE has funded an international group of speculative fiction authors and artists—from all seven continents—to help give them a "social license to operate."[36]

In watching their introductory and celebratory video, "We are XPRIZE," one is struck by its global, totalizing visual scope; the way in which it frames competition and its financial rewards as the way to incentivize technology and the future; the focus on "big" and "grand challenges"; the narrative that capitalism is something that "solves" problems but industry is lagging and thus needs a technological fix (as they claim—"problems that the markets have failed to solve"). Like the invocation of the *Nautilus*, XPRIZE employs a narrative of adventure on a journey to the future as well as to the deep oceans and outer space; and their film emphasizes going "deep into the *imagination*," to extract ideas and transform them into a techno-utopia of "hopefulness," "added value," and a STEM future, where petrocapitalism still reigns but is slightly cleaner.[37] The organization emphasizes oil cleanup technologies because they do not imagine a future outside of petrocapitalism.

In its efforts to commodify the depth imagination, XPRIZE funded *Current Futures*, a title that cleverly plays with the ontology of fluidity as well as time. The perspective in this interactive website is submarine, with the waves gently moving over the text that reads:

> Inspired by the awarding of the Shell Ocean Discovery XPRIZE and in celebration of World Oceans Day, XPRIZE partnered with 18 sci-fi authors and 18 artists, with contributions from all seven continents, to create an anthology of original short stories in a future when technology has helped unlock the secrets of the ocean. The series is a "deep dive" into how some of today's most promising innovations might positively impact the ocean in the future, meant to remind us about the mystery and majesty of the ocean, and the critical need for discovery and stewardship.[38]

Feminist scholars such as Carolyn Merchant and Val Plumwood have long challenged the narrative of the way in which nonhuman nature is rendered as female gendered space, waiting passively for the penetration of masculine technology and capital to "unlock [its] secrets," a trope that has long been associated with oceanic "wilderness."[39] The language of the prize draws upon a long western tradition of representing the ocean in terms of the sublime that is simultaneously "mystery" as well as a site for conquest through discovery and techno-capitalism.

Since their partnership with Shell in the wake of the Deepwater Horizon extraction disaster,[40] XPRIZE has been particularly focused on oil spill technologies and their goal to map the entire seabed by 2030 with an interest in the new industries of the blue economy and so that Shell can explore new blue frontiers of extraction.[41] As Kara Keeling has documented, since the 1970s Shell has been invested in the concepts of exploration, innovation, imagination, speculation, interdisciplinarity, and a "future scenarios initiative" of storytelling to shore up its global network of extractive ecological disaster zones.[42] This "critical need for discovery" is part of the contemporary scramble for new submarine minerals as much as establishing a "social license to operate" through the popular genre of science/speculative fiction, providing an extractive imaginary that plumbs the depths of the seas.

Mining Cultural Capital

Who is behind this particular effort to mine cultural capital and promote an extractive imaginary? XPRIZE was established in 2011 by Royal Dutch Shell;[43] its current investors include the major venture capitalists associated with the neoliberalization of education, health, the subject, and the global commons. They include Amazon, Elon Musk, Google, the military and aerospace conglomerate Northrup Grumman, health insurance companies like Anthem and Blue Cross, and transnational mining corporations like Tata Steel.[44] To those who've read Arundhati Roy's *Capitalism: A Ghost Story*, the story of philanthropic colonialism will be a familiar one. In that book, she details the history of the way in which corporate foundations such as Rockefeller and Carnegie created institutions to ensure their "global corporate governance" after World War II through the establishment of the Council on Foreign Relations (CFR) which, thanks to further support by the Ford Foundation, created a global structure that undergirded the creation of the United Nations and appointed nearly all the presidents of the World Bank since 1946. From that trajectory she concludes that "corporate philanthropy has turned out to be the most visionary business of all time."[45] As she has demonstrated, these foundations—working with the CIA—not only generated enormous profits out of postcolonial nations' debt but also restructured academic disciplines in international and area studies.[46] Jane Mayer's research has similarly unearthed a decades-long campaign by ultra right-wing American plutocrats to undermine the liberalism of universities, think tanks, government agencies, and philanthropy.[47]

In reflecting on the ways in which non-governmental organizations (NGOs) have restructured activism and labor into non-intersectional fragments, Roy grimly concludes that "funding has fragmented solidarity in ways that repression never could."[48] This is the result of the privatization of everything, including ideas. It's this *innovative* side to global capitalism that we need to consider more carefully before we rush to adopt terms like the "blue humanities" amidst this drive for an extractive blue economy. Roy reminds us that global elites "can adapt and constantly innovate[,] ... are capable of quick thinking and immense tactical cunning," and, as we'll see here, capable of harnessing the creative imagination. In fact, we might conclude that these extractive industries are dependent on it.

Let's submerge further by taking a closer look at the XPRIZE *Current Futures: A Sci-Fi Ocean Anthology* and its goals. While the collection includes predominantly American writers and is in English, it does seek to be global in scope, including emergent and established authors from the Caribbean, Africa, UK, Australia, India, and China. One of the authors kindly shared with me the list of criteria sent to the writers:

1. Stories should be between 2,500 and 3,500 words;
2. Stories should be original and unpublished;
3. Stories should take place in the 2030–2050 timeframe, far enough out for significant technological developments to have occurred, but still relatively accessible;
4. Each writer should choose from the below list of focal areas (or submit her/his own), to ensure a diversity of stories;
5. The underlying tone should be that of techno-optimism, in line with XPRIZE's vision of the future of our oceans to be healthy, valued, and understood.

The commodification of the imagination is not a new story of course, as we might see this as a kind of patronage system that supported artists and writers under the regime of dynasties all over the world. But in this context the new dynasty is a neoliberal techno-optimism as the sovereign of the future.

Turning to XPRIZE's assigned focus areas we can see how the extractive industries are imagining the future:

1. Exploration of Shipwrecks or other human artifacts;
2. New Energy Sources;
3. Environmental DNA (eDNA), Metagenomics;
4. Advanced Communications, Acoustics, Interspecies Communications;
5. Discovery of new species/lifeforms;
6. Discovery of new landscapes;
7. Eco-Tourism;
8. Terraforming, Underwater Human Habitats;
9. Advanced Conservation or Restoration Techniques;
10. Ocean Data, A.I.;
11. AUVs, Robotic Exploration, Transportation;
12. Advanced Imaging, Sensors, Tagging, Monitoring;
13. Clean-up Technologies. [49]

There is a long history of colonial tropes of transoceanic expansion that are harnessed here and in the techno-optimistic architects of the blue economy. If we think back to the World Bank's definition of the blue economy, we see many of the same features: the ocean is imagined as a space of extraction of both energy and protein; the ocean is a space of "maritime transport," particularly submersibles such as Autonomous Underwater Vehicles (AUVs); the ocean is a space of tourism and a blue-green frontier.[50] We know that in the practice of storytelling, movement across space produces narrative. Thus, travel across and beneath the seas provides the possibility of narrating adventure and the "discovery of new species." The "discovery of new landscapes" is an established colonial plot device as it has been for centuries of maritime fiction.[51] This sense of wonder at the discovery of nonhuman nature is then commodified through practices like ecotourism.

The repetition of the term "discovery" here frames the alterity of the ocean as optimistically subject to human technology and the sublime. While feminist materialists argue—compellingly, I believe—for the importance of oceanic submergence, the haptic, and the sensory encounter with our nonhuman others,[52] surprisingly XPRIZE also emphasizes the ways in which other senses are integral to knowing the ocean through acoustics, sensing, AI, advanced knowing, and interspecies communications. One of the hallmarks of the field of the environmental humanities is its multispecies theories and imaginaries; yet here we see XPRIZE is poised to tap into the ways in which contemporary sf writers provide a depth imagination of our nonhuman others that benefits an extractive imaginary and practice.

Building on the foundational work of Martin Rudwick, the geographer John Childs reminds us of the complexity and cunning of extractive industries.

He argues: "The deep sea's liveliness and its material properties may actually be very well recognized by those very actors who seek to exploit it."[53] This reflects a larger capacity for mining companies to reflect and develop what Rudwick terms "geognosy" (or perhaps in this context, aquagnosy) in thinking in complex, three-dimensional, and specifically visual ways with maps, surfaces, and depth.[54] This is evident in XPRIZE's description of its project as well as the way in which the website adopts a visual volumetric in which one submerges deeper with each story. While the homepage has an image of "wet matter" in gentle motion (placing the viewer undersea), we also note that each story has a commissioned art piece that in almost all of the cases, domesticates the alterity of the sea through familial and maternal images.

As we shift our focus to the stories themselves, I want to bring forward a few important arguments about sf that will help us unpack the way in which this genre and its techno-utopian futures emerge in an era of speculative finance and inform the extractive imaginary. The Frankfurt School thinkers argued that cultural production is structured by the commodity form and capitalism itself. Building upon this work and commenting on the ontological flatness of sf in general,[55] Fredric Jameson has pointed out the limitations of the genre as a whole, declaring that "our imaginations are hostages to our own mode of production ... at best Utopia can serve the negative purpose of making us more aware of our mental and ideological imprisonment ... and that therefore the best Utopias are those that fail the most comprehensively."[56] Certainly there are failures in this XPRIZE anthology but they fail precisely in ways that illuminate the imaginative bankruptcy of neoliberal capital: a new era that is loosened from the commodity forms that concerned the Frankfurt School and its theorizations of culture. In our contemporary context, speculation is as much genre as capital's new (blue) spatial fix.

A recent issue of *The Centennial Review* seeks to unpack the economic totality of "immaterial financialization" and speculative futures in both economic and narrative terms. What the editors term "extractive speculations" in relation to venture capital I adopt here to think through the extractive as material and interpretive practice in speculative fiction.[57] There is a vital body of scholarship exploring "fictitious capital"[58] and the way in which neoliberal capitalism financializes the subject through economies of debt (mortgages and loans) as well as speculation and risk (insurance). Building upon this work, scholars are examining how the narrative structures and imaginaries of speculative fiction are often entangled and informed by speculative finance.[59] The majority of scholarship on sf has read these generally utopian texts against the grain of the homogenizing reach of global capital, as resistant texts to the relentless competitive individualism of neoliberalism and toward more community-based modes of knowing and being.[60] The current trend is to argue that progressive sf imaginaries make the everyday violence of finance capital visible.[61] But we must ask: is rendering something visible another way to make it available for consumption?

I am brought to this question through the critique of green and blue capital-ism, which seeks to render ecological damage visible to capital so that it becomes valuable and then a source of investment for "blue growth."[62] My question is how to disentangle what becomes visible—be it newly discovered deep-sea creatures or the mapping of sea floor vents—from commodification. The work linking speculative fiction and capital also raises for me second, more generic question—*are utopian, alternative visions of the future the only way to imagine ourselves outside of neoliberal, extractive regimes of capital?* Are there no other genres that might also do this work?

In the remaining space I have I would like to bring these questions about speculation in relation to a few of the stories of the anthology, and examine how they speak to what XPRIZE calls—without irony—an "innovation pipeline." Not coincidentally, Shell has long used the concept of innovation as critical to its extractive imaginary:

> Innovation is the reason why we are able to drill for oil miles under the ocean, turn gas into liquid and transport it from the desert to cities, and unlock new sources of energy such as biofuels from plants. Rising demand for energy, together with the need to reduce carbon dioxide emissions, makes the role of innovation even more important.[63]

Innovation is a through line of the *Current Futures* anthology, which is the origi-nary mechanism of techno-optimism. Although there is a wide range of queer, feminist, non-binary, and more-than-human protagonists and imaginaries in the anthology, all of the stories feature techno-optimism in the wake of severe cli-mate change. Techno-optimism is an eco-modernist conceit that human inge-nuity will solve the ecological crisis caused by racial capitalism.[64] By extension, the narrative of techno-optimism highlights and even re-entrenches a nature/culture divide. Some of the sf authors in the anthology attempt to dissolve this division by narrating experiences of enchantment and the sublime, as their human characters become awe-struck by the beauty of submarine life. In those stories, the visual consumption of the ocean and its creatures through submers-ible technologies or by gazing through aquarium glass creates a sense of won-der and commitment to conservation. Nevertheless, the species barrier generally remains intact and most of the narratives remain anthropocentric.[65]

Collectively, the authors of *Current Futures* document the perils of ocean acidification and warming, sea-level rise, animal extinction, devastation of coral reefs, increasing hurricanes, the expansion of plastic waste and/or oil spills in the ocean, AI technologies designed for toxic cleanup, and address poverty, famine, and environmental refugees caused by the Anthropocene. While the frame of techno-optimism provides a spatial and technological "fix," it also leads to the creation of some extremely competitive and individualistic protagonists. Many of the anthology's protagonists are young women or non-binary, who are vying to win technology grants and financial support from elite white men, who reside on "super yachts" or elaborate crypts in Paris. These men become the audience to whom the protagonists need to "pitch" ideas or technologies ranging from robot fish and sensor webs to Subjective Behavioral Immersion (SBI) suits. In Brenda Cooper's short story, for instance, money gleams from the opening paragraph where the chief CEO scientist of an ocean preservation foundation "signed approvals for so much money she could have fed all of Washington state for a year."[66] This anthology has a remarkable presence of foundation leaders and launches, financial investors, and lavish investors' parties, which are not particularly compelling nor do they warrant much plot movement outside of a pedagogical one in which one character "pitches" her product. In the words of Lauren Beukes's character, "The (investors) want guarantees ... telling them what they want to hear. It's all compromise." In fact, in Deborah Biancotti's story of industrial coral farming, one protagonist is labeled by another a "corporate patsy." Read allegorically, we might see these gender and power relations in terms of the largely female authors' relationships with their XPRIZE benefactors. Nevertheless, it's troubling because as broad and experimental as the sf genre can be, the diversity of the protagonists and their worlds has not ruptured the suture to the neoliberal operators of extractive capital.[67]

Since the general tone of *Current Futures* is post-apocalyptic, the stories imagine adaptation and innovation as ways to navigate the future, terms that are critical to circulatory capitalism and its extractive practices. In the words of Elizabeth Bear's character, "The rising sea can't be stopped, but its force can be shifted." Many of the narratives are framed as futuristic detective stories in which the young female characters need to demonstrate their scientific reason to solve ecological crises and win the support of the wealthy male investment class, which is always rendered as a cosmopolitan, transnational elite. Kaushik Sunder Rajan's description of the "venture science"[68] of neoliberal regimes is literalized in many of these stories—truth is given "truthiness" (to borrow from Stephen Colbert) because the venture-capital-funded scientist is projecting the possibilities of technology into the future, which cannot be known. By writing them as "current futures," the authors of the anthology provide *anticipatory evidence* of the imagination for extractive capital.[69]

There's a maxim attributed to both Fredric Jameson and Slavoj Žižek that it's easier to imagine the ends of the earth than the ends of capitalism. Narratively, the sf writers of capitalism's ends find their spatial fix in the ocean and in the depth imaginary. While most of the stories do not render a future outside of capitalism, others directly imagine its blue spatial fix. In Biancotti's story of industrial coral farming and its potential pirating, the narrative resolution comes to rest on the realization that "trading was how the Blue Economy worked." Her venture scientist protagonist seeks to assist an unfunded coral conservationist by trading a meal for her genetically altered coral as a way to circumvent her corporate overlords, but it does not question genetically modified organism (GMO) technologies or the trade in forms of life. The critique only goes so far in that "the people funding the grants" decide who trades commodities. In Gwyneth Jones's story, a student protagonist finds a way to harvest animal intelligence to produce "sustainable" abyssal plains mining, and determines "trade is the breath of life." It's hard to place these particular stories that use the trade in life or naturalize seabed extractivism as progressive sf imaginaries because they trade, narratively speaking, in neoliberal individualism and extractivism.

In fact, the extractive imaginary in many of these stories harvests data and the "mysteries of the ocean" for circulatory capital. Madeleine Ashby's protagonist "took a deep breath, feeling the data pouring in all around her. It felt like the secrets of the sea were speaking to her." She then shares these secrets with her boss, the head of an elite transnational organization, who decides he will use it to "help (him) decide some future investments." So while sf as a genre has often been attributed with resistance to transnational extractive regimes and in positing liberatory ecological and multispecies speculative futures, in this collection the stories are not necessarily even environmental. For example, Cooper's foundation director is extremely dismissive of the environmental movement and the critique of the corporate abuse of science, remarking that the "greatest environmental cliché is *Save the Whales*." When a young journalist complains that "science made plastic and atom bombs and gasoline. Science stole everything from my generation," she is dismissed as using "such old, stale talking points."

While I'm particularly sympathetic to the critic's desire for spaces of revelation, critique, and enchantment in speculative fiction, it was a challenge for me as a reader to sympathize with many of these individualistic, flat characters who function as problematic allegories of the Anthropocene. I've argued elsewhere that allegory as a formal device is critical for interpreting the multiscalar crises of the Anthropocene and, following Walter Benjamin, have demonstrated that it represents a way of reading the disjuncture between weather and climate, human and the planet. In that book, I drew from work that is complicating the very human-nature binaries that the Anthropocene enacts by turning to relational ontologies, interspecies relations, and what I termed "sea ontologies," which are about the merger and dissolving of self into "wet matter."[70] I argued that our partial understanding of global climate change produces new economies of speculation, and that sea-level rise, our most globally visible manifestation of climate change, contributes to the production of new generic forms. Fluidity and mutability are hallmarks of the oceanic imaginary—these concepts of transformation are also integral to allegory as a form because it is about the metamorphosis of the subject and, eventually, reader. In a later piece co-written with Tatiana Flores, we argued for the importance of "submerged" visions,[71] inspired by the work of Stacy Alaimo who wrote:

> Submersing ourselves, descending rather than transcending, is essential lest our tendencies toward Human exceptionalism prevent us from recognizing that, like our hermaphroditic, aquatic evolutionary ancestor, we dwell

within and as part of a dynamic, intra-active, emergent, material world that demands new forms of ethical thought and practice.[72]

I've traced this ethical, more-than-human engagement in arts and fiction, but I have to admit it gave me pause to see the ways in which XPRIZE was looking to encourage writers to submerge themselves and to explore submerged perspectives and "interspecies communications" to suit extractivist aims. There are many stories in the collection that are about oceanic submersion but *not* about generic or subjective transformation; in fact, submersion does not necessarily transform genre nor does it dissolve the competitive, individualistic aims of the protagonists. For example in Sheila Finch's story, which focuses on dolphin communication, the protagonist is writing a dissertation that "would be cutting edge, and she wasn't going to be easily thwarted." When she learns her neural implants allow her to communicate with cetaceans, "she imagined stunning her doctoral committee with her results" rather than the deeper ontological meanings of what that multispecies communication might do to transform both her and her non-human interlocutor. When she does communicate telepathically with an octopus she declares that her "scientific training prevailed" and she dismisses her earlier attachment to the Hawaiian concept of multispecies `ohana, or family, as "childish." This current of the extractive imaginary might be better accounted for in a deeper engagement with the claims about sf as a genre and the imagined futures of the blue humanities.

I'll conclude my chapter with some possibilities that our sf authors have provided to read neoliberalism against the grain, creating spaces and bodies that are less subject to petro-commodification and extraction. There are three stories of merger and submersion that are profoundly transformative of both genre and the subject that open possibilities of alterity that challenge the extractivist imaginary through affect, the body, and ontology. These are the stories that, against the grain of neoliberal extraction narratives, imagine (sea) ontologies that deepen narrative possibilities. As Astrida Neimanis observes, "Our watery relations within ... a more-than-human hydrocommons ... [can] present a challenge to anthropocentrism, and the privileging of the human as the sole or primary site of embodiment."[73] For instance, Beukes's story, "Her Seal Skin Coat," challenges the "new golden age of exploration" invoked by James Cameron. While her protagonist travels to Antarctica to work with technologies that allow one to merge into the body of Weddell seals, she critiques her wealthy benefactor—and likely XPRIZE—by having her protagonist remark "you're paying me so you can play at being an explorer." The immersion tank that tourists enter in their Antarctic journey claims to give one access "inside the mind of a Weddell Seal" but her character knows "it's still *your* mind inside the body of a Weddell Seal. And isn't that the problem?"

In a later experience termed "dysmorphia" the protagonist Maia becomes renamed when she attempts to become "one with the ocean" through the immersion tank. Because she lives and breathes seals for 14 to 16 hours a day she develops a "ghost sense of fish or favorite octopus in her *other mouth.*" At the ending of the story the seal she is embodying is killed by an orca, a physically and emotionally wrenching experience that she allegorizes as her relationship to her white male benefactor. There is no collectivity or utopia to be found except in her return to the tank which allows her this immersion that is provided by— and takes her outside of—the neoliberal narrative of masculinist extraction and discovery. In this sense, the story suggests that "watery embodiment presents a challenge to three related humanist understandings of corporeality: discrete individualism, anthropocentrism, and phallogocentrism,"[74] even as neoliberal technologies may provide the materiality or structure.

Malka Older also imagines technologies to merge human and animal consciousness, not in the service of techno-optimism or extraction but like Beukes, to register empathy and the capacity to feel nonhuman pain. In her story, "octovision" enables the sharing of octopus memory with humans, and the once thriving coral reef that has since died is recorded, felt, and grieved rather than commodified. In Catherynne M. Valente's story, the only one featuring a non-human narrator—a pregnant orca—human and whale memory merge and the story dissolves realism into a poetic, lyric reflection of intergenerational memory of underwater life rather than isolated individualism and achievement. In these stories, the authors foreground the intimacy and care that is possible between human and nonhuman, engaging what Bachelard termed "the dialectics of immensity and depth," producing a "depth imagination" that inscribes a multi-scalar "concordance of world immensity with intimate depth of being."[75] Perhaps that has not been commodified. To return to the Sylvia Earle epigraph that opens this chapter, these particular stories remind us that the extraction of "our existence" is dependent on intimacy and species being with our nonhuman others.

Acknowledgments

This chapter was written and presented on the unceded territories of the Gabrielino Tongva, where we are currently experiencing the deadly after-effects of yet another offshore oil spill. My thanks to the coordinators of the Cambridge Theory, Criticism, and Culture Seminar for sponsoring my first iteration of this paper in December 2020; my thanks to Irus Braverman and our Laws of the Sea group for their support and engagement, including Andreas Philippopoulos-Mihalopoulos for his helpful feedback on an earlier draft.

Notes

1 "5 Questions for Dr. Sylvia Chapman," Conservation Law Foundation, 2015, https://www.clf.org/conservation-matters-articles/5-questions-for-dr-sylvia-earle/.
2 Gaston Bachelard, *The Poetics of Space: The Classic Look at How We Experience Intimate Places* (New York: Penguin Books, 2014), 188.

3 While there is now a large body of work, an early example of the use of the term "blue humanities" can be found here: https://www.neh.gov/humanities/2013/mayjune/feature/the-blue-humanities. That neoliberalism in academia is a post-Reagan and Thatcher development is nicely outlined here: https://www.aaup.org/article/tyranny-neoliberalism-american-academic-profession#.YWTVO9rMJPY.

4 Bachelard, *Poetics*, 208.

5 Stuart Elden, "Secure the Volume: Vertical Geopolitics and the Depth of Power," *Political Geography* 34 (2013): 35–51.

6 John Hannigan, *The Geopolitics of Deep Oceans* (Cambridge: Polity Press, 2016) and Surabhi Ranganathan, "Ocean Floor Grab: International Law and the Making of an Extractive Imaginary," *European Journal of International Law* 30, no. 2 (2019): 573–600. For the connection between UNCLOS and the oceanic humanities see Elizabeth DeLoughrey, *Routes and Roots: Navigating Caribbean and Pacific Island Literatures* (Honolulu: University of Hawai'i Press, 2007).

7 DeLoughrey, *Routes and Roots*; Peter Linebaugh and Marcus Rediker, *The Many-Headed Hydra* (Brooklyn: Verso Books, 2000); Paul Gilroy, *The Black Atlantic* (Cambridge: Harvard University Press, 1993).

8 DeLoughrey, *Routes and Roots*.

9 Ivan Illich, *H2O and the Waters of Forgetfulness* (Berkeley: Heyday Books, 1985), 43; DeLoughrey, *Routes and Roots*.

10 Edward LiPuma and Benjamin Lee, *Financial Derivatives and the Globalization of Risk* (Durham: Duke University Press, 2004), 15.

11 Zygmunt Bauman, *Liquid Modernity* (Cambridge: Polity Press, 2000), discussed in DeLoughrey, *Routes and Roots*, 56–57.

12 Philip Steinberg and Kimberley Peters, "Wet Ontologies, Fluid Spaces: Giving Depth to Volume through Oceanic Thinking," *Environment and Planning D: Society and Space* 33, no. 2 (2015): 247–264.

13 Stacy Alaimo, "Oceanic Origins, Plastic Activism, and New Materialism at Sea," in *Exposed: Environmental Politics and Pleasures in Posthuman Times,* eds. Serenella Iovino and Serpil Oppermann (Bloomington: Indiana University Press, 2014), 186–203; Astrida Neimanis, *Bodies of Water: Posthuman Feminine Phenomenology* (London: Bloomsbury Publishing, 2017); Elizabeth DeLoughrey and Tatiana Flores, "Submerged Bodies: The Tidalectics of Representability and the Sea in Caribbean Art," *Environmental Humanities* 12, no. 1 (2020): 132–166; Alexis Pauline Gumbs, *Undrowned: Black Feminist Lessons from Marine Mammals* (Chico: AK Press, 2020).

14 Arvid Pardo, *Common Heritage: Selected Papers on Oceans and World Order, 1967–1974,* ed. Elisabeth Mann Borgese (Malta: Malta University Press, 1975); DeLoughrey, *Routes and Roots*; Zoe W. Brent et al., "The Blue Fix: Unmasking the Politics behind the Promise of Blue Growth," *Transnational Institute* (2018): 3–19; Ranganathan, "Ocean Floor Grab." For a critique of how Pardo prepared the seabed as a space for mining see John Childs, "Performing 'Blue Degrowth': Critiquing Seabed Mining in Papua New Guinea Through Creative Practice," *Sustainability Science* 15, no. 1 (2019): 117–129.

15 The terms "blue humanities" and "blue economy" seem to have arisen simultaneously around 2009 and have been greatly expanded as concepts since then.

16 Christopher L. Connery, "The Oceanic Feeling and the Regional Imaginary," in *Global/Local: Cultural Production and the Transnational Imaginary*, eds. Wimal Dissanayake and Rob Wilson (Durham: Duke University Press, 1996), 286–287.

17 Ibid.

18 David Harvey, *The Condition of Postmodernity: An Enquiry into the Origins of Cultural Change* (Oxford: Blackwell Publishing, 1990), 183.

19 LiPuma and Lee, *Financial Derivatives*, 14.

20 Ibid., 13–14.

21 Brent, "The Blue Fix," 3, 5.

22 Ibid., 16.

23 My thanks to Phil Steinberg for calling my attention to this competition. See https://www.isa.org.jm/world-oceans-day.

24 I explore this in more depth in a special issue of the journal *English Language Notes* on hydropower: "Toward a Critical Ocean Studies for the Anthropocene," *English Language Notes* 57, no. 1 (2019): 22–36.

25 Alexander Campbell, "Extractive Poetics: Marine Energies in Scottish Literature," *Humanities* 8, no. 1 (2019): 7.

26 Anna Lowenhaupt Tsing, "Natural Resources and Capitalist Frontiers," *Economic and Political Weekly* 38, no. 48 (2003): 5100–5106, 5102.

27 See R. Carver et al., "A Critical Social Perspective on Deep Sea Mining: Lessons from the Emergent Industry in Japan," *Ocean and Coastal Management* 193 (2020): 1–10.

28 Convention on the Law of the Sea, Article 136, December 10, 1982, 1833 UNTS 397 [UNCLOS]: "[T]he Area and its resources are the common heritage of mankind." "The Area" is "the sea-bed and ocean floor, and the subsoil thereof, beyond the limits of national jurisdiction," and its "resources" are "all solid, liquid or gaseous mineral resources in situ in the Area at or beneath the sea-bed, including polymetallic nodules." UNCLOS art. 1, 133a. See also John Childs, "Greening the Blue? Corporate Strategies for Legitimising Deep Sea Mining," *Political Geography* 74 (2019): 1–12; Childs, "Performing 'Blue Degrowth'"; John Childs, "Extraction in Four Dimensions: Time, Space and the Emerging Geo(-)politics of Deep-Sea Mining," *Geopolitics* (2020): 189–213.

29 See Carver, "A Critical Social Perspective."

30 Liam Campling and Alejandro Colás, "Capitalism and the Sea: Sovereignty, Territory and Appropriation in the Global Ocean," *Environment and Planning D: Society and Space* 35, no. 4 (2017): 776–794; Ranganathan, "Ocean Floor Grab."

31 Colin Filer and Jennifer Gabriel, "How Could Nautilus Minerals Get a Social Licence to Operate the World's First Deep Sea Mine?" *Marine Policy* 95 (2019): 394–400.

32 Carver et. al., "A Critical Social Perspective," 3.

33 In the scholarship I've consulted the term is first used in John Childs and Julie Hearn, "'New' Nations: Resource-Based Development Imaginaries in Ghana and Ecuador," *Third World Quarterly* 38, no. 4 (2016): 844–861, 847. See also Campbell, "Extractive Poetics," 16; Ranganathan, "Ocean Floor Grab."

34 "Current Futures: A Sci-Fi Ocean Anthology," XPRIZE, 2021, https://go.xprize.org/oceanstories/.

35 See ibid.

36 The emphasis on the seven continents is given by Chanda Gonzales-Mowrer (Vice President, Prize Operations) in the 2019 award ceremony recorded here: https://www.youtube.com/watch?v=gB35nTmiX7w&t=6s&ab_channel=XPRIZE.

37 "A Global Future Positive Movement," XPRIZE, 2021, https://www.xprize.org/about/mission.

38 XPRIZE, "Current Futures."

39 Carolyn Merchant, *The Death of Nature* (New York: Harper & Row Publishers, 1980) and Val Plumwood, *Feminism and the Mastery of Nature* (New York: Routledge, 1993).

40 "Information on Concept Paper for Partnerships Dialogue of the Ocean Conference," XPRIZE, accessed October 15, 2021, https://sustainabledevelopment.un.org/content/documents/13700XPRIZE2.pdf.

41 Many of the websites outlining the partnerships between Shell and XPRIZE have disappeared since I started writing this essay in 2020. This one outlines the 2015 ocean floor mapping prize partnership: https://techcrunch.com/2019/05/31/teams-autonomously-mapping-the-depths-take-home-millions-in-ocean-discovery-xprize/.

42 Kara Keeling, *Queer Times, Black Futures* (New York: New York University Press, 2019).

43 Braden Kelly, "Exploring for Innovation by Land, Sea, and Space," Customer Think, October 15, 2011, https://customerthink.com/exploring_for_innovation_by_land _sea_and_space/.

44 "Prize Sponsors," XPRIZE, accessed October 15, 2021, https://www.xprize.org/ about/benefactors/sponsors.

45 Arundhati Roy, *Capitalism: A Ghost Story* (Chicago: Haymarket Books, 2014), 25.

46 Ibid., 49.

47 Jane Mayer, *Dark Money: The Hidden History of the Billionaires Behind the Rise of the Radical Right* (New York: Doubleday, 2016).

48 Ibid., 37. See also the discussions about "philanthrocapitalism" in David Rieff, "Philanthrocapitalism: A Self-Love Story," *Nation*, October 1, 2015, https://www .thenation.com/article/archive/philanthrocapitalism-a-self-love-story/; and William Easterly, *The Tyranny of Experts: Economists, Dictators, and the Forgotten Rights of the Poor* (New York: Basic Books, 2014). A larger body of work on this topic includes Robert F. Arnove, ed., *Philanthropy and Cultural Imperialism: The Foundations at Home and Abroad* (Bloomington: Indiana University Press, 1982); Anand Giridharadas, *Winners Take All: The Elite Charade of Changing the World* (Visalia: Vintage, 2019); Mark Dowie, *American Foundations: An Investigative History* (Cambridge: MIT Press, 2002); Inderjeet Parmar, *Foundations of the American Century: The Ford, Carnegie, and Rockefeller Foundations in the Rise of American Power* (New York: Columbia University Press, 2015). Many thanks to Nizan Shaked for suggesting the list of texts here.

49 The criteria were shared with me by an XPRIZE author who prefers to remain anonymous.

50 See Stefan Helmreich, *Alien Ocean: Anthropological Voyages in Microbial Seas* (Berkeley: University of California Press, 2009).

51 DeLoughrey, *Routes and Roots.*

52 See Stacy Alaimo, "New Materialisms, Old Humanisms, Or, Following the Submersible," *Nordic Journal of Feminist and Gender Research* 19, no. 4 (2011): 280–284; Eva Hayward, "Fingeryeyes: Impressions of Cup Corals," *Cultural Anthropology* 25, no. 4 (2012): 577–599; DeLoughrey and Flores, "Submerged Bodies."

53 John Childs, "Extraction in Four Dimensions."

54 Martin J. S. Rudwick, *Bursting the Limits of Time: The Reconstruction of Geohistory in the Age of Revolution* (Chicago: University of Chicago Press, 2005).

55 "I have said this negatively in connection with contemporary utopias: that their shallowness is not the mark of their failure of imagination, but rather very precisely their political function on the formal level—namely, to bring the reader up short against the atrophy of the utopian imagination and of the political vision in our own society." Fredric Jameson, *Archaeologies of the Future: The Desire Called Utopia and Other Science Fictions* (New York: Verso, 2005), 6.

56 Jameson, *Archaeologies of the Future*, xi. See also Fredric Jameson, "Reification and Utopia in Mass Culture," *Social Text* no. 1 (1979): 130–148.

57 David M. Higgins and Hugh C. O'Connell, "Introduction: Speculative Finance/ Speculative Fiction," *New Centennial Review* 7 (2019): 1–9.

58 See Sheryl Vint, "Promissory Futures: Reality and Imagination in Finance and Fiction," *New Centennial Review* 19, no. 1 (2019): 11–36.

59 See Jameson *Archaeologies of the Future*; Vint, "Promissory Futures"; Aimee Bhang, *Migrant Futures: Decolonizing Speculation in Financial Times* (Durham: Duke University Press, 2018).

60 See Vint, "Promissory Futures," 12.

61 See ibid., 18.

62 Brent, "The Blue Fix," 7.

63 "Energy and Innovation," Shell, 2021, https://www.shell.us/energy-and-innovation .html#vanity-HR0cHM6Ly93d3cuc2hlbGwudXMvZW5lcmd5LWFuZC1pbm5v dmF0aW9uL29jZWFuLWlubm92YXRpb24uaHRtbA. See also Keeling, *Queer Times*, 13.

64 John Asafu-Adjaye et. al, "An Ecomodernist Manifesto," Ecomodernism, 2015, http://www.ecomodernism.org/; (This was popularized by a 2015 manifesto written primarily by men working at the neoliberal Breakthrough Institute).

65 While there are a few stories that imagine human/nonhuman mergers, the only story told from a nonhuman point of view is "The Seething Sea Sufficeth Us" by Catherynne M. Valente. See https://go.xprize.org/oceanstories/.

66 All *Current Futures: A Sci-fi Ocean Anthology* citations can be found on the website: https://go.xprize.org/oceanstories/.

67 There are of course exceptions—Vandana Singh and Catherynne M. Valente are the most experimental with form and narrative, whereas Nalo Hopkinson challenges colonial tourism through its natural decay; Lauren Beukes makes a powerful commentary on the exploitation of women and marine mammals. See https://go.xprize .org/oceanstories/.

68 Quoted in Vint, "Promissory Futures," 27. Kaushik Sunder Rajan, *Biocapital: The Constitution of Postgenomic Life* (Durham: Duke University Press, 2006).

69 This is part of a larger argument about the rise of capital in relation to speculation and transnational labor/slavery. See Ian Baucom, *Specters of the Atlantic: Finance Capital, Slavery, and the Philosophy of History* (Durham: Duke University Press, 2005), 31; Keeling, *Queer Times.*

70 Elizabeth DeLoughrey, *Allegories of the Anthropocene* (Durham: Duke University Press, 2019).

71 DeLoughrey and Flores, "Submerged Bodies."

72 Alaimo, "New Materialisms," 280–284.

73 Neimanis, *Bodies of Water,* 2.

74 Ibid., 3.

75 Bachelard, *Poetics,* 210, 188–189.

7

NAVIGATING THE STRUCTURAL COHERENCE OF SEA ICE

Philip Steinberg, Greta Ferloni, Claudio Aporta, Gavin Bridge, Aldo Chircop, Kate Coddington, Stuart Elden, Stephanie C. Kane, Timo Koivurova, Jessica Shadian, and Anna Stammler-Gossmann

Introduction

In November 2017, Baffinland Iron Mines Corp., operator of the Mary River Mine in Canada's Arctic territory of Nunavut, announced that it was amending the expansion plan filed with the Nunavut Planning Commission. Although residents of Pond Inlet (Mittimatalik), the predominantly Inuit community closest to the mine, had varying views regarding the proposed expansion, one component of the plan had few if any supporters: a proposal for icebreaking vessels to retrieve ore from the mine's loading facility at Milne Inlet on Eclipse Sound during the winter season. The company had already reduced the proposed frequency of winter shipping in response to community opposition. However, even the latest version of the proposal, which called for a maximum of two vessels each year between December and February, was unacceptable to residents of Pond Inlet, which also fronts Eclipse Sound. As Joe Enook, then the region's representative to the Nunavut Legislative Assembly and generally a supporter of the mine, noted, disturbing the winter sea ice would jeopardize local residents' ability to travel and hunt. "Eclipse Sound is our grocery store," Enook said, explaining his opposition. "[With the winter shipping proposal] there was a potential for disruption."[1]

As the Pond Inlet residents' intransigence suggests, breaking sea ice,[2] although usually conducted with the singular objective of enabling maritime navigation, can have myriad negative environmental and economic impacts, on land as well as at sea, from disrupting algal blooms that are at the base of the food chain to upending the lifeways of Indigenous peoples. As such, it would seem to be an

FIGURE 7.1 US Coast Guard's "Healy" (the Coast Guard's icebreaker) in pack ice. Public domain.

DOI: 10.4324/9781003205173-8

activity suitable for environmental regulation, potentially employing environmental impact assessments, cost-benefit analyses, and other regulatory tools. And yet, although icebreaking is fundamentally an act of environmental violence, it is conceptualized legally as a freedom of navigation, essentially the same as a ship gliding over what is idealized as a formless, featureless surface. Thus, although it would appear that managing the impact of ice breaking poses a seemingly simple practical regulatory problem, approaching this problem in a way that values sea ice's structural coherence, and thereby affirms Indigenous peoples' rights to self-determination in governance of their landscapes and seascapes, must necessarily challenge underpinning ideas about surfaces, volumes, structures, and movements of and in ocean-space that are inherent to Western conceptions of mobility, time, and territory.

To address this challenge, this chapter proceeds in four sections. In the first section, we review the role of sea ice in northern economies and ecologies as well as the potential impact of icebreaking. Following this, the second section considers and rejects the argument that barriers to the regulation of icebreaking are specifically *legal*. In fact, frameworks and precedents exist for regulating ocean uses (including navigation) to protect environments and Indigenous livelihoods, and these could be applied to limit the right to break ice, especially when there are communities of interest that have a shared concern for maintaining sea ice as a predictable space with structural integrity. Therefore, we suggest in the third section that, in the absence of overriding legal or political obstacles, the fundamental barrier to adopting sea ice protections that acknowledge Indigenous perspectives and claims instead rests in the ways that Western legal reasoning conceives of the spaces across which vessels move as lifeless, formless, and frictionless surfaces. We therefore turn to the geophilosophical (or ontological) challenges posed when the ocean, including in its frozen state, is understood not as a surface to be crossed but as a lively space of intersecting mobilities, interdependencies, and transformations. Finally, the fourth section situates our brief consideration of icebreaking within a broader literature in marine planning that explores how thinking from an oceanic perspective can challenge the limits of law and territory, and how a legal approach to icebreaking can suggest new modalities for understanding and governing the ocean.

Mobilities on/of Sea Ice

As Joe Enook reminds us in his opposition to icebreaking in Eclipse Sound, sea ice is foundational for regional ecologies and economies across much of the Arctic. Sea ice is never just "frozen water," as expressed in the hundreds of local names used to distinguish sea ice types.[3] It is always in a process of becoming and dissolution across space and time, undergoing continuous structural alterations through snow accumulation, lead formation, wind advection, brine rejection, and countless other ice processes.[4] The underside of sea ice, particularly in marginal zones, hosts algal communities that provide the base energy for some of the world's richest marine ecosystems.[5] Conversely, the upward-facing surface of ice provides crucial denning and feeding grounds for a range of species from seals to polar bears.[6]

Sea ice, in its fascinating complexity, is a fundamental aspect of lives and livelihoods for Indigenous peoples throughout much of the Arctic. During winter months, sea ice provides a stable hunting platform for Inuit whose diet largely depends on marine ecosystems. Shorefast ice (stationary ice extending from shore, usually fixed by sections of thicker ice that are grounded on the seabed) allows hunters to follow whales, seals, polar bears, and other game far out into what would be summertime open water.[7] Sea ice acts as a "highway" that connects communities to each other; in some cases it is the only route between settlements.[8] Reindeer herders in parts of Russia use sea ice to move their herds to summer pastures, circumventing rivers that have already melted.[9] In Alaska, sea ice provides protection from winter storms that cause coastal erosion and claim vital infrastructure, homes, and lives.[10]

Additionally, for the Inuit in particular, sea ice is central to a traditional culture that is characterized by a deep attachment to and respect for the ocean (including when frozen) as well as land.[11] Being able to use sea ice to provide for one's family and community contributes to wellbeing as part of what it means to live a fulfilling life.[12] As a demanding environment, sea ice teaches important lessons of patience, endurance, courage, and good judgment.[13] Arctic Indigenous

FIGURE 7.2 Children playing on sea ice, near the settlement of Igloolik, Nunavut, Canada. Photo by Claudio Aporta. Used with permission.

peoples know "all possible facets of sea ice":[14] its numerous forms but also its numerous material, cultural, and spiritual functions.[15]

The value of sea ice lies not just in its *quantity* but also in its *quality*. For traveling, sea ice must be thick, strong, and smooth so that hunters and herders (and their equipment) can easily move by dog sled or snowmobile without falling through the ice and risking hypothermia or drowning.[16] Multi-year ice (sea ice that has survived at least one summer melt season) provides additional stability and is a source of fresh drinking water, an essential resource when hunting for long periods away from shore.[17] In marginal zones where hunters catch whales and seals as they break through the ice to breathe, sea ice must be the perfect balance between a breakable ceiling for the animals and a sturdy platform for the hunters. When the quality of sea ice is degraded, knowledge that had been accumulated over millennia loses its relevance, or needs adapting, reducing hunters', herders', and travelers' ability to interpret the icescape, its opportunities, and its dangers.[18]

While climate change is partially responsible for destabilizing the qualities of sea ice that sustain Indigenous lifeways in the Arctic,[19] disturbance by icebreaking vessels also plays an important role. Break-off events, where large sections of sea ice separate from shorefast ice, pose a significant danger, as they can lead hunters and herders (and their herds) to fall into the water or be carried away on a broken-off floe.[20] Sea ice disturbance increases the likelihood of such events and makes them more difficult to predict.[21] Icebreaking can also impact the trails used by snowmobiles and sleds, as well as the migration patterns of land mammals, such as caribou.[22] While in some cases it might be possible to cross the ice as soon as one hour after a ship has passed, the ice will refreeze as a rubble mess that hunters might need to axe their way through, and snowmobiles risk getting stuck or breaking down.[23] Potentially fatal delays can result if, when returning from a hunt for instance, one finds that a vessel has cut through the ice trail being used for the return journey, or that it has separated ice floes that were previously close enough to step across.[24]

Additionally, shipping vessels and icebreakers are loud, potentially scaring polar bears and caribou (overall, from the region, but also specifically during a hunt), and increasing animal deaths from collisions with passing vessels.[25] Icebreakers emit noise from bubbling systems which blow pressurized air underwater to push ice away and their propellers make sharp, intermittent ramming noises when stuck in ice; these noises mask cetacean inter-species communication, possibly causing behavioral and physiological changes that affect their well-being, reproduction, and migration.[26] Icebreaker activity also creates waves that can flood and freeze the openings to polar bear dens and seal breathing holes.[27] Waves can also pose a danger to hunters, as their small boats are unable to cross ship wakes safely.[28]

Furthermore, when ice re-forms after disruption by a passing vessel, new, unpredictable variables are added that may confound the calculations of experienced hunters. Usually, hunters' judgments of the strength, thickness, and structural integrity of sea ice are based on close monitoring of weather and

sea ice conditions prior to and during time out on the ice. However, icebreaking causes the ice to fracture and refreeze in erratic ways, creating unpredictable conditions.[29] Finally, by hindering sea ice formation and speeding sea ice
breakup, icebreaking can shorten the period during which animals can use the
ice for breeding and migrating, further impacting livelihoods in both animal and
human communities.[30]

It is unsurprising, then, that when the Arctic Corridors and Northern Voices
project held meetings with 13 Inuit communities to gather perspectives and recommendations regarding Low Impact Shipping Corridors in Canada's North,
residents consistently referenced the threats posed by icebreaking.[31] Many communities voiced concerns about how icebreaking activities disrupt animals and
their habitats and jumble ice trails and routes, endangering communities' and
hunters' lives and livelihoods. They proposed areas where icebreaking should not
happen, others where noise should be kept to a minimum, and suggested better
communication with shipping traffic control to be able to plan around ships and
their routes. Such concerns are nothing new, nor are they restricted to Canada.
In 1975, a group of Greenlandic hunters physically blockaded an icebreaker *en
route* to a mining site at Marmorilik, as it was disrupting their hunting practices
in the Uummannaq Fjord.[32] Negotiations there and then on the ice edge resulted
in an agreed route that would be less disruptive to hunting.

At one level, the differences between Indigenous peoples, who share sea ice
environments with nonhuman inhabitants, and shippers, who seek to cross the

FIGURE 7.3 Tourist boat surrounded by icebergs, Iceland. Photo by Anna Stammler-
Gossmann. Used with permission.

ocean's surface, appear insurmountable: Shippers perceive ice as an obstacle and aspire to navigation on an ice-free ocean while Indigenous peoples perceive a continuum by which land, frozen water, and liquid water are all spaces that enable and constitute the web of Indigenous and animal livelihoods. However, both groups have an interest in understanding conditions in an environment that changes both seasonally and over the long term. Indeed, although the Arctic Corridors and Northern Voices project identified some Indigenous concerns that stemmed from Inuit ways of valuing, thinking of, and using ocean-space that are largely foreign to Western thinking (e.g., the value of ice as a "highway" of hunting trails), other concerns voiced by Indigenous peoples were likely shared by shippers (e.g., concern over lack of accurate navigational charts), or were in broad alignment with the environmental priorities that already underpin marine management (e.g., protection of breeding grounds).[33] A survey of cumulative effects of marine shipping conducted by Transport Canada and an initiative organized by the ICE LAW Project and the Company of Master Mariners of Canada resulted in similar findings.[34] In particular, the ICE LAW Project initiative found significant points of overlap between "Western" and "Indigenous" interests, once one turns away from viewing ice as a legal abstraction and instead focuses on encounters with sea ice in its materiality.[35]

To summarize, Indigenous communities and shipping companies share an understanding that watery spaces, in their multiple frozen states, require addressing on their own terms. The challenge for lawmakers (at international and national scales) is thus to manage navigation in a manner that safeguards fragile Arctic ecosystems, protecting both the biota that thrive above and below the sea ice surface and the lives and livelihoods of Indigenous peoples. As we argue below, this requires attentiveness not just to sea ice's multiple functions and uses, but also to its underlying structural coherence.

Law, Navigation, and Ice-Covered Waters

By any measure, a central feature of the United Nations Convention on the Law of the Sea (UNCLOS) is its commitment to safeguarding freedoms of navigation. In addition to asserting that "Every State, whether coastal or land-locked, has the right to sail ships flying its flag on the high seas" (Article 90), UNCLOS extends navigation rights to other areas of the ocean, including exclusive economic zones (EEZs), territorial seas, international straits, and archipelagic waters.[36] However freedom of navigation is not absolute. Numerous articles in UNCLOS mandate that international navigation rights must be exercised with due regard to the rights of coastal states in zones of national jurisdiction (e.g., Articles 19, 21(4), 39, 40, 41(7), 43(4), 53(11), 58(3), and 60(7)) and navigation in ice-covered EEZs must accommodate heightened environmental vulnerability (Article 234). Even on the high seas the right to navigation is balanced with rights of overflight, laying submarine cables, constructing artificial islands, fishing, and conducting scientific research (Article 87). Furthermore, state practice has evolved to permit the establishment of temporary zones on the high seas where shipping (and fishing) are prohibited to allow for weapons testing.[37] This suggests that Article 87's list of enumerated rights is not exhaustive and potentially could be extended, with further restrictions placed on navigation.

All of this is to suggest that, hypothetically, current legal instruments could be employed to balance the right to navigation in ice-covered waters (including the right to engage in icebreaking) with other interests.[38] Furthermore, as Indigenous activists have noted, when international laws that guarantee the rights of states are balanced with those such as the United Nations Declaration on the Rights of Indigenous Peoples (UNDRIP) that guarantee the rights of Indigenous peoples to maintain their livelihoods, cultures, and collective identities through control of ancestral lands, waters, and the resources contained therein, these protections can be extended to cover preservation of the environment that enables the maintenance of Indigenous peoples' lifeways. This is precisely the argument made by former Inuit Circumpolar Council (ICC) Chair Sheila Watt-Cloutier in *The Right to be Cold*: "I believe the campaigns to link climate change to human rights protection—efforts that acknowledge our shared humanity and our shared future—are the most effective way to bring about lasting change."[39] Additionally, whether in international law or domestic legislation and practice, the recognition of Indigenous peoples' rights to their environment and the appreciation of nature as multi-faceted and dynamic often are inseparable from each other: Indigenous self-determination involves the articulation and implementation of Indigenous perspectives on the environment, and vice versa, a point that is acknowledged in UNDRIP (Article 5) and by Indigenous leaders such as ICC Chair Dalee Sambo Dorough.[40] Although some of the best documented examples of Indigenous perspectives on the marine environment being incorporated into state-led planning have occurred in Aoteaora/New Zealand,[41] initiatives in the Arctic also stand out. For instance, the Agreement to Protect High Seas Fisheries in the Central Arctic Ocean links recognition of Indigenous knowledges and perspectives on the environment, protection of Indigenous rights and interests, and empowerment of Indigenous peoples in participation and decision-making.[42] A further example can be seen in the Pikialasorsuaq Commission, a trans-boundary, ICC-led marine management initiative endorsed by the governments of Canada and Greenland to develop a management regime for the Pikialasorsuaq Polynya, an exceptionally biologically productive area of open water surrounded by sea ice that spans the two countries' EEZs.[43]

Notwithstanding these examples, however, efforts at securing the integrity of sea ice by balancing navigation rights with those of communities that derive other values from frozen ocean environments may be limited as a means for conserving the environment, while respecting Indigenous self-determination. Not all Indigenous people are in equivalent positions regarding their relationship to either the state or the environment which suggests that different situations may require targeted approaches. For example, Watt-Cloutier's embrace of a human rights agenda (mobilized through UNDRIP) for preserving Inuit culture and, by extension, the environment that sustains that culture, has been challenged by some legal scholars who question whether Indigenous interconnections with the environment can be adequately articulated through instruments that see all rights as derived from the anthropocentric, universalist notion of individual human rights.[44] Turning to Arctic Canada, Todd, for instance, argues that protection and recognition of Indigenous culture requires not simply securing the right to culture through Western law but recognition of Inuit law which, in the case of coastal Inuvialuit, is based on an understanding of human-fish entanglements that sits outside the Western tradition.[45]

Despite these limits, this review points to the existence of legal mechanisms that potentially could protect sea ice amidst its entanglements with individuals, communities, environments, biogeophysical processes, and the lives of more-than-human entities. Additionally, the presence of shared interests among a diversity of groups suggests that viable regulations should be politically feasible. Nonetheless, we note that sea ice, as a material form to be structurally preserved, remains beyond the scope of legal regulation. We therefore suggest that the fundamental obstacle to the implementation of effective sea ice protection is *ontological*. That is, in order to protect the structural integrity of sea ice a shift is needed in the way that territory and oceans are understood in the Western geographic imaginary.

Rethinking Ocean Territories

Land and sea are often counterposed as binaries in Western political and legal thought: the former understood as capable of being transformed, developed, and bounded (i.e., "territorialized") and the latter as immune to these social exertions, a featureless space of frictionless flows and untethered resources, capacious in its liquidity. This binary can be found in, for example, Hugo Grotius's 17th century *Mare Liberum* or Carl Schmitt's 20th century works *Land and Sea* and *The Nomos of the Earth*.[46] In fact, recent scholarship has explored how modern notions of both land and sea, although seemingly in opposition to each other, are grounded in a common understanding of territory, in which the reduction of space to fixed points, with relative resource values, in relative location to each other on an inert, two-dimensional field is conceptually divided from the experience of engaging with and strategizing movement through the planet's biogeophysical materiality.[47] The same logic that isolates points on land as places to be developed and bounded, distinct from the features of terrain that join these points, facilitates the construction of the ocean as ideally the opposite. In a modern, point-based ontology, the ocean is understood as a space where distance can be annihilated through mobility (since the ocean is understood as having no terrain) and where the ocean's points exist solely as mathematical abstractions, freed from the differentiating power of nature that makes land suitable for development and enclosure.[48] As Schmitt puts it, the ocean has no "character," no places, and hence no potential for transformation or territorialization.[49] In the history of state formation on land, the ocean has played the role of "constitutive other," the obverse side of territory: a formless, placeless, liquid environment "outside the lines," an essential space of unmanageable fluidity across which one navigates to traverse between land-based territories or into which one descends to extract "free" resources that are brought back to develop land.[50]

Having inherited this idea of the ocean as a fluid essence, however, jurists and regulators have struggled at the margins of its liquidity, because the ocean is not simply a formless surface or a placeless, voluminous depth that exists in determinate opposition to the territories of society. At the most obvious level, the geophysical binary between land and ocean that underpins the geopolitical binary between territory and non-territory breaks down at the coast, where boundaries between land and sea are often indistinct and mobile, not only in tidal zones and estuaries but also due to the subjective nature of sea-level calculations and long-term trajectories of climate change.[51] UNCLOS makes little effort to accommodate the dynamic nature of oceanic systems or the complex nature of ocean-land-human-animal-atmosphere interfaces, as is acknowledged both by those who view this disassociation of the hard boundaries of law from the vicissitudes of geography as a weakness in the system and by those who see it as a strength.[52] Even when UNCLOS attempts to accommodate geographic dynamism, complexity, and indeterminacy, the "fixes" implemented fail to account for the ways that watery spaces are used and experienced. For instance, although UNCLOS acknowledges that a line that consistently follows the low-water mark may not always be the best means for distinguishing land from ocean, its alternative, permitting the drawing of straight baselines "where the coastline is deeply indented and cut into, or if there is a fringe of islands along the coast in its immediate vicinity" (Article 7), only secondarily considers the interests or practices of coastal communities.[53] Likewise, UNCLOS's regime for archipelagic states (Articles 46–54) permits states, under certain conditions, to draw straight lines that designate areas of ocean as internal water, but this fails to reflect the ways that people and other biota surrounding islands inhabit the ocean-spaces that interweave with islands.[54]

Just as the construction of the ocean as an idealized non-territory of formless liquidity is challenged by murky distinctions between land and sea, it is also challenged by the presence of solid land at the bottom of the ocean.[55] Beyond the territorial sea (and its seabed), ocean law, both pre- and post-UNCLOS, has distinguished the seabed from the waters above, rendering the ocean floor, but not the water column, suitable for point-based investments (for oil and gas drilling) and bounded enclosure (for seabed mining).[56] This separation of the marine environment into distinct strata with unique and differentiated territorial properties has implications for the regulation of deep-sea mining, as it could lead to environmental impact assessments that inadequately account for the ways in which "harms" and "losses" extend across both space and time.[57] More broadly, this reification of a binary between land (even if submerged) as a

FIGURE 7.4   Fishing trawlers in port, Kirkenes, Norway. Photo by Anna Stammler-Gossmann. Used with permission.

"character-full" space of bounded, developable places (i.e., "territory") and the ocean as a formless liquid abstraction (i.e., "non-territory") limits one's ability to extend insights about the ocean's turbulent materiality to the greater hydrosphere.[58] It constrains the Western legal tradition's comprehension of oceanic features, like waves or currents, as simultaneously forces *and* objects; as entities that simultaneously occur in place, move across space, *and* constitute place; as unique entities *and* analytic categories; as metonyms that both reflect *and* shape the conceptual foundations that are used to understand livelihoods that largely occur beyond ocean-space's geographic limits.[59]

Returning to the specific question of regulating icebreaking, the idealized binary between territorial land and formless ocean directly challenges any effort to preserve the structural coherence of sea ice: how can one preserve *form* in a space that is legally constructed as formless? Despite the advances made in the collaborative governance of Arctic waters discussed above, efforts to preserve sea ice's structural integrity through recognition of its value have been necessarily limited by the overarching ontology applied to the ocean. Indeed, the one article in UNCLOS that acknowledges that seawater can ever take non-liquid form— Article 234, which gives states the right to impose additional environmental protections in ice-covered areas of their EEZs—constructs sea ice solely as a hazard, a potential source of disruption to the formless surface idealized by Western navigators, not as a valued form to be preserved for its specific capacities.[60] It would be difficult to align Article 234's perspective with one that acknowledges sea ice's existence as a dynamic object at the intersection of biological, geophysical, and cultural processes, let alone one that affirms sea ice's many functions and affordances for more-than-human ecologies and climate systems.

The power of the territory-non-territory binary and the difficulties encountered when one attempts to apply it to sea ice are evidenced in Canadian justifications for defining the waters of the Arctic archipelago as its historic internal waters. Secretary of State Joe Clark's address to Parliament when Canada declared straight baselines around the archipelago—affirming that "from time immemorial Canada's Inuit people have used and occupied the ice as they have used and occupied the land"—flips sea ice to the other side of the binary. However, it rather misses the point raised earlier in this chapter about the distinct ways in which northern peoples incorporate sea, ice, and land into their lifeways in ways that mimic neither the Western sense of land as territory nor its idealized maritime negation.[61]

Finally, efforts to incorporate and govern sea ice as territory have been confounded not just by its presence in the ocean, which is legally designated as a space beyond territory, but also by its indeterminate and dynamic properties. Sea ice is constantly moving as well as melting and freezing, it exists in varying concentrations that change rapidly over time, and it generally has indistinct borders, and this has led to inconsistency and uncertainty in the development and implementation of ice-sensitive regulations.[62] Each of these properties not only makes sea ice a difficult environment to regulate; it also makes it difficult to conceive how, from the perspective of Western law, we might protect its material integrity as a spatial object.

Managing Dynamic Ocean-Space

UNCLOS and, more broadly, the laws and regulations of the sea, are fundamentally spatial. After defining the ocean as a juridical space (the area of earth's surface and subsurface beyond the limits of internal waters), UNCLOS defines the contours of state power in subsidiary areal zones (territorial seas, contiguous zones, EEZs, the high seas), horizontal strata (surface, water column, seabed), and features (rocks, islands, low-tide elevations, archipelagos). Marine planning initiatives then work within this spatial framework to define the spaces within which management can be applied (the marine protected area, the regional fishery management organization zone, et cetera) in order to govern specific uses. Some scholars have described this process as one of marine territorialization, as areas of the ocean's surface, water column, and floor are bounded and allocated, a process by which land-based ontological assumptions and spatial planning mechanisms are applied to the ocean.[63]

However, building on the understanding of territory as a political technology, we take issue with this characterization.[64] Rather than seeing territory simply as a bounded space, this approach seeks to analyze the making and remaking of territory, comprehending territory as a process rather than as an outcome. It seeks to explore the practices or techniques—such as cartography, surveying, and population management, as well as legal instruments and abstractions—and their relation to the places that are measured and controlled. The abstractive measurement and control of territory gives one aspect of its political-legal form but masks the complexity of the dynamic nature of territory, through the forms of its terrain, understood as the material surfaces and depths encountered and affected by moving bodies.

When management lines are drawn at sea the concept of terrain is elided as the ocean's dynamic materiality is reduced to an atemporal abstraction.[65] Returning to Schmitt, when abstract lines are drawn in the ocean with little regard to its geophysical dynamics (its connections to land, atmosphere, and distant seas), its changes in form, or the mobilities of its (nonhuman and human) inhabitants, the ocean is perceived as a space without "character,"[66] a point frequently noted by critics of hard-bordered spatial management tools.[67] And if the ocean—even the managed, governed, conserved ocean—is seen as having no terrain, or no "character," then it likewise has no places, no features, no form. In such an environment, preserving sea ice as an oceanic feature that serves specific functions makes no more sense than conserving an individual wave or water molecule. In the eyes of the modern planner or jurist, the ocean's parts, disaggregated into points, are never reaggregated through the practice of terrain into meaningful entities. Instead, they are stranded as ephemeral elements adrift in ocean-space, to be managed rationally and spatially through calculative linear abstractions. Sea ice, an unacknowledged and unacknowledgeable feature, is simply allowed to melt away.

Could the spatial nature of ocean governance be mobilized not to constrain possibilities but to open new alternatives? Numerous scholars have pointed to the ocean as a site of potential legal innovation. For Mann Borgese, the ocean's global value as a space that connects the world's economies and ecologies, its local meanings as an arena of livelihoods, and its political status largely outside state territorial boundaries can be mobilized through law to bring new ethics of care, stewardship, and self-determination to governance.[68] Van Dyke et al. extend this agenda, calling for the norm of "freedom of the seas" to be replaced with one of "freedom *for* the seas," wherein, instead of understanding the ocean as a space of individual, protected rights, the ocean is understood as a socionatural space that joins a diversity of biogeophysical (including human) functions, services, and interconnections.[69] From such a perspective the value of the ocean's forms—its waves, its currents, its ice, et cetera—would lie not just in their present functions (as a hunting surface, as a global climate regulator, et cetera) but in the meanings that have been ascribed to and derived from them over millennia by the interspecies web of inhabitants who engage the ocean environment.

The challenge, then, is to reterritorialize the ocean through new understandings of terrain that can be applied to the sea, recognizing the "character" that the ocean already has and that is continually being reproduced through biogeophysical processes and human interventions. If earlier attempts to theorize terrain and its relation to territory emphasized the political-strategic aspects of its control,[70] more recent work has stressed the way the complex materiality of terrain helps to

FIGURE 7.5 Snowmobile tracks from land to sea ice, off Melville Peninsula, toward Igloolik, Nunavut, Canada. Photo by Claudio Aporta. Used with permission.

ground and add depth to our understanding of territory.[71] Thinking through terrain to grasp the materiality of territory forces an analysis of the relation between land and sea, in complex and dynamic environments, and ultimately collapses any straight-forward binary division.

Yet this is not an easy task. Implementing "territory beyond *terra*" presents a range of challenges,[72] and to use the concept of terrain to understand spaces which were previously seen solely as water (whether liquid or frozen) may be particularly problematic.[73] Critical marine planners have shown, however, that when we replace the hard boundaries of the marine protected area with an understanding of ocean-space as existing within flows—flows of histories, data, knowledges, and practices, as well as water and biota—we develop new perspectives that shed light on the processes through which terrain is encountered and enacted.[74] Conversely, when we reorient ourselves toward oceanic terrains, by listening to those who engage the ocean as a material space, new planning mechanisms (and, potentially, legal institutions) emerge for the ocean environment.

Of course, people who actually encounter the ocean have long understood the ocean as terrain, and there may even be common perspectives held by divergent users. As the Arctic Corridors and ICE LAW Projects both found, sea ice users with seemingly diametric interests in the integrity of ice—the Inuit hunter who wishes to preserve frozen hunting trails and the Coast Guard officer who wishes to maintain liquid shipping routes—likely still have more in common with each other than does either with the drafters of UNCLOS who largely ignored ice's presence. This suggests that in the ocean (as elsewhere), it is crucial that law be developed and implemented by those who experience a space in its multiplicity. To be sure, differences in interest remain even among users with shared concerns and perspectives, and there may be no singular perspective on sea ice that joins them together. However, as Squire proposes, a "pluriversal" understanding of terrain, including oceanic terrain, may well be tenable.[75]

To conclude, the fundamental obstacle to development and implementation of a comprehensive legal regime that protects the integrity of frozen ocean environments from icebreaking is neither strictly legal nor political: It is *ontological*. The challenge is to understand the ocean not as a formless surface—the antithesis of land-based territory—but as ice/water terrain, with character and form, with history and affordances. Only then, we argue, can one develop a regime to protect sea ice from acts of environmental violence that would undermine its structural integrity and socioecological functions.

Acknowledgments

This chapter is the result of discussions and workshops held by the Project on Indeterminate and Changing Environments: Law, the Anthropocene, and the World (the ICE LAW Project), a three-year, interdisciplinary project that ended in July 2019 and that sought to "investigate the potential for a legal framework that acknowledges the complex geophysical environment in the world's frozen regions and explore the impact that an ice-sensitive legal system would have on topics ranging from the everyday activities of Arctic residents to the territorial foundations of the modern state" (http://icelawproject.weebly .com). The ICE LAW Project was funded by the Leverhulme Trust (Grant IN-2015-033).

Notes

1 Lisa Gregoire, "Nunavut Mining Company Takes Icebreaking off the Table," *Nunatsiaq News*, November 6, 2017, https://nunatsiaq.com/stories/article/65674nuna-vut_mining_company_takes_ice_breaking_off_the_table/. Since 2017, opposition has strengthened. Prior to hearings in February 2021, Pond Inlet's mayor took a stand against the expansion, which also called for building a railroad from the mine to the Milne Inlet dock. As the hearings concluded, local hunters protested by blockading the company's airstrip for seven days. Randi Beers, "Mary River Mine Protesters Announce End to Blockade," *Nunatsiaq News*, February 11, 2021, https://nunatsiaq .com/stories/article/mary-river-mine-protesters-announce-end-to-blockade/.
2 Due to the importance of various sea ice stages to Indigenous activities and animal habitats, we take "icebreaking" to mean all forms of sea ice disruption to facilitate navigation by ships, including where icebreakers are not required. This ranges from breaking routes through solid pack ice to pushing already broken up ice floes further apart.
3 Igor Krupnik et al., eds., *SIKU: Knowing Our Ice—Documenting Inuit Sea Ice Knowledge and Use* (Dordrecht: Springer, 2010).
4 David N. Thomas and Gerhard S. Dieckmann, eds., *Sea Ice: An Introduction to Its Physics, Chemistry, Biology and Geology* (New York: Blackwell, 2008); Claudio

Aporta, "Markers in Space and Time: Reflections on the Nature of Place Names as Events in the Inuit Approach to the Territory," in *Marking the Land: Hunter-Gatherer Creation of Meaning within Their Surroundings*, eds. Robert Whallon and William Lovis (Abingdon: Routledge, 2016), 67–88.

5   L.A. Codispoti et al., "Synthesis of Primary Production in the Arctic Ocean: III. Nitrate and Phosphate Based Estimates of Net Community Production," *Progress in Oceanography* 110 (2013): 126–150; Victoria Hill et al., "Synthesis of Integrated Primary Production in the Arctic Ocean: II. In Situ and Remotely Sensed Estimates," *Progress in Oceanography* 110 (2013): 107–125; P. Matrai et al., "Synthesis of Primary Production in the Arctic Ocean: I. Surface Waters, 1954–2007," *Progress in Oceanography* 110 (2013): 93–106.

6   A.M. Pagano et al., "High-Energy, High-Fat Lifestyle Challenges an Arctic Apex Predator, the Polar Bear," *Science* 359, no. 6375 (2018): 568–572.

7   John C. George et al., "Observations on Shorefast Ice Dynamics in Arctic Alaska and the Responses of the Iñupiat Hunting Community," *Arctic* 57, no. 4 (2004): 363–374.

8   "The Sea Ice Is Our Highway: An Inuit Perspective on Transportation in the Arctic," Inuit Circumpolar Council Canada, 2008, https://secureservercdn.net/104.238.71 .250/hh3.0e7.myftpupload.com/wp-content/uploads/2019/01/20080423_iccamsa _finalpdfprint.pdf; Claudio Aporta, "Routes, Trails and Tracks: Trail Breaking Among the Inuit of Igloolik," *Études/Inuit/Studies* 28, no. 2 (2004): 9–38.

9   Anna Stammler-Gossmann, "'Translating' Vulnerability at the Community Level: Case Study from the Russian North," in *Community Adaptation and Vulnerability in Arctic Regions* (Heidelberg: Springer Netherlands, 2010), 131–162.

10   Zhanpei Fang et al., "Reduced Sea Ice Protection Period Increases Storm Exposure in Kivalina, Alaska," *Arctic Science* 4, no. 4 (2018): 525–537.

11   Krupnik et al., *SIKU: Knowing Our Ice.*

12   Agata Durkalec et al., "Climate Change Influences on Environment as a Determinant of Indigenous Health: Relationships to Place, Sea Ice, and Health in an Inuit Community," *Social Science and Medicine* 136–137 (2015): 17–26.

13   Sheila Watt-Cloutier, *The Right to Be Cold: One Woman's Story of Protecting Her Culture, the Arctic and the Whole Planet* (Minneapolis: University of Minnesota Press, 2015).

14   Stammler-Gossmann, "'Translating' Vulnerability," 141.

15   H.P. Huntington et al., "Sea Ice Is Our Beautiful Garden: Indigenous Perspectives on Sea Ice in the Arctic," in *Sea Ice*, ed. David N. Thomas (Hoboken: Wiley, 2017), 583–799.

16   Aipilik Inuksuk, "On the Nature of Sea Ice Around Igloolik," *Canadian Geographer* 55, no. 1 (2011): 36–41; Stammler-Gossmann, "'Translating' Vulnerability"; Gita J. Laidler et al., "Travelling and Hunting in a Changing Arctic: Assessing Inuit Vulnerability to Sea Ice Change in Igloolik, Nunavut," *Climatic Change* 94 (2009): 363–397.

17   George et al., "Observations on Shorefast Ice."

18   Natalie Carter et al., "Arctic Corridors and Northern Voices: Governing Marine Transportation in the Canadian Arctic—Pond Inlet Nunavut," 2018, https://www .arcticcorridors.ca/reports/.

19   Josefino C. Comiso, "Large Decadal Decline of the Arctic Multiyear Ice Cover," *Journal of Climate* 25, no. 4 (2012): 1176–1193; J.C. Stroeve et al., "Changes in Arctic Melt Season and Implications for Sea Ice Loss," *Geophysical Research Letters* 41, no. 4 (2014): 1216–1225.

20   Inuksuk, "On the Nature"; Laidler et al., "Travelling and Hunting"; Stammler-Gossmann, "'Translating' Vulnerability."

21   Karen E. Kelley and Gita J. Ljubicic, "Policies and Practicalities of Shipping in Arctic Waters: Inuit Perspectives from Cape Dorset, Nunavut," *Polar Geography* 35, no. 1 (2012): 19–49.

22   Natalie Carter et al., "Arctic Corridors and Northern Voices: Governing Marine Transportation in the Canadian Arctic—Ulukhaktok Northwest Territories," 2018, https://www.arcticcorridors.ca/reports/.

23 Inuit Circumpolar Council Canada, "The Sea Ice Never Stops—Circumpolar Inuit Reflections on Sea Ice Use and Shipping in Inuit Nunaat," accessed August 15, 2021, https://www.inuitcircumpolar.com/project/the-sea-ice-never-stops-circumpolar -inuit-reflections-on-sea-ice-use-and-shipping-in-inuit-nunaat/.

24 Jackie Dawson et al., "Infusing Inuit and Local Knowledge into the Low Impact Shipping Corridors: An Adaptation to Increased Shipping Activity and Climate Change in Arctic Canada," *Environmental Science and Policy* 105 (2020): 19–36; Inuit Circumpolar Council Canada, "The Sea Ice Never Stops."

25 Susan C. Wilson et al., "Assessment of Impacts and Potential Mitigation for Icebreaking Vessels Transiting Pupping Areas of an Ice-Breeding Seal," *Biological Conservation* 214 (2017): 213–222.

26 Christine Erbe and David M. Farmer, "Zones of Impact Around Icebreakers Affecting Beluga Whales in the Beaufort Sea," *Journal of the Acoustical Society of America* 108, no. 3 (2000): 1332–1340.

27 Natalie Carter et al., "Arctic Corridors and Northern Voices: Governing Marine Transportation in the Canadian Arctic – Sachs Harbour Northwest Territories," 2018, https://www.arcticcorridors.ca/reports/.

28 Carter et al., "Pond Inlet Nunavut."

29 Hajo Eicken, "Indigenous Knowledge and Sea Ice Science: What Can We Learn from Indigenous Ice Users?" in *SIKU: Knowing Our Ice: Documenting Inuit Sea Ice Knowledge and Use*, eds. Igor Krupnik et al. (Dordrecht: Springer, 2010), 357–376; Claudio Aporta, "The Sea, the Land, the Coast, and the Winds: Understanding Inuit Sea Ice Use in Context," in *SIKU: Knowing Our Ice: Documenting Inuit Sea Ice Knowledge and Use*, eds. Igor Krupnik et al. (Dordrecht: Springer, 2010), 163–180.

30 Kelley and Ljubicic, "Policies and Practicalities."

31 Dawson et al., "Infusing Inuit."

32 Jens Dahl, "Mining and Local Communities: A Short Comparison of Mining in the Eastern Canadian Arctic (Nanisivik/Arctic Bay) and Greenland (Marmorilik/ Uummannaq)," *Études/Inuit/Studies* 8, no. 2 (1984): 145–157.

33 Dawson et al., "Infusing Inuit."

34 Claudio Aporta et al., "Shipping Corridors through the Inuit Homeland," *Limn* 10 (2018); "Qaqqaliaq: Going to the Hilltop to Scan," ICE LAW Project, 2017, https:// icelawproject.weebly.com/uploads/1/2/6/4/126423318/april_2017_mm_qaqqaliaq _printed-layout.pdf; "The Cumulative Effects of Marine Shipping: What We've Heard so Far," Transport Canada, 2018, https://letstalktransportation.ca/cems; Leah Beveridge, "Inuit Nunangat and the Northwest Passage: An Exploration of Inuit and Arctic Shipping Conceptualizations of and Relationships with Arctic Marine Space in Canada," in *Governance of Arctic Shipping: Rethinking Risk, Human Impacts and Regulation*, eds. Aldo Chircop et al. (Berlin: Springer, 2020), 137–149.

35 Aporta et al., "Shipping Corridors."

36 Tommy T.B. Koh, "A Constitution for the Oceans: Remarks at the Final Session of the Conference at Montego Bay," in *United Nations Convention on the Law of the Sea* (New York: United Nations, 1983), xxxiii–xxxvii.

37 Jon M. Van Dyke, "Military Exclusion and Warning Zones on the High Seas," *Marine Policy* 15, no. 3 (1991): 147–169.

38 Ilker K. Basaran and Hayat Cemre Cakioglu, "Marginal Ice Zone: Profit vs Protection," in *Arctic Yearbook 2020*, eds. Lassi Heininen et al. (Akureyri: Arctic Portal, 2020), 132–137.

39 Watt-Cloutier, *The Right to Be Cold*, xxiii.

40 Dalee Sambo Dorough, "The Rights, Interests and Role of the Arctic Council Permanent Participants," in *Governance of Arctic Shipping: Balancing Rights and Interests of Arctic States and User States*, eds. Robert C. Beckman et al. (Leiden: Brill, 2017), 68–103.

41 John Reid and Matthew Rout, "The Implementation of Ecosystem-Based Management in New Zealand—A Māori Perspective," *Marine Policy* 117 (2020): 103889.

42 "Agreement to Prevent Unregulated High Seas Fisheries in the Central Arctic Ocean," Government of Canada, 2018, https://www.dfo-mpo.gc.ca/international/agreement-accord-eng.htm.

43 Aporta et al., "Shipping Corridors"; ICE LAW Project, "Qaqqaliaq"; Pikialasorsuaq Commission, "An Inuit Strategy for the Future of Pikialasorsuaq," accessed August 15, 2021, http://pikialasorsuaq.org/kl/; see also Aldo Chircop et al., "Is There a Relationship between UNDRIP and UNCLOS?" *Ocean Yearbook* 33, no. 1 (2019): 90–130. For further Canadian examples, see Sheila Watt-Cloutier, "Everything Is Connected: How Marine Spatial Planning Links Environment, Economy, Sustainability, Human Rights and Leadership in the Twenty-First Century Arctic," *Northern Public Affairs Magazine*, 2016, http://www.northernpublicaffairs.ca/index/everything-is-connected-how-marine-spatial-planning-links-environment-economy-sustainability-human-rights-leadership-in-the-twenty-first-century-arctic/; "Tide Turning on Marine Co-Governance in Canada," West Coast Environmental Law, August 8, 2017, https://www.wcel.org/blog/tide-turning-marine-co-governance-in-canada.

44 Karen Engle, "On Fragile Architecture: The UN Declaration on the Rights of Indigenous Peoples in the Context of Human Rights," *European Journal of International Law* 22, no. 1 (2011): 141–163.

45 Zoe Todd, "Fish Pluralities: Human-Animal Relations and Sites of Engagement in Paulatuuq, Arctic Canada," *Études/Inuit/Studies* 38, no. 1–2 (2014): 217–238.

46 Hugo Grotius, *The Freedom of the Seas; or, The Right Which Belongs to the Dutch to Take Part in the East Indian Trade* [1608] (New York: Oxford University Press, 1916); Carl Schmitt, *The Nomos of the Earth in the International Law of Jus Publicum Europaeum* [1950], trans. G.L. Ulmen (New York: Telos Press, 2006); Carl Schmitt, *Land and Sea: A World-Historical Meditation* [1942], trans. Samuel Garrett Zeitlin (New York: Telos Press, 2015).

47 Stuart Elden, "Missing the Point: Globalization, Deterritorialization and the Space of the World," *Transactions of the Institute of British Geographers* 30, no. 1 (2005): 8–19; Stuart Elden, "Land, Terrain, Territory," *Progress in Human Geography* 34, no. 6 (2010): 799–817; Stuart Elden, *The Birth of Territory* (Chicago: University of Chicago Press, 2013).

48 Philip E. Steinberg, *The Social Construction of the Ocean* (Cambridge: Cambridge University Press, 2001).

49 Schmitt, *The Nomos of the Earth*.

50 Under UNCLOS, states are recognized as possessing territorial seas (out to 12 nautical miles). However these waters, as well as their continental shelves, are granted territorial status because they are considered contiguous extensions of land, not because the ocean itself is understood as having territorial properties. See Jones and Ranganathan, this volume; Steinberg, *The Social Construction*; Philip E. Steinberg, "Sovereignty, Territory, and the Mapping of Mobility: A View from the Outside," *Annals of the Association of American Geographers* 99, no. 3 (2009): 467–495.

51 Katherine G. Sammler, "The Rising Politics of Sea Level: Demarcating Territory in a Vertically Relative World," *Territory, Politics, Governance* 8, no. 5 (2020): 604–620.

52 Clive Schofield and David Freestone, "Islands Awash Amidst Rising Seas: Sea Level Rise and Insular Status Under the Law of the Sea," *International Journal of Marine and Coastal Law* 34, no. 3 (2019): 391–414; Kate Purcell, *Geographical Change and the Law of the Sea* (Oxford: Oxford University Press, 2019).

53 UNCLOS acknowledges that "economic interests … [as] evidenced by long usage" may be cited to justify straight baselines (Article 7(5)), but only when the geometric conditions have already been met.

54 Carter, *Decolonising Governance: Archipelagic Thinking* (London: Routledge, 2019).
55 John Childs, "Extraction in Four Dimensions: Time, Space and the Emerging Geo(-) Politics of Deep-Sea Mining," *Geopolitics* 25, no. 1 (2020): 189–213.
56 Marta Conde et al., "Mining Questions of 'What' and 'Who': Deepening Discussions of the Seabed for Future Policy and Governance," under review, used with permission; Jones and Ranganathan, this volume; Surabhi Ranganathan, "Ocean Floor Grab: International Law and the Making of an Extractive Imaginary," *European Journal of International Law* 30, no. 2 (2019): 573–600.
57 Lisa A. Levin et al., "Defining 'Serious Harm' to the Marine Environment in the Context of Deep-Seabed Mining," *Marine Policy* 74 (2016): 245–259; Holly J. Niner et al., "Deep-Sea Mining with No Net Loss of Biodiversity—An Impossible Aim," *Frontiers in Marine Science* (2018).
58 Jessica Lehman et al., "Turbulent Waters in Three Parts," *Theory & Event* 24, no. 1 (2021): 192–219; Kimberley Peters and Philip E. Steinberg, "The Ocean in Excess: Towards a More-than-Wet Ontology," *Dialogues in Human Geography* 9, no. 3 (2019): 293–307; Philip E. Steinberg and Kimberley Peters, "Wet Ontologies, Fluid Spaces: Giving Depth to Volume through Oceanic Thinking," *Environment and Planning D: Society and Space* 33, no. 2 (2015): 247–264.
59 Stefan Helmreich, "Nature/Culture/Seawater," *American Anthropologist* 113, no. 1 (2011): 132-144; Stefan Helmreich, "Waves: An Anthropology of Scientific Things— Transcript of the Lewis Henry Morgan Lecture given on October 22, 2014," *HAU: Journal of Ethnographic Theory*, 4 (2014): 265–284.
60 Claudio Aporta, "Shifting Perspectives on Shifting Ice: Documenting and Representing Inuit Use of the Sea Ice," *Canadian Geographer* 55, no. 1 (2011): 6–19.
61 Clark's speech is reprinted in Franklyn Griffiths, ed., *Politics of the Northwest Passage* (Kingston & Montreal: McGill-Queen's University Press, 1987): 269–273.
62 Ingrid Bay-Larsen et al., "Mapping Ice in the Norwegian Arctic–on the Edge between Science and Policy," *Landscape Research* 46, no. 2 (2020): 167–181; Philip E. Steinberg et al., "Edges and Flows: Exploring Legal Materialities and Biophysical Politics of Sea Ice," in *Blue Legalities*, eds. Irus Braverman and Elizabeth R. Johnson (Durham: Duke University Press, 2020), 85–106; Siri Veland and Amanda H. Lynch, "Arctic Ice Edge Narratives: Scale, Discourse and Ontological Security," *Area* 49, no. 1 (2017): 9–17.
63 Noella J. Gray, "Charted Waters? Tracking the Production of Conservation Territories on the High Seas," *International Social Science Journal* 68, nos. 229–230 (2018): 257–272; Kimberley Peters, "The Territories of Governance: Unpacking the Ontologies and Geophilosophies of Fixed to Flexible Ocean Management, and Beyond," *Philosophical Transactions of the Royal Society B: Biological Sciences* 375, no. 1814 (2020): 20190458.
64 Elden, *The Birth of Territory.*
65 Purcell, *Geographical Change.*
66 Schmitt, *The Nomos of the Earth.*
67 Robert W. Duck, "Marine Spatial Planning: Managing a Dynamic Environment," *Journal of Environmental Policy and Planning* 14, no. 1 (2012): 67–79; Stephen Jay, "The Shifting Sea: From Soft Space to Lively Space," *Journal of Environmental Policy and Planning* 20, no. 4 (2018): 450–467; Sue Kidd and Geraint Ellis, "From the Land to Sea and Back Again? Using Terrestrial Planning to Understand the Process of Marine Spatial Planning," *Journal of Environmental Policy and Planning* 14, no. 1 (2012): 49–66; Sara M. Maxwell et al., "Dynamic Ocean Management: Defining and Conceptualizing Real-Time Management of the Ocean," *Marine Policy* 58 (2015): 42–50; see also Steinberg and Peters, "Wet Ontologies."
68 Elizabeth Mann Borgese, *The Oceanic Circle: Governing the Seas as a Global Resource* (Tokyo: United Nations University Press, 1998).
69 Jon M. Van Dyke et al., eds., *Freedom for the Seas in the 21st Century: Ocean Governance and Environmental Harmony* (Washington, D.C.: Island Press, 1993).

70 Elden, "Land, Terrain, Territory."

71 Stuart Elden, "Legal Terrain: The Political Materiality of Territory," *London Review of International Law* 5, no. 2 (2017): 199–224; Stuart Elden, "Terrain, Politics, History," *Dialogues in Human Geography* 11, no. 2 (2021): 170–189; Gastón Gordillo, "Terrain as Insurgent Weapon: An Affective Geometry of Warfare in the Mountains of Afghanistan," *Political Geography* 64 (2018): 53–62; Gastón Gordillo, "The Power of Terrain: The Affective Materiality of Planet Earth in the Age of Revolution," *Dialogues in Human Geography* (2021).

72 Elden, "Legal Terrain"; Kimberley Peters et al., *Territory beyond Terra* (London: Rowman & Littlefield, 2018).

73 Elden, "Terrain, Politics, History," 8.

74 Leslie Acton et al., "What Is the Sargasso Sea? The Problem of Fixing Space in a Fluid Ocean," *Political Geography* 68 (2019): 86–100; Noëlle Boucquey et al., "The Ontological Politics of Marine Spatial Planning: Assembling the Ocean and Shaping the Capacities of 'Community' and 'Environment,'" *Geoforum* 75 (2016): 1–11; Noëlle Boucquey et al., "Ocean Data Portals: Performing a New Infrastructure for Ocean Governance," *Environment and Planning D: Society and Space* 37, no. 3 (2019): 484–503; Luke Fairbanks et al., "Assembling Enclosure: Reading Marine Spatial Planning for Alternatives," *Annals of the American Association of Geographers* 108, no. 1 (2018): 144–161; Peters, "The Territories of Governance."

75 Rachael Squire, "Where Theories of Terrain Might Land: Towards 'Pluriversal' Engagements with Terrain," *Dialogues in Human Geography* (2021); Gordillo, "The Power of Terrain."

8
UNCLOS AS A GEOPOLITICAL CHOKEPOINT

Locked Down, Locked In, Locked Out

Elspeth Probyn

> The law of the sea ... offers a microcosm of the opening and foreclosure of international law's rulemaking protocols.
> —Surabhi Ranganathan, "Decolonization and International Law"[1]

The turbulent times of Covid-19 continue to reveal with force the multifaceted nature of the inequalities and inequity of those locked down, and those locked out. From the perspective of the ocean, I examine how Covid-19 has laid bare structural chokepoints. I attend to the situation of stranded seafarers unprotected by any state, and I note those whose livelihoods are being ravaged by the mobility of distant-water fisheries (e.g., those that fish outside of their 200-nautical-mile exclusive economic zone [EEZ]).

For some, it may come as a surprise that these situations are enabled by one of the most wide-ranging and ambitious pieces of legal thinking of the 20th century. Indeed, the United Nations Convention on the Law of the Sea ("UNCLOS"[2]) is one of the most radical if deeply uneven legal documents ever produced about the ocean. Phillip Allot's view from 1992 (a couple of years before UNCLOS was in full force) was that UNCLOS contained "a half-formed new structural uniqueness, full of painful ambiguities and exciting possibilities."[3] It tried to envision the ocean as a global commons but became a conduit for enclosing the more-than-human marine environment. As I argue, certain articles enabled conceptual and practical chokepoints, disrupting the circulation of its vaunted quest for harnessing the ocean for humanity. And one of its greatest occlusions

FIGURE 8.1 Commercial container ship, "Ever Given," stuck in the Suez Canal in March 2021. Image is about 2.64 kilometers wide. Processed by Pierre Marcuse. Licensed under the Creative Commons Attribution 2.0 Generic.

DOI: 10.4324/9781003205173-9

is the lack of recognition of Indigenous rights to their traditional sea countries. As Charlie Watts, an Inuk Senator, states about Inuit Rights in Canada, "the UNCLOS system does not provide a mechanism to ensure the participation of Indigenous peoples in any matters relating to the law of the sea, even when these directly affect Indigenous peoples' rights."[4]

In this chapter, I explore how certain articles enshrined in the Law of the Sea, as well as the underlying tensions at the time of its final articulation in 1982, haunt us. The very question of that "us" compels my argument. Forged within a certain understanding of humanity, and of humanitarianism, the language of UNCLOS contains blind spots in its in/occlusions. The "structure of feeling"[5] that fed UNCLOS's formulation of humanity is today certainly no longer in play. As Ayça Çubukçu queries in her recent review of Achille Mbembe's latest book on Afropolitics and decolonization,[6] can we envision "a humanist invitation to live up to humanity?"[7] However, to hope for better we must better understand the historical present. This is where the analyses of UNCLOS's "opening and foreclosure" continue to be critical. What spheres—cultural, economic, humanitarian, social, geopolitical—are being brought together or forced apart, reworked, or occluded through the legacy of the law?

My argument proceeds in four parts. The first part positions the strange locked down nation I inhabit, and through the optic of cruise ships and cargo fleets I examine the plight of seafarers locked out at sea during Covid-19. The second part examines certain regulatory terms that emerged alongside and through UNCLOS, which I argue allow for the evasion of the protection of human and more-than-human lives.[8] The third part examines the philosophical and political underpinnings of UNCLOS. In the fourth part, I explore the notion of legal and marine geopolitical chokepoints, before finally turning in my conclusion to a reflection on whether UNCLOS as a juridical system of opening and enclosure may prompt us to re-evaluate the legacies of maritime law. I also raise the crucial issue of how Indigenous people were and continue to be locked out of their traditional sea countries.

Seafarers Stuck at Sea

The hackneyed phrase, "we're all in this together" continues to be countered by the dizzying ways in which we are not. At the height of the Covid-19 crisis, there were over a million seafarers caught at sea. On land, some hoarded toilet paper and mastered sourdough, while some were under considerable hardship caused by socio-economic pressures, and others working on the ocean were denied basic human rights. A report from the International Transport Workers' Federation frames seafarers as "out of sight, out of mind."[9] But it is because of them that commodities circulate—more than 90 percent of global trade is shipped around the world. They enable our everyday consumption (of tea and spices or now, plasma TVs), and are indeed out of the minds of most. Seafarers have been variously described as "social marginals within their home societies

... away from home ... harbingers of drunkenness and disorder."[10] Conversely, in colonial Southeast Asia, "the sea was integral to a 'home place.'" Relegated by ethic difference, seafarers were designated as "sea gypsies" with no home*land*.[11]

With Covid-19, the feeling of being trapped inside has become a widespread effect of being trapped, one that elsewhere I typify as "cleithrophobia."[12] All those images of people's faces pressed against windows to see their loved ones wave on the other side. Since March 2020 in Australia, it is illegal for citizens and permanent residents to leave or enter their island home. The Commonwealth budget delivered on May 12, 2021 came with the bald statement that international borders would not be opened until at least mid-year 2022. At the time of writing (May 2021), those trying to enter Australian from India, including Indian Australian citizens and residents, face a six-year jail term and fines of up to $AUD66.600. As one ex-pat Australian reporter stuck in the United States writes: "There are up to 40,000 Australians around the world registered with the Department of Foreign Affairs who identify as 'stranded'—that is, they desperately want to return home, but they can't."[13]

I think of the irony of White men who proclaimed this land as terra- and aquanullius free to be conquered, who now in the figure of the present prime minister, Scott Morrison, won't let anyone leave or return. And the awful weight of White history reminds me of how the Indigenous people of this land have been locked down since the arrival of Cook. As Bradley Moggridge, an Indigenous hydrogeologist from the Kamilaroi Nation, succinctly puts it: "we've been locked up on missions and reserves, losing our language and stories; locked out of country and we still don't have the keys to our country."[14] It was, after all, only in 2021 that the Australian government condescended to change the words of the Australian National Anthem from "For we are young and free" to "For we are one and free." The oldest continual civilization in the world does not, however, enjoy freedom from everyday harassment and the trauma of past and present injustice.

This is to say that I write from a very parochial place. An island nation that has benefited hugely from globalization now pulls down the iron shutters—as Lester says, we are the new "Hermit Kingdom."[13] Australia has become very insular with a "[literally superficial] view of connections."[15] It isn't by chance, although it is ironic, that Australia has locked down its own citizens and residents. As Itamar Mann argues, Australia's offshore refugee detention system is one "of cruelty by design."[16] Those who live on the island "know" at some level—some fiercely opposed, others in favor, and still more who seemingly don't care—of the atrocity that several different governments of both political parties[17] have inflicted on refugees. It was John Howard's Coalition government that in 2001 first introduced "offshore processing" in Nauru and Papua New Guinea. This is a reality that I cannot further explore here but it remains a significant fact of Australian life[18]—and one that other governments such as the UK's have been tempted to replicate.

In his ethnography of cargo ship crew, Ben-Yehoyada spatializes how "One's social worth was tied to one's ability to get off the ship."[19] Is it too far a bow to draw to consider how some Australian citizens and permanent residents took for granted their ability to get off the island—their social worth measured in business trips, overseas holidays and cruises?

And what of cruise ships—those floating gin palaces—which during the height of the crisis entrapped people in different ways. At one level, they are enormous spaces of privilege that criss-cross the seas. A publicity line from one of the largest cruise companies states:

> From the moment you step aboard, we want you to feel welcomed and right at home. And with attentive service from a friendly staff that knows what hospitality means, you'll find your Princess® ship truly is your home away from home.[20]

Those homes soon became unheimlich as they were proven to be floating containers of the virus. In Australia, the most notorious case was a US-headquartered Carnival ship operating under the flag of Bermuda. On March 19, 2020, the ship slunk into Sydney Harbour in the early morning with over 660 infected people on board. Passengers disembarked although there were no results from the very few swab tests conducted. They immediately flew off to their various homes across Australia and the world. Amy Dale writes that "the cruise cluster, which is believed to have originated from an infected crew member distributing food and drinks, has been responsible for at least 20 deaths."[21] Twenty-one of the overall 908 deaths thus far recorded in Australia came from one ship.

Then New South Wales Police Commissioner Mick Fuller had a stern message to all cruise ship operators: "They don't pay taxes in Australia; they don't park their boats in Australia … time to go home."[22] Natalie Klein writes, "In early April 2020, it was estimated that 15,000 crew were stranded on 18 cruise ships around the Australian coast with concerns that coronavirus would take hold and spread."[23] On April 23, the *Ruby Princess* left for the Philippines where it joined a huge, stilled flotilla: mid-year, Manila Bay was the world's biggest "parking lot" for cruise ships, with many thousands of crew still on board.[24] Freya Higgins-Desbiolles notes that "[a]s the cruise ships became stranded around the world as ports closed to them, the question of exactly where home for them was, as they operated under FoC [Flags of Convenience], began to be discussed."[25] "Cruise companies choose to use a FoC as part of their economic model, helping their business gain profits by helping them avoid stringent economic, social and environmental regulations."[26]

For cruise ship workers, being locked away at sea for a long period was an extraordinary situation. They normally would change crew when the cruise reaches its destination. However, for the seafarers on cargo ships incredibly long hours and year-long periods at sea are normal. It is also normal to be abandoned "in calculated economic decisions by ship owners … to stop paying for the upkeep of the crew and the ship."[27] State and international government responses to Covid-19, were exacerbated by shipping companies' cavalier attitudes. As one crew member stated: "'We just want to request … on humanity ground [sic] please release us,' said Gaurav Singh, 29, an officer on the Anastasia, where several crew members are suicidal after waiting for about five months."[28] The living conditions in their tiny, shared cabins are horrendous, with one crew member saying he feels like he is "in prison, with a bunch of very grumpy men."[29] Another trapped ship member said, "We are simultaneously always leaving and never leaving."[28]

Locking Up the More-than-Human Ocean: EEZ, MSY & TAC, ITQ, FoC

The question of "home" is vexed because of the complicated system of vessel flags, especially "open registry" versus "closed registry." Under UNCLOS, the latter represents a "genuine link" between the ship and the state under which flag it sails, in the French "un lien substantiel": "a substantial link" hinting at some possibility of substantiating that connection. Conversely, the former relies on weak claims: "it may be that a ship has no physical connection with its flag State. Indeed, it may never visit its notional 'home port,' or even find it possible to do so, given that some open registries, such as those of Mongolia or Bolivia, are based in land-locked States."[30]

Flying the flag of whatever state is cheapest and most blind to human rights and environmental abuse, ships shuttle commodities around the world with a brutal efficiency. When, however, they were stopped, caught by Covid-19, some of the realities of their labor came to light. Strangely enough, while cruise ships and cargo ships were refused entry to ports, distant-water fisheries industries continued to operate illegal, unreported, and unregulated (IUU) fishing in coastal waters of the Global South.[31] These huge enterprises rarely need to land as the fish is frozen and processed at sea and transhipped to smaller vessels. This is not necessarily an innocent manoeuver: it allows for illegal fish to go unnoticed.

I want to step back a moment to consider how international regulations in the last century allowed for this strange situation. The alphabet soup in the subheading above includes some of the most important ways in which the ocean was, and continues to be, framed. They were measures propelled by the realization that the seemingly inexhaustible supply of fish in the sea was in fact not the case. It dawned on scientists and regulators that fish could be finite.

The acronyms mentioned above: EEZ (Exclusive Economic Zone), MSY (Maximum Sustainable Yield) & TAC (Total Allowable Catch), ITQ (Individual Transferable Quota), and FoC (Flags of Convenience) are interlinked. Following the Truman Proclamation of September 28, 1945, the United States asserted exclusive jurisdiction beyond their traditional territorial seas. The machinery of UNCLOS eventuated in the 1982 decision to make EEZs applicable for all coastal countries. UNCLOS ruled that a state's EEZ is an area beyond and adjacent to the territorial sea, extending seaward to a distance of no more than 200 nautical miles out from its coastal baseline. This would lead to states realizing that they could protect "their" fish through fisheries science.

The Maximum Sustainable Yield (MSY) was first enshrined in American fisheries policy as early as the 1950s. Its author Wilbert Chapman, who comes across as an ardent American fish nationalist, argued that "there was no time to waste in staking an American claim to high-seas fish."[32] As Carmel Finley and Naomi Oreskes bluntly state,

> US policy was designed to draw the seas—in particular the Pacific—under US influence and control. Thus, while not a physical enclosure, in the sense of fencing in a commons, it was, for all intents and purposes, a 'political' enclosure.[33]

MSY soon became widely seen as the vehicle that would arrive at the holy grail: sustainable fisheries. It is ensconced in Total Allowable Catch (TAC), which dictates how many fish of different species can be taken out of the sea. How much is being fished is measured against estimates of the reproduction of fish stocks. But of course, counting fish is not an exact science. As one fisheries regulator I interviewed wryly acknowledged, "fish have tails" and no regard for manmade lines in the sea. As Finley and Oreskes put it, "MSY is an example of the proverbial three-legged stool. It began as policy, it was declared to be science, and then it was enshrined in law."[33]

These two moves to lock in the ocean were accompanied by another move that many see as the ultimate privatizing of the ocean.[34] Individual Transferable Quotas (ITQs) were progressively introduced in numerous fishing nations in the early 1980s. If it was hard to count fish, ITQs promised to render fish as private property to be rented or sold. To take a striking Australian instance, in 1984 the highly lucrative Southern Blue Tuna fishery was divided up by ITQ. Based on the historical catch of boats, owners were given a yearly quota to fish. The hope was that this would be a check to "the race to fish" as everyone was assured of their portion. However, due to a number of factors (mainly very high interest rates), a majority of boat owners sold out, and the fate of tuna ended up in the hands of less than 20 boat owners (from a previous fleet of over 200).[35]

These interlinked mechanisms profoundly altered the ocean, effectively enclosing it in different ways—parceling it up into privatized enclaves, and rendering fish as livestock. One could say that it ended the ocean as global commons. It also strangely divorced fish from their marine habitat—the former to be counted and allocated to owners, the latter fenced off. As Liam Campling and Alejandro Colás put it, capitalism at sea "reshaped coastlines and reconfigured marine ecosystems."[36]

I've already discussed the final acronym—FoC—which, as we will see, combined with another of UNCLOS's articles, continues to plague the more-than-human marinescape. As I will describe shortly, UNCLOS tried to evenly distribute the sea to developing nations and even to noncoastal states. Article

91 states, "Ships have the nationality of the State whose flag they are entitled to fly. There must exist a genuine link between the State and the ship."[37] As I flagged above, there has been much discussion over decades as to the interpretation of what a "genuine link" entails. While one might think that UNCLOS would have been clearer in its wording, Surabhi Ranganathan notes "UNCLOS is not simply permissive but suggestive in how prohibitions of FoCs may be evaded."[38]

If we can simplify the question of national flags, it seems that UNCLOS sought to extend rights to the ocean for all countries. Article 62(1), however, applies only to coastal states. It requires "coastal states to promote optimum utilisation in their EEZs."[39] As Parzival Copes states, this "essentially … constitute[s] a commandment that 'thou shalt not waste fish,' imposing a moral obligation on the coastal state to be reasonable in sharing resources in excess of its own capacity to utilize them."[40] While state oceanic boundaries were laid out, new "moral obligations" to far-flung nations were imposed.

Together these articles form the condition of possibility for the millions of seafarers left in precarious situations, and the countless small fishers who struggle to find fish in their national EEZ waters overexploited by the distant-water fishing industries. These long-distance fleets often engage in illegal fishing and operate under ever-changing or even multiple flags. This has produced no-go areas for local fishers whose inshore fish stocks are overexploited, or it forces them further out to sea at great risk in their small boats (see Florian Grisel, this volume, on small-scale fishers in France). West Africa has become a global hub of illegal fishing, losing an estimated $1.3 billion annually to the trade, according to a report from the Africa Progress Panel.[41] China's distant-water fishing fleet reports only an estimated eight percent of its catch.[42] Lest this seem to point solely at China, other major distant-water fisheries are from Spain, South Korea, and Russia—and privately-owned industrial fishing fleets are handsomely subsidized by their governments.

UNCLOS: Humanitarian Chaos

How did this come about? How did a document dedicated to resolving much of the world's economic and political divisions allow for this situation? Reading the history of the lead up to and implementation of the Law of the Sea Convention is heart-rending. As Ranganathan asks in 2021, looking back at the long and tortured story of the three UNCLOS, what "alternative political geographies, economic imaginaries and epistemic approaches were highlighted in the process?"[43]

Let us pause on the sheer breadth of what was hoped for. One of the central figures in the epic tale of UNCLOS was Elisabeth Mann Borgese. The daughter of the exiled German writer Thomas Mann, Borgese's gendered vision for the oceans encapsulated the hopes of post-World War II.[44] As Behnam, a diplomat with the UN, central to the UN Conference on Trade and Development (UNCTAD), and a close colleague of Borgese, writes:

A new paradigm for the ocean was in construction—no longer mare liberum, no longer mare clausum—but the common heritage of mankind. Elisabeth's love for the ocean was surpassed only by her commitment to peace and the well-being of humankind. She saw in the making of the constitution of the ocean through the Third United Nations Conference on the Law of the Sea (UNCLOS III), the making of a new world order.[45]

Following her death in 2002, the tributes to her vision and political tenacity were copious. But equally the titles of some of the articles reveal the disappointment that promises of a new order were betrayed. Behnam entitles one article, "The unfulfilled promise of the seventies,"[46] and another, "Twilight of the flag state control."[47] Others were more wishful: "Peace and the Law of the Sea."[48] Reading accounts of the times, what is striking is the sheer amount of activity involved in trying to come up with documents that would be purpose-fit for the vision of the ocean for all humanity. Interwoven throughout was the question of how to best harness the economic possibilities of the ocean for developing countries. It is remarkable to consider the geopolitical changes taking place. As Behnam puts it: "The period 1947 to 1964 witnessed the birth and the struggle for existence of some 75 new and developing States."[49] He continues:

> Trade was envisaged as the engine of growth and development for fledging economies and a vehicle by which developing countries could integrate themselves into a world economy so as to acquire the capacity to accumulate wealth and the capacity to deal with the kaleidoscope of development problems.[49]

And trade meant transport, and more precisely the ocean as a medium of transport—as it always has been but now in a different key. Behnam was clear-sighted about the Global North's motivation despite the 1970s exuberance to include the whole of mankind in a teleological passage to economic salvation. From within the bowels of UNCTAD, Behnam reports that "[a] rhetorical question was being bandied about in the corridors and smoke-filled rooms of UNCTAD: 'if developing countries cannot develop in the field of shipping, then where can they?'"[50] But as the developing nations increasingly used their majority votes to put their demands to the fore, Behnam notes how the goodwill of the 1970s began to evaporate. In Behnam's estimation, that "goodwill" was carried by "the remnants of a guilty conscience for colonial domination."[51] Writing from the perspective of the American military, Mark Rosen is even more blunt:

> the LOS Convention was negotiated during the height of the Cold War, in which there were basically three competing factions: (a) major maritime states such as the United States and the USSR, which wanted broad rights to ocean access; (b) the developing countries that made up the G-77 [the

now 134 strong UN collation of developing nations], which were mostly concerned with gaining access to marine resources and revenues commensurate with their population size; and (c) coastal states, which were interested in being able to exclusively exploit and protect their coastal resources and being able to hold the navies of the major maritime powers at arm's length.[52]

Now as oceans are filled with ships hiding under and changing flags of state from one moment to the next, and as coastal nations of the Global South gain precious little in fishing access fees, it is harder to cheer at these at times well-meaning sentiments. And to wonder at the political stakes. For instance, Harry Pitt-Scott relates how "The drafting of Liberia's Maritime Code ... was checked and approved by the American Overseas Tanker Corporation, Standard Oil, and ESSO (ExxonMobil), who wished to use the Liberia flag to weaken ship workers' unions and undermine the European shipping nations."[53]

UNCLOS as Chokepoint

What I want to explore now is how the ocean is a minefield of chokepoints, both geophysical and jurisdictional, and test out whether the concept of the law as chokepoint is useful. Donald Rothwell notes how marine chokepoints have traditionally been described:

> The law of the sea and maritime security has often placed emphasis upon so called "choke points," that is those navigation routes which either due to their geographical location or strategic significance are navigation routes through which large volumes of shipping pass and as a result the legal regime regulating that passage and the geopolitical factors within those waters take on particular significance to the international community.[54]

Chokepoints trouble the still dominant ideas about the fluidity of the ocean, whether celebratory or not. Allan Sekula, the famed filmmaker and writer, describes how under "the world's increasingly grotesque 'connectedness,' the hidden merciless grinding away beneath the slick superficial liquidity of markets," global capitalism rules.[55] That connectedness has increasingly become unstuck by economic and (il)legal chokepoints. The myth of the smooth connectedness of globalization across liquid expanses has been torn. Sometimes the chokepoints occur when geographical narrowing meets the vast size of super Panamax cargo ships. For instance, in March 2021 the 400 meters long (1,300 feet) *Ever Given* got stuck in the Suez Canal—blocking the passage of hundreds of cargo ships. The Suez, an engineering feat opened in 1869, is simply not wide or deep enough for the ever-growing size of cargo ships. Built to facilitate passage between the Mediterranean Sea to the Red Sea through the Isthmus of Suez, the canal also produces restrictions and immobility.

While there is not room here to properly discuss it, the 1990s return of piracy on a large, and international scale brought renewed attention to the precarity of ships traveling through the chokepoints of narrow straits. For instance, it was estimated that US$13 to $16 billion per year was lost to piracy "concentrated in the waters between the Red Sea and the Indian Ocean, off the Somali Coast, the Strait of Malacca and Singapore."[56] In the case of Somalia, no longer a functioning state, several argued that the "pirates" were "fishers" whose income was destroyed by distant-water fleets who through intensive IUU left fishers with no fish. As Stig Hansen reports, "Speaking on 20 November 2011, President Farole of Puntland reiterated what has become one of the most repeated explanations of Somali piracy: 'The piracy started when fishermen defended themselves against illegal fishers.'"[57] This view has been contested by many.[58] However, as one former fisher turned pirate or "protector of the sea" put it, "why would anyone go back to catching tuna when you can catch an oil tanker?"[59]

Jatin Dua's ethnography of Somali piracy[27] is a compelling account of the delicately intertwined histories of power in that region. More recently, he and colleagues in anthropology have turned to the notion of chokepoint to broaden its conceptual scope. As Ashley Carsh et al. argue, the concept of the chokepoint exposes "the underside of global circulation—the situated processes through which deterritorilized flows are channeled, diverted and bogged-down in the murky, sticky particularities of localities."[60] While this is somewhat obvious in the case of natural chokepoints, more widely they state that "the chokepoint is a useful analytic for examining the operative—and often generative—interplay of circulation and constriction in the contemporary world."[61]

What is particularly useful about their reconceptualizing of chokepoints is that they free it from a focus on immobility. Chokepoints do not only stop movement, they also operate relationally and temporally: "chokepoints are different things for different people at different times."[62] Carsh et al. argue generatively how "chokepoints are not only good to control or pass through, but they are also—to use Levi-Strauss's timeworn phrase—good to think with."[63] They turn to a particularly interesting question: "how do we think about something ... that is both a concept and a thing in the world?"[64]

Can we think of UNCLOS itself as a chokepoint? UNCLOS conceptually created and in practice produced new "things of the world," new borders, new ways of trying to measure the ocean's immensity, and new ways of functioning on and in the ocean. Article 62(1), whereby "coastal states to promote optimum utilisation in their EEZs," constitutes a chokepoint in that it lays open coastal nations to the push to fully exploit their resources while it encloses them to the predations of long-distance fishing fleets. In terms of a mechanism to control circulation, UNCLOS tries to get around a "natural" chokepoint of landlocked states with no access to marine resources by bringing them into the coastal more-than-human family. For instance, Article 91 gifts flags to any and all states whether they have a coast or not. Carsh et al.'s description below seems to apply to UNCLOS's action in the world:

> What emerges in and around the chokepoint, then, are high-stakes inter-plays and tensions between circulation and regulation, local and remote forces, and human and nonhuman agencies, often with unexpected and far-ranging effects.[65]

In its conceptual reach, chokepoint allows us to consider the law as a mobile force that remakes oceanic and terraqueous arrangements (see Henry Jones's chapter in this volume for an argument about the co-constitution of law and geography). The importance of chokepoint as both a thing in the world and as a concept, as Elizabeth Dunn points out, is that it forges "geopolitics based on the control of circulation rather than the control of territory."[66] Most obviously UNCLOS pro-duces a construction of marine EEZs based on previous imperial conquests. As we've seen, it forms the conditions of possibility for different regimes of labor as seafarers are locked down on stilled ships. Through regulations that EEZs allow for, fish become enclosed as private property. In short, UNCLOS has reformed the relations between land and sea, between flow and constriction, opening and closing, and between nations and people.

Conclusion: Accounting for Occlusion

In this chapter, following Ranganathan's argument about oceanic opening and foreclosure cited in my epigraph, I examined UNCLOS as a juridical system of opening and enclosure and foreclosure understood as a conceptual chokepoint that allows for flow and blockage. I focused on the unintended consequences of certain articles in UNCLOS that have resulted in a free-for-all for vessels, owned in one country, often run by companies in another, manned by crew mainly from the Global South although overseen by officers from the North. The more-than-human is devastated by unregulated fishing practices that ravage the marine environment and deplete the fish stock of developing coastal nations, depriving them of precious sources of protein. Instead of a set of interlocking parts, UNCLOS can be seen as a kaleidoscope or "a telescope for exploring relationships and disjunctures across multiple spatial and temporal scales."[63] We can palpably feel those temporal and spatial disjunctures underlying the accounts I have related of UNCLOS's deep commitment to, and the equally profound betrayal of, the humanitarian conviction of some players in the last century. That legacy has produced our present ocean: piecemeal watery parcels fought over through ocean grabbing, illegal fishing, or home to the ultimate in late-capitalist formations such as seasteading.[67]

Perhaps most heinously, UNCLOS and the deliberations and discussion behind it were deeply mired in colonial blindness. Blindness to the imperial histories of colonization resulted in very differently sized EEZs. As Peter Nolan astutely points out, the maritime imperial nations such as France and Britain were given the EEZs of their considerable colonies.[68] For instance, the British Indian Ocean Territory with a land mass of 60 square kilometers has an EEZ of 639,000 square kilometers (compared to China's total of 900,000 square kilometers). The British are free to do as they wish to "their" territory even though it was not terra- or aqua-nullius. The Chagossian People, who had lived on the islands since the 1790s, were completely and utterly removed from Diego Garcia, the largest island, so that the British could allow the US military to set up a large air and naval base.

In Australia, the Indigenous coastal people were, and still largely remain, locked out of their sea countries. The deep materiality and history of this form of lockout is hard for non-Indigenous people to fathom. From 1770 to 1829, 6,363,000 square kilometers of Australia's EEZ was locked down to Indigenous coastal people. Without treaty, taking back those unceded marine areas has been a long and painful fight. The year UNCLOS was enacted in 1982 was also when Eddie Koiko Mabo and his fellow Meriam kinspeople first started their challenge of legal fictions of terra- and aqua-nullius in Australia. On June 3, 1992, the High Court of Australia held that the Meriam possessed the traditional ownership of the lands of Mer, which lead the passing of the Native Title Act of 1993, providing the framework for all Australian Indigenous people to make claims of native title.[69] On July 31, 2008, Australia's High Court granted traditional owners exclusive native title rights to the intertidal zone in the Blue Mud Case, which is to say the area between high and low water marks including river mouths and estuaries. This gave Indigenous sea-country people control over fishing rights in that zone. Now the Northern Territory Aboriginal Land Council grants these licenses to recreational fishers, and commercial fishers cannot fish in the areas covered by the Aboriginal Land Rights (NT) Act.[70]

These are history-breaking legal judgments for Australian Indigenous people (in Aotearoa/New Zealand, Iwi/Māori now own about 50 percent of quota). They are unfortunately ongoing as each case has to be brought by individual Indigenous land councils to the courts. In addition to the immense hard work that this legal work entails, around the world, and especially in the Global South, there is a fine web of organizations, communities, NGOs, and different para- and governmental bodies that have over the decades sometimes used UNCLOS, and sometimes not, to bring about forms of oceanic justice. These range from the at times aggressive attitude of groups like Sea Shepherd who starting in 2014 chased the Bandit 6 for two years until they caught them. These were six notorious illegal fishing vessels (four of which were owned by a Spanish company, Vidal Armadores that still sails illegal fleet under various flags, including North Korea) who were plundering toothfish in the southern seas. In a productive and long-term move, in 1982 eight tiny Pacific island countries decided to do something against the foreign fishing fleets that caught much of the world's skipjack tuna but for which they received little recompense. They formed the Parties to the Nauru Agreement[71] and over the years they have rendered their fishing practices sustainable, and have forced foreign fishing fleets to properly pay them for being able to fish in their combined EEZ. Here we see some of the hopes of UNCLOS realized: developing nations banding together to protect their marine resources and ways of life.

Alongside these examples, countless "soft laws" have enabled better outcomes for the oceans, marine life, and seafarers. Others, such as WorldFish and the Gender in Aquaculture and Fisheries Section of the Asian Fisheries Society, have turned to trying to count what goes uncounted—refugees, women's work, discards—and organizations like the International Marine Organization, and the different national and international bodies of marine unions have pursued the objective of ensuring human and more-than-human marine rights. These bodies are crucial in accounting for the occlusions of UNCLOS.

Acknowledgments

My thanks to the interlocutors from whom I learned much in the process of realizing this book, and to our editor Irus Braverman, and especially to Surabhi Ranganathan for her help when I so needed it. Thanks, too, to Brett Nielson for his eagle-eyed remarks.

Notes

1 Surabhi Ranganathan, "Decolonization and International Law: Putting the Ocean on the Map," *Journal of the History of International Law* 23, no. 1 (2021): 161–183.
2 While the acronym UNCLOS (United Nations Convention(s) of the Law of the Sea) refers to the three UN Conventions I 1956–1958; II 1960; III 1973–1982), and LOSC (Law of the Sea Convention) is generally preferred by legal specialists, I use UNCLOS here as I am interested in the processes and the underlying philosophies that yielded the law.
3 Philip Allott, "Mare Nostrum: A New International Law of the Sea," *American Journal of International Law* 86, no. 4 (1992): 764–787. Also cited in Joanna Mossop, "Can We Make the Oceans Greener: The Successes and Failures of UNCLOS as an Environmental Treaty," *Victoria University Wellington Law Review* 49 (2018): 573–594.
4 Charlie Watts, "Inuit Rights to the Arctic," *Law Now*, May 7, 2015, https://www.lawnow.org/inuit-rights-to-the-arctic/.
5 The cultural studies scholar Raymond Williams (2015) uses this term to focus on the different ways of representation (from government documents to popular culture) that emerge at any one time in history to become part of dominant culture. Raymond Williams, *Politics and Letters: Interviews with New Left Review* (London: Verso Trade, 2015).
6 Achille Mbembe, *Out of the Dark Night* (New York: Columbia University Press, 2021).
7 Ayça Çubukçu, "Book Review. *Out of the Dark Night: Essays on Decolonization* by Achille Mbembe," *LSE Review of Books* (April 29, 2021).
8 While I am mainly concerned in this section with how the oceans and fish became privatized through regulation, my understanding is framed by a more-than-human perspective that focuses on the interrelation of marine animals and environments and humans. See Elspeth Probyn, "The Cultural Politics of Fish: A More-than-Human

Habitus of Cultural Consumption," *Cultural Politics* 10, no. 3 (2016): 287–299; Elspeth Probyn, *Eating the Ocean* (Durham: Duke University Press, 2016).

9 International Transport Workers' Federation, *Out of Sight, Out of Mind: Seafarers, Fishers & Human Rights* (London: International Transport Workers' Federation, 2006).

10 Stephen Davies, "The Parallel Worlds of Seafarers: Connections and Disconnections on the Hong Kong Waterfront (1841–1970)," in *Meeting Place Encounters Across Cultures in Hong Kong, 1841–1984*, eds. Elizabeth Sinn and Christopher Munn (Hong Kong: University of Hong Kong Press, 2017), 131.

11 Jennifer Gaynor, "Maritime Ideologies and Ethnic Anomalies: Sea Space and the Structure of Subalternity in the Southeast Asian Littoral," in *Seascapes: Maritime Histories, Littoral Cultures, and Transoceanic Exchanges*, eds. Jerry H. Bentley et al. (Honolulu: University of Hawai'i Press, 2007), 53–70.

12 Elspeth Probyn, "Cleithrophobia, and the Oceanic Unheimlich" (Keynote, The Return of the Gothic Marine Conference, James Cook University, 2021).

13 Amelia Lester, "My Country, the World's New Hermit Kingdom," *Foreign Policy*, May 12, 2021, https://www.crikey.com.au/2021/05/12/my-country-the-worlds-new-hermit-kingdom/?success=q5jpqd.

14 Bradley Moggridge, Talk given at the launch of Emily O'Gorman's *Wetlands in a Dry Country* (Seattle: University of Washington Press, 2021) (Macquarie University, Zoom, July 15, 2021).

15 Naor Ben-Yehoyada, *The Mediterranean Incarnate: Region Formation between Sicily and Tunisia since World War II* (Chicago: University of Chicago Press, 2017), 230.

16 Itamar Mann, "Attack by Design: Australia's Offshore Detention System and the Literature of Atrocity" *European Journal of International Law* 32, no. 1 (2021): 309–326.

17 The two dominant political parties in Australia are the Labor Party and the Coalition Party, which is the more conservative, and over the last two decades has become harder right-wing.

18 See for instance, Tania Penovic and Azadeh Dastyari, "Boatloads of Incongruity: The Evolution of Australia's Offshore Processing Regime," *Australian Journal of Human Rights* 13, no. 1 (2007): 33–61; Alexander Reilly, Gabrielle Appleby, and Rebecca Laforgia, "'To Watch, to Never Look Away': The Public's Responsibility for Australia's Offshore Processing of Asylum Seekers," *Alternative Law Journal* 39, no. 3 (2014): 163–166; Madeline Gleeson, *Offshore: Behind the Wire on Manus and Nauru* (Sydney: NewSouth, 2016); Adrienne Millbank, "World's Worst or World's Best Practice? European Reactions to Australia's Refugee Policy," *People and Place* 12, no. 4 (2004): 28–37.

19 Ben-Yehoyada, *The Mediterranean Incarnate*, 171.

20 "Ruby Princess Cruise Ship," Princess Cruises, accessed July 15, 2021, https://www.princess.com/ships-and-experience/ships/ru-ruby-princess.

21 Amy Dale, "COVID-19: All Out to Sea," *LSJ: Law Society of NSW Journal* no. 66 (2021): 40–43.

22 Mark Reddie, "First Ruby Princess Crew Members Disembark After Coronavirus Isolation, Hundreds Still Left on Board," *ABC News*, April 22, 2020, www.abc.net.au/news/2020-04-21/some-crew-on-coronavirus-cruise-ship-ruby-princess-taken-off/12167856.

23 Natalie Klein, "Explainer: What Are Australia's Obligations to Cruise Ships Off Its Coast Under International Law?" *The Conversation*, April 2, 2020, https://theconversation.com/explainer-what-are-australias-obligations-to-cruise-ships-off-its-coast-under-international-law-135428.

24 Candace Sutton, "Cruise Ship Staff Still Adrift After 110 Days and Counting," *News Com*, July 2, 2020, https://www.news.com.au/travel/travel-updates/health-safety/cruise-ship-staff-still-adrift-after-110-days-and-counting/news-story/fb7b81b83c7807420133096c2088feca.

25 Freya Higgins-Desbiolles, "Socialising Tourism for Social and Ecological Justice After COVID-19," *Tourism Geographies* 1, no. 14 (2020): 1–12, 7.

26 Ibid., 6.

27 Jatin Dua, *Captured at Sea: Piracy and Protection in the Indian Ocean* (San Francisco: University of California Press, 2019), 500.

28 Raul Dancel et al., "Stuck at Sea: Seafarers Who Kept International Trade Humming Despite Covid-19," *Straits Times*, November 15, 2020, https://www.straitstimes.com /world/stuck-at-sea-seafarers-who-kept-international-trade-humming-despite -covid-19.

29 Ibid.

30 Stuart Kaye, "Port Access and Assistance to Cruise Ships During the COVID-19 Pandemic," *Australian Law Journal* 94, no. 6 (2020): 420–426.

31 Sally Yozell and Amanda Shaver, *Shining a Light: The Need for Transparency Across Distant Water Fishing*, Stimson Center Report, 2019, https://www.stimson.org/2019 /shining-light-need-transparency-across-distant-water-fishing/.

32 Cited in Carmel Finley and Naomi Oreskes, "Maximum Sustained Yield: A Policy Disguised as Science," *ICES Journal of Marine Science* 70, no. 2 (2013): 245–250, 247.

33 Ibid., 248.

34 Gísli Pálsson and Agnar Helgason, "Figuring Fish and Measuring Men: The Individual Transferable Quota System in the Icelandic Cod Fishery," *Ocean & Coastal Management*, 28, nos. 1–3 (1995): 117–146.

35 Elspeth Probyn, *Eating the Ocean* (Durham: Duke University Press, 2016).

36 Liam Campling and Alejandro Colás, *Capitalism and the Sea: The Maritime Factor in the Making of the Modern World* (London: Verso, 2021).

37 David Anderson, "Article 19: Nationality of Ships" (presentation, March 2020), https://wwwcdn.imo.org/localresources/en/OurWork/Legal/Documents/ IMLIWMUSYMPOSIUM/1%20Negotiating%20history%20of%20article%2091 %20of%20UNCLOS.pdf.

38 Surabhi Ranganathan, e-mail message to author, May 5, 2021.

39 Mossop, "Can We Make the Oceans Greener," 577.

40 Parzival Copes, "The Impact of UNCLOS III on Management of the World's Fisheries," *Marine Policy* 5, no. 3 (1981): 217–228.

41 Edward McAlister et al., "African Migrants Turn to Deadly Ocean Route as Options Narrow," *World News*, 2018, https://www.reuters.com/article/us-europe- migrants-africa/african-migrants-turn-to-deadly-ocean-route-asoptions-narrow- idUSKBN1O210J.

42 Andrew Jacobs, "China's Appetite Pushes Fisheries to the Brink," *New York Times*, April 30, 2017, https://www.nytimes.com/2017/04/30/world/asia/chi- nasappetite-pushes-fisheries-to-the-brink.html?_r=0; Tabitha Grace Mallory, "China's Distant Water Fishing Industry: Evolving Policies and Implications," *Marine Policy* 38 (2013): 99–108.

43 Ranganathan, "Decolonization," 161.

44 Irini Papanicolopulu, ed., *Gender and the Law of the Sea* (Leiden: Brill Nijhoff, 2019).

45 Awni Behnam, "A Personal Reflection by the IOI Honorary President," International Ocean Institute, accessed July 15, 2021, https://www.ioinst.org/elisabeth-mann -borgese/a-personal-reflection-by-the-ioi-honorary-president/.

46 Awni Behnam, "The Unfulfilled Promise of the Seventies: Shipping and Developing Countries," *Ocean Yearbook Online* 18, no. 1 (2004): 453–487.

47 Awni Behnam and Peter Faust, "Twilight of Flag State Control," *Ocean Yearbook Online* 17, no. 1 (2003): 167–192.

48 Agustin Blanco-Bazan, "Peace and the Law of the Sea," *Ocean Yearbook Online* 18, no. 1 (2004): 88–97.

49 Awni Behnam, "The Unfulfilled Promise of the Seventies," 455.

50 Ibid., 475.

51 Ibid., 476.

52 Mark E. Rosen, "Challenges to Public Order and the Seas," Center for Naval Analyses, 2014, https://apps.dtic.mil/sti/pdfs/ADA597094.pdf, 5.

53 Harry Pitt Scott, "Offshore Mysteries, Narrative Infrastructure: Oil, Noir, and the World-Ocean," *Humanities 9*, no. 3 (2020): 5.

54 Donald R. Rothwell, "Arctic Ocean Choke Points and the Law of the Sea" (ANU College of Law Research Paper No. 10-81, Canadian Council of International Law Annual Meeting, Ottawa, October 28, 2010), 4.

55 Allan Sekula, *Between the Net and the Deep Blue Sea (Rethinking the Trace in Photographs)* (Cambridge: The MIT Press, 2002), 7.

56 Rashid U. Sumaila and Mahamudu Bawumia, "Fisheries, Ecosystem Justice and Piracy: A Case Study of Somalia," *Fisheries Research* 157 (2014): 154–163.

57 Stig Jarle Hansen, "Debunking the Piracy Myth: How Illegal Fishing Really Interacts with Piracy in East Africa," *RUSI Journal* 156, no. 6 (2011): 26–31.

58 Axel Klein, "The Moral Economy of Somali Piracy–Organised Criminal Business or Subsistence Activity?" *Global Policy* 4, no. 1 (2013): 94–100; Awet T. Weldemichael, "Maritime Corporate Terrorism and its Consequences in the Western Indian Ocean: Illegal Fishing, Waste Dumping and Piracy in Twenty-First-Century Somalia," *Journal of the Indian Ocean Region* 8, no. 2 (2012): 110–126.

59 Dua, *Captured at Sea*, 59.

60 Ashley Carsh et al., "Chokepoints: Anthropologies of the Constricted Contemporary," *Ethnos* (2020): 1–11, 4.

61 Ibid., 2.

62 Ibid., 7.

63 Ibid., 9.

64 Ibid., 5.

65 Ibid., 3.

66 Elizabeth Cullen Dunn, "Warfare and Warfarin: Chokepoints, Clotting and Vascular Geopolitics," *Ethnos* (2020), https://doi.org/10.1080/00141844.2020.1764602.

67 Surabhi Ranganathan, "Seasteads, Land-Grabs and International Law," *Leiden Journal of International Law* 32, no. 2 (2019): 205–214; Alexander Mawyer, "Floating Islands, Frontiers, and Other Boundary Objects on the Edge of Oceania's Futurity," *Pacific Affairs* 94, no. 1 (2021): 123-144.

68 Peter Nolan, "Imperial Archipelagos: China, Western Colonialism and the Law of the Sea, *New Left Review* (2013): 77–95.

69 "Mabo Decision," National Museum of Australia, accessed August 20, 2021, https://www.nma.gov.au/defining-moments/resources/mabo-decision.

70 "Blue Mud Bay High Court Decision," Creative Spirits, November 29, 2020, https://www.creativespirits.info/aboriginalculture/land/blue-mud-bay-high-court-decision.

71 "Parties to the Nauru Statement," accessed July 15, 2021, https://www.pnatuna.com/.

9

FROM EXTENDED URBANIZATION TO OCEAN GENTRIFICATION

Miami's River Port and the Precarious Geographies of Haitian Shipping

Jeffrey S. Kahn

During the 1970s and 1980s, the US federal district court in Miami served as a testing ground for a series of legal confrontations over the rights of Haitian asylum seekers arriving by sea.[1] More than three decades after these litigation struggles began, I found myself in the chambers of one of the judges who had issued several key rulings on matters related to the procedures to be used in assessing Haitian asylum applications. I had arranged the meeting with the hope that the Judge might share some of his personal recollections from his time presiding over these lawsuits. As we sat together, he set the mood for his account of the cases that had come before him by explaining that prior to the arrival of Haitians and Cubans, the city had been nothing more than "a sleepy little town on the banks of the Miami River." With this choice of narrative frame, the Judge had selected a particular temporality for his tale, one that evoked a break from a bygone provincial urbanity. At its heart was the image of a modest, now-long-displaced settlement perched on the shores of a slow-moving estuary. In terms of demographics, this vision of the Miami of old was the antithesis of its contemporary, multiethnic incarnation. In terms of materiality, one could almost imagine the gentle current of the river as the embodiment of a fondly remembered southern languor.

The Judge's comforting depiction of his hometown seemed to arise from a sense of nostalgia and loss around not just the Miami of the past, but, more specifically, the Miami River that once was. The authors of a 1992 Grand Jury report concerning the fate of the by-then polluted waterway evoked parallel sentiments when they recalled that "Miami" was a Seminole word meaning "sweetwater"—a frequently repeated, but incorrect, bit of local folklore—and

FIGURE 9.1 Wooden Haitian freighter under tow by US Coast Guard Vessel during the 1980s. Plastic containers and bicycles can be seen tied to the roof of the combined wheel and cabin house. Courtesy of the US Coast Guard Historian's Office, Washington, DC.

DOI: 10.4324/9781003205173-10

that one of the city's early founding families, the Brickells, had "built their home and store by [the river's] clear water and reveled in its beautiful palm and mangrove lined banks." The Brickells' neighbor on the north bank, Julia Tuttle, wrote of her desire to transform the "wildness" of the river's edge, its "tangled mass of vine, bush, trees, and rock," into a "prosperous country."[2] Miami's history, however, is a bit more tumultuous, a bit more blood-soaked, than this southern pastoral—or subtropical frontier—poetics suggests. One need only look to the open terrorism of the Ku Klux Klan in Miami's Colored Town and the brutal fighting of the Seminole wars to recognize that this image of a placid, riverine backwater surrounded by a verdant terra nullius hides a good deal.[3]

In some ways, though, the Judge was correct. Haitians and Cubans would remake the city along with its eponymous river. By the 1980s, the waterway had become one of Florida's major shipping centers. Hundreds of vessels steamed across the Caribbean to dock at its scattered terminals, which were officially renamed the Port of Miami River in 1986—unceremoniously, as it turns out, to satisfy a Coast Guard regulation related to bilge pump outs.[4] For Haitians struggling to establish themselves in a city deeply hostile to their presence, the river became a key site in what was an emerging maritime economy of breakbulk shipping that would connect provincial Haitian ports to this newly established outpost of the diaspora.[5]

In this chapter, I explore the Haiti trade of the Miami River as a shifting geography of possibility that emerged under conditions of near impossibility. It is a geography of mobile, primarily Black subjects whose racialized ways of sustaining life under conditions of extreme precarity have been subjected to a host of regulatory measures designed to manage, surveil, and constrain. Despite these very real obstacles, Haitian seafarers have managed to produce riverine and maritime space through their economic improvisations, their material circulation, and their often mundane, signifying visibility within the commercial waterway of a city that would have preferred to keep them hidden away. In this sense, they are part of a longer tradition of Black seafarers whose contributions to the making of the spatiality of the Atlantic world are too often erased in favor of portrayals in which they appear as mere passive objects of White geographic agency—a critical observation Catherine McKittrick and Clyde Woods have made with regard to Black space-producing capacities more generally.[6]

I argue that by attending to Haitian shippers' and seafarers' concrete itineraries, their ascribed value within a symbolic economy of urban "decay" and gentrifying "revitalization," and their encounters with regulatory and commercial interventions designed to manage how they shape the geographies they inhabit, one can see the multiple ways in which the fashioning of riverine urban space and maritime "hinterland" space are intertwined. Here, I draw on Neil Brenner and Christopher Schmid's concept of "planetary urbanization"—the extension of urbanization processes beyond the traditionally imagined urban core to what have long been imagined as extra-urban "hinterlands" or "wildernesses"—and

Nancy Couling's broadening of the concept to include "ocean urbanization."[7] As Couling notes, if we are to take seriously processes of extended urbanization, we must also recognize that the increasingly dense maritime traffic (whether shipping or fishing) and the more fixed sea-based extractive industries that support global capitalism are as much a part of urbanized society as land-based activities (an idea I will explore further in the pages ahead). The lens of "ocean urbanization" is particularly useful because it brings more conventional urban sites—the metropolitan port, for example—and the sea spaces to which they are connected into a single frame. This allows one to discern the ways the regulation of urban ports transforms not only the city but also the sea and the maritime worlds that exist beyond metropolitan cores.

To reveal the particular interconnections of port and sea in Miami requires attention to how Black, subaltern geographies become enmeshed with and influence the often more visible—or at least, more-often-acknowledged—spatialities of formal urban orders and their processes of transformation. Not the least of these are the racialized codifications of urban space so central to the question of gentrification and the population displacements that are its *sine qua non*. In this chapter, I place the urbanized space of Miami's river port in the same frame as commercial sea lanes, asking not just whether ocean urbanization exists but what particular kinds of racializing urban processes extend to the oceans. More specifically, I explore whether the gentrification of the Miami River shaped who belonged on the maritime routes between Haiti and South Florida, and, if so, whether these reverberations developed as an element of ocean urbanization that we might call *ocean gentrification*.[8]

Miami's "Savage Slot": The River Port

When one thinks "Miami," one envisions beaches and bays, neon-accented art deco hotels, and gleaming skylines. It is, in other words, a metropolis synonymous with shorelines and high-end properties. Or, that is what comes to mind when one imagines the postcard version of the "Magic City," long heralded as a salubrious (but also hedonistic) American Riviera.[9] If the white sands of South Beach and the warm waters of Biscayne Bay stand in for Miami's luxury resort aesthetic, then another body of water, the Miami River, has done the same for the city's multiethnic laboring class and its underworld economies. For decades, this 5.5-mile working river, with its multiple, scattered, private shipping terminals catering to the shallow draft ports of the Caribbean, has been derided as a crime-ridden "demimonde" and a blight on Miami's urban core.[10] The stigmatization of the river has been a thorn in the side of those select few who have yearned for its redevelopment with pricey but profitable condominiums and mega-yacht marinas. And yet, for many, the river is almost invisible. "Most of Dade's citizenry," a 1991 Miami-Dade County Grand Jury charged with assessing the state of the river acknowledged, "barely notice the Miami River" at all.[11] Cutting through the heart of the city, the river hides in plain sight.

For those who do care to notice, the association of the river with toxicity, racialized alterity, and criminality became firmly entrenched during the 1980s and 1990s. In 1992, the aforementioned Grand Jury report declared that "benign neglect and planned desecration" had transformed the once pristine waters of the river into a "cesspool."[12] Intentional and unintentional dumping of raw sewage, industrial refuse, contaminated bilge water, and other pollutants had blanketed its silt-filled bottom "with a layer of toxic waste."[13] One journalist described the material detritus of "industrial boatyards, fuel docks, scrap iron yards, and a hash of small eyesore worksheds" as further defiling "the water and a riverbed that has not been dredged for half a century."[14] Robert Parks, former head of the committee that served as a forum for river issues in the 1980s and 1990s, conveyed the degree of pollution thus: "If you dug up the sediment and touched a match to it, it would burn."[15]

Intertwined with the river's reputation for physical contamination was also a sense that the waterway materialized a racialized difference that gave a titillating edge to Miami's status as a troublingly porous contact zone with the United States' neighbors to the south—Haiti and Cuba in particular. This status as a peculiar border metropolis—one without a proximate land border—marked the city as not fully of the normatively White, sanitary US body politic.[16] In other words, there was another sort of "pollution" at play here. One profile of the river described it as "reminiscent of the settings of Joseph Conrad's novels—subtropical, multilingual, milling with activity, and largely unfettered, with dark subtexts."[17] Another also referred to it as a "Conradian waterway" and likened the Coast Guard Captain who took responsibility for safety regulations at the river port in the mid-1990s to "Lord Jim," one of Conrad's White protagonists engulfed by "savage" worlds.[18] Adding further "color" were the almost obligatory descriptions of the bloated remains of animal sacrifice—allusions to the demonized rituals of Miami's Haitian and Cuban populations—jettisoned human corpses, and discarded arms shipments.[19] When night falls, one journalist proclaimed, the river begins to look like "a modern day Barbary Coast."[20] Instead of North African corsairs, however, Haiti's *Tonton Makout*, the secret police of the ousted Duvalier dynasty, were rumored to lurk on the river, having supposedly turned their sights from human smuggling to overseeing the drug shipments that fueled Miami's urban degradation and violence.[21]

In many ways, the Miami River is an embodiment of what anthropologist Michel-Rolph Trouillot has called the "Savage slot" of the West's own "geography of imagination"—a foil that, he argues, the West requires as the condition of possibility for its own utopian self-conceptualization.[22] The river's particular "Savage" inflection has much to do with existing US popular templates of Haitian alterity—themselves crafted from the sense that Haiti, a nation born of the largest successful slave revolt in history, has exhibited a purer form of Black Africanity than is perceived to exist elsewhere in the Americas.[23] These templates, of course, are not fixed; they have shifted, shedding and accreting associations over time. As tens of thousands of Haitian asylum seekers made their way by sea to Miami during the 1970s and 1980s, for instance, a noxious brew of racializing stereotypes that associated Haitians with AIDS, criminality, and "voodoo" rites combined to sharpen the anxieties around Haitian migration and an emerging border panic.[24] This fraught moment surrounding the robustness of American sovereignty was sparked, in large part, by the departure of more than one hundred and twenty thousand Cubans from the port of Mariel and their arrival in South Florida during the spring of 1980.[25] Much of the concerns related to the predominantly light-skinned Cubans, however, would be borne by Haitians, a pattern of displaced fear that would repeat itself in various forms in the years ahead. For reasons that will become clear momentarily, visions of an exceptional Haitian alterity would also be transposed to the river and Haitian-run river commerce.[26]

Tested by unprecedented litigation campaigns concerning the treatment of Haitian asylum seekers, scrambling to respond to the aftermath of the Mariel boatlift—as the Cuban exodus was dubbed—and grappling with a surge in coeval boat arrivals from Haiti, the administration of President Ronald Reagan launched an offshore border enforcement program initially dubbed Haitian Migrant Interdiction Operations (HMIO) in the autumn of 1981. HMIO, a collaboration between the US Coast Guard and the US Immigration and Naturalization Service (INS), was an effort to move the asylum processing regime of South Florida out onto the decks of Coast Guard vessels patrolling the Windward Passage, thereby permitting INS officials a freer hand in adjudicating Haitian migrant claims. It was a juridical "spatial fix" of sorts that relied

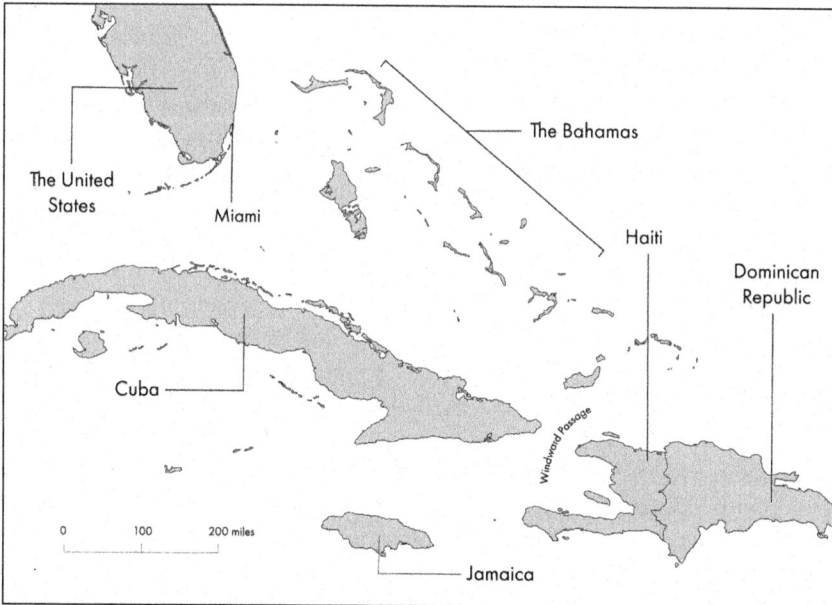

FIGURE 9.2   The northern Caribbean. Drawn by author.

on existing notions of ocean exceptionalism and extraterritoriality as cover for the creation of a new regime that masqueraded as asylum screening but served more as a means to achieve virtually across-the-board repatriations.[27] Resting on this jurisdictional arbitrage, the program effectively cut off migration aboard wooden sailing vessels from Haiti, but only after the aforementioned litigation campaigns had *prevented* the INS from emptying South Florida of the 50,000 to 70,000 Haitians who had made it to US shores.[28] Stigmatized by the press and the rhetoric of local and state officials, the Haitian community that took hold in North Miami would become a collective stock character in evolving, albeit standardized, narratives of Miami's threatening tropical alterity.[29]

By the 1980s, Haitians trading between Miami and Haiti's ports had emerged as players on the Miami River, creating new, visible maritime networks of exchange. At the same time, the Haitian vessels and the Haitian trade more generally became iconic of certain visible aspects of river commerce in ways that paralleled the stereotypes around the larger Haitian community, fueling popular imaginaries of what journalists would proclaim as the waterway's Conradian mystique.

Regulating the Port, Regulating the Sea

Just as the Coast Guard began intercepting the wooden vessels used to transport Haitian asylum seekers by sea, more and more Haitian boat owners entered the *commercial* trade routes between Haiti and South Florida. Cohorts of steel and wooden *motorized* freighters started to move between provincial ports and the small-scale terminals of the Miami River, their owners eager to connect the burgeoning diaspora community to markets back home.[30] Many aspects of the actual transportation dimensions of this trade were run by Haitians for Haitians, and it provided a variety of goods, from bulk food shipments to secondhand bicycles, that otherwise would have been difficult to access for consumers in Haiti's provinces. And yet the relative openness of the trade routes—including their accessibility to low cost, by US shipping standards, wooden freighters—was not to last. Just as Haitian asylum seekers traveling by sea became targets of draconian exclusion policies and, eventually, Coast Guard maritime patrols beginning in the early 1980s, so, too, would the Haitian freighters on the Miami River draw the attention of regulators seeking to control and restrict their mobility.

By the 1990s, vessels moving between Haiti and Miami River terminals had become associated with the smuggling of people and illicit drugs as well as a junkyard aesthetic, itself an effect of the large quantities of secondhand goods lashed to their decks as they exited the river. It was neither immigration nor drug enforcement actions, however, that actually began squeezing Haitian shippers out of South Florida. In May of 1994, the US Coast Guard's 7th district initiated Operation Safety Net, a program designed to enforce minimum safety standards for freight vessels under 500 gross tons, of which 238 were calling in the Port of Miami River at the time.[31] The resulting regulatory changes would push dozens of Haitian shippers out of Miami.

Operation Safety Net began with an interim inspection program that focused, according to the Coast Guard, on "firefighting, lifesaving, and crew require-ments" for the smaller freighters of the Miami River.[32] Some Haitian vessels were detained pending remediation of code deficiencies and all were warned that the applicable regulations would be fully implemented by July of 1997 with no exceptions.[33] The idea behind the phased approach was to allow vessel own-ers to bring their freighters up to code or to acquire sufficient funds to buy new vessels—in the case of the owners of wooden freighters, this meant purchasing steel-hulled replacements.[34]

In 1995, US Coast Guard Captain David Miller took responsibility for Operation Safety Net, intensifying inspections and continuing the detention of vessels in Miami. In the 1995–1996 fiscal year, Coast Guard officials on the river carried out five times the number of inspections and issued five times the number of citations compared to the previous year.[35] For many of the Haitian merchants most vulnerable to the new inspection regime, the writing was on the wall.

From the beginning, it was inevitable that the wooden freighters would even-tually succumb to the new regulations. In 1991, a Miami-Dade Grand Jury esti-mated there were between 25 and 35 such vessels calling on the Miami River. On the eve of the deadline for implementation of the freighter restrictions in 1997, it appears that number may have remained steady or possibly increased despite the difficulties that an Organization of American States trade embargo on Haiti had imposed on the river fleet between 1991 and 1994.[36] A single ship-ping terminal on the river, for instance, claimed that 62 Haitian vessels used its facility in 1997, the majority of them wooden freighters.[37] By the following year, they were gone.[38]

In the end, the Coast Guard did not deploy existing US regulations to block the Haitian freighters on the river—initially they had intended to apply provisions of Title 46 of the Code of Federal Regulations—but instead turned to the newly created Code of Safety for Caribbean Cargo Ships as a substitute standard.[39] This safety code, which applied to vessels under 500 gross tons, was adopted as part of the final negotiations surrounding the signing of the Caribbean Memorandum on Port State Control in 1996.[40] The Memorandum was one of many similar agreements modeled after the 1982 Paris Memorandum on Port State Control that the International Maritime Organization had been encouraging states to adopt as a means of dealing with inadequacies in the regime of vessel owner and flag state responsibility insofar as vessel safety was concerned.[41] Given that it cov-ered the same vessels as Operation Safety Net, the code provided a way for the Coast Guard to soften the edge of its enforcement actions: rather than imposing US regulations on foreign-flagged vessels from the Caribbean, officials could simply apply the safety standards adopted by states in the region served by the river freighters. US unilateralism was replaced with an ostensibly collaborative North-South internationalism that involved borrowing from Caribbean interna-tional law for the purpose of transforming the Miami River. Notably Haiti and the Dominican Republic, two of the top three destinations for freighters on the river, were not party to the Memorandum.[42]

Given that each vessel was itself a small-scale business operation, the Coast Guard's pivot to a more tough-minded approach to the river effectively shuttered dozens of Haitian-owned businesses that operated, in part, in Miami. At the same time, the removal of the wooden freighters effectively refashioned the aesthetics of this urban riverscape by decreasing the number of Haitian-operated vessels bearing what was perceived as "junk" and possibly "stolen" cargo through the heart of downtown Miami (more on this shortly). While the Coast Guard emphasized congressional mandates unrelated to Miami politics as the basis for its enforcement actions, others suspected something else was afoot. Since the 1980s, local hostility toward Haitian shipping on the river had been palpable, and not because of vessel safety issues. Many connected the new Coast Guard regulations with existing anti-Haitian sentiment, surmising that the impetus behind Operation Safety Net could not be entirely separated from the "redevelopment" goals of real-estate firms seeking to expel Black, immigrant shippers and the terminals that served them from the river all in an effort to transform the waterfront into a site for highly profitable new-build gentrification.

Gentrification on the River

In 1987, the *New York Times* ran a story with the attention-grabbing headline, "Miami's 6-Mile River: Pollution, Aliens, and Drugs." The tone was typical for coverage of the waterway, even if the outlet, with its national audience, was not. Despite its use of well-worn tropes of savage alterity and romantic visions of Miami's "flowering past"—and a pronounced amnesia with regard to its histories of imperial warfare and racial segregation—the piece captured how the specter of gentrification loomed over the river as far back as the 1980s. Jon Nordheimer, the article's author, wrote,

> those who see charm in the workaday world of the river, the paint-laden smell of the boat sheds and the decaying charm of faded bungalows of the side canals that branch off it mutter darkly that progress means gentrification, trendy riverside cafes with expense account customers and limited access to the public. They see an end to use of the river as a colorful port for the rust bucket tramp freighters, some as long as 200 feet, that squeeze down it laden with foodstuffs, machine parts, clothes, and just about anything else that can be lashed down to their decks.[43]

There were, in Nordheimer's framing, "two rivers," each with a set of temporally marked modes of existence—on the one hand, there was the gritty, rough and tumble world of river commerce (itself imagined as a holdover from an earlier time), and, on the other hand, there were the real-estate developments and new-build gentrification projects of a "revitalized" river-to-come.[44]

Others have noted something more specific working beneath the surface of this presumed attempt to attract real-estate capital to the riverfront. Under the veneer of general concern over crime, pollution, and the overall look of the working river was a more targeted agenda. In the 1980s, the rise of the Haitian freighter trade drew the attention of local leaders and federal agencies who began cracking down on stolen goods "smuggled" out of the river port. Because the Haitian freighters transported cargo both in their holds and strapped to their decks—often with piles of bicycles, lawn furniture, and plastic jugs visible to anyone standing on the river's edge—and were made of wood (often brightly painted), they gave a certain feel to the river, one not appreciated by all. Captain David Miller, who brought Operation Safety Net to full throttle in 1995, described his first encounter with the wooden vessels this way:

> I thought I was back in the 1900s [sic]. ... Here were all these wooden-hull coastal freighters. Then you came to a small shipping terminal and you would see a sign nailed to a tree, "This boat bound for Port-au-Prince on this date." Everywhere else this is done over computers, but here we were in the last century.[45]

Despite the fact that the cargo carried by wooden vessels made up a fraction of the total tonnage hauled out of the river each year, the Haitian freighters took center stage in conversations about the waterway's makeover and became prime targets for law enforcement. Here one finds unmistakable echoes of the previous decade during which Haitians, who made up a minuscule percentage of overall undocumented migration, had become highly publicized targets of new immigration control programs.[46] As aesthetic focal points and material embodiments of a Black, Caribbean, make-do precarity, the Haitian trade also seemed to absorb public anxieties, in this instance those concerning the criminality and pollution that plagued the river.[47] And just as HMIO had closed off migration routes out of Haiti, so would the new Coast Guard regulations block a good portion of the Haitian fleet from the river and, by extension, the waterways between Haiti and South Florida.

Back in 1990, several years before the Coast Guard would turn in earnest toward regulating river traffic, this sense that the Haitian vessels were being specifically targeted as part of gentrification efforts was already widespread. Customs officers and local Miami police raided dozens of freighters, confiscating "hot" bicycles—some of them supposedly bore the stickers from a police impound lot near the river, while, in other cases, the Haitian captains simply had no paperwork to prove ownership, which is hardly surprising for an informal economy of this type.[48] While policing on the river was meant to impact the drug trade from Latin America more generally, there seemed to be an almost farcical obsession with Haitian freighters and bicycles.

Clearly perturbed by having to participate in these operations, one Customs official was quoted as saying that raids by federal agencies and local police "shared the same goal—to run the Haitian boats off the river and make it a more attractive place for business."[49] This same official indicated his irritation at doing work that, essentially, was geared toward "supporting local real-estate development," especially given that Haitians "'have a right to ship out of the river'" as much as anyone else.[50]

The sense that law enforcement actions were tied to the commercial interests of Miami's well-heeled citizens did not arise out of thin air. In the 1980s, the Miami River Coordinating Committee (MRCC) served as a forum for the articulation of river-focused redevelopment plans, although it was by no means unified in this regard.[51] Aspirations for a waterfront of greenways, parks, and towering condominium complexes had, in other words, been in the atmosphere for decades. Operation Safety Net seemed perfectly tailored to complement the goals of those members of the MRCC seeking river redevelopment and to achieve what some of the earlier law enforcement operations targeting petty smuggling had not—that is, getting the unsightly wooden Haitian freighters off the river.

In the wake of Operation Safety Net, the Florida legislature created the Miami River Commission, which in turn launched projects to draft new urban infill and dredging plans.[52] The multi-million dollar improvements, funded in large part through federal programs, led to selective rezoning of commercially zoned properties for residential building projects on the river. It also provoked a series of legal battles between the river's trade association and city commissioners who had declared the working river "dead."[53] The rumblings about gentrification from the late 1980s had proved prescient.

By the late 1990s and early 2000s, the explicit interweaving of redevelopment goals and policing was unmistakable. Enforcement programs with names like Operation River Sweep and Operation River Walk, in many ways progeny of the operations of the 1980s (one of the earlier iterations was called Operation River Watch), brought local, state, and federal agencies together to target the steel-hulled Haitian freighters that remained on the river. Operation River Walk, created at the behest of Florida Governor Jeb Bush, mobilized multiple state and federal agencies to police the river freighters with the explicit goal of "[s]pur[ring] economic development along the river corridor."[54] As before, Haitian vessels seemed to bear the burden of this policing, at least in public rhetoric surrounding the operation.[55] In this new era of river investment, there was no longer any need to conceal the intersection of policing and real-estate development interests.

River Urbanization/River Gentrification, Ocean Urbanization/Ocean Gentrification

As scholars of transnationalism began to argue decades ago, diasporic subjects (though they would not have used this term) often do not remain locked in place with regard to identity or mobility once they have left their countries of birth, but instead maintain ties and may continue to circulate across juridically segmented, nationally defined spaces.[56] The network of Haitian shipping routes extending between Haiti's ports and the Miami River are a material testament to such transnational connection between diasporic Haitians and communities back "home." At the same time as this recognition of transnational subjectivity and practice emerged, however, so too did a widespread association of global interconnectivity with liquid metaphors of frictionless planetary flows.[57] And yet the maritime mobility of the wood-hulled freighters of the 1980s and 1990s, and even a significant number of the larger, steel-hulled ships of today, does not conform to the poetics of smooth oceanic circulation.[58] Much of the Haiti trade was labor-intensive, breakbulk shipping (carried out almost exclusively by Haitians), not highly mechanized, computerized intermodal transport. As I have already described, it proved vulnerable to regulatory whims and the territorial sovereignties that legitimize them—a terrain that hardly resembles visions of seamless global integration. Moreover, the breakbulk sectors of the Haiti trade involve a multiplicity of senders and receivers (as opposed to a handful of large, and largely homogeneous, shipments), many of whom are not hyper-mobile, "flexible citizens"[59] circulating in the borderless world that 1990s-era metaphors of flow and declarations of waning sovereignty proclaimed—to the contrary, they are frequently locked into place as a result of tightened immigration controls and uneven, racialized geographies of violence and precarity.[60]

Across the steep value gradients of these bordered geographies, mundane, often discarded objects become valuable commodities or media for satisfying obligations of reciprocity and care.[61] A dozen mattresses, a sack or two of rice, a tightly wrapped cardboard carton of secondhand clothing, some car batteries: these are just some of the cargo the stevedores load up in Miami and unload from many of the ships in Haiti's ports. River commerce, in other words, was not exclusively defined by massive, single-commodity bulk shipments. Exchanges of modest, when disaggregated, amounts of goods, and, as such, the maintenance of intimate family connections, small-scale retail operations, and, often enough, a combination of the two (although there are certainly "bigger" entrepreneurs involved in the trade with Haiti as well, but that is another story) were also central to the trade. These exchanges have unfolded across a half dozen or so provincial ports, interweaving the urban and rural peripheries (from the viewpoint of the capital, Port-au-Prince) of what is often perceived as the most peripheral nation in the Caribbean (from the viewpoint of the United States and, also, many of Haiti's neighbors) with the urban core and ethnic enclaves of one of the largest metropolitan areas in the United States.

The supply chains and circulation of goods that tie these sites together suggest a type of expansive interconnectivity that urban geographers Neil Brenner and Christopher Schmid have characterized as an outgrowth of processes of "planetary urbanization."[62] Drawing on Henri Lefebvre's provocative hypothesis that "society has become completely urbanized,"[63] Brenner and Schmid argue for a shift in focus from constrained understandings of "urban form"—a concern with urban cores, for instance—to "urbanization processes" that are far more extended in scope.[64] Interpreting their position generously allows for an acknowledgment of the distinctiveness of certain sites of "concentrated urbanization"[65] while also foregrounding the more expansive contours of far-flung, "poly-nucleated" urbanized regions and the ways spaces conventionally imagined as the antithesis of the urban—ocean "wildernesses," for example—are "increasingly interconnected with the rhythms" of these urbanization processes.[66] Within this view, even "transoceanic shipping lanes" can be reimagined as part of the "urban fabric" because of the ways they "enmesh" nodes of planetary political-economic activity.[67]

Oceans have long been envisioned as interstitial voids, and, as such, have escaped the "terracentric" focus of much humanistic and social science inquiry, something that has started to change of late with a host of so-called "oceanic turns" in law, history, literature, geography, and anthropology, among other disciplines.[68] One of my own contributions to this literature explores the production of space at sea by drawing a parallel between the pedestrian itineraries that Michel de Certeau identifies as actualizing urban spaces and the repetitive sea voyages that materialize networks of oceanic interconnectivity.[69] For de Certeau, the act of walking in the city signified; it had a material semiotic enunciatory capacity that he likened to language's phatic function. The phatic, a key concept within sociolinguistics, refers to the effect of utterances aimed less at conveying or eliciting information than at establishing interpersonal connection. Take, for example, the formulaic "how are you?" greeting, which is designed less to discover an interlocutor's state of being and more to ritualistically create or reinforce a social tie.[70] Although de Certeau is vague with regard to the interplay between phatic effects and human mobility, his argument suggests that movement, like an uttered "how are you?," signifies in a way that also leads to connection. Visible movement sends a message, in other words, much like a bird's song—an aural example of the phatic—sends a message. Bodies-in-motion implicitly exclaim "I am here" among a chorus of "I am here[s]." In this way, the existence of visible, bodies-in-motion can draw other bodies into what de Certeau calls a "mobile organicity" in which a multiplicity of itineraries in time form a spatial symphony of varying intensity that literally makes the city. There is, in other words, a nonverbal, material semiotic dimension to wandering urban corridors, one that actualizes but also "manipulates spatial organizations."[71] For de Certeau, the "phatic topoi" such "pedestrian enunciation" generates are *of* the city's built environment (they are constrained by it) but they do not succumb entirely to its panoptic features.[72] Rather, these forms of circulation produce and give life to a partially improvised urban environment.

I have argued elsewhere that something similar to de Certeau's phatic mobility also unfolds at sea, wherein the repetitive voyaging of, for example, Haitian freighters in the waters between Haiti, The Bahamas, Cuba, and South Florida produces maritime space.[73] Such passageways are not voids. They are deeply familiar highways, the contours and density of which are etched in the minds of those who work them and those who surveil them.[74] Moreover, much like de Certeau's "pedestrian enunciation," they fashion new geographies through oceangoing enunciation, and they do so with stylistic particularity.[75] As such, they "cannot be reduced to their graphic trail"—that is, they cannot be reduced to the courses they chart in the sea. For Haiti's wood-hulled freighters, this "style" was evinced in their architecture, their surfaces, the goods they hauled and lashed to their decks, and the racialized bodies of those who hauled them.[76] As Haitian captains charted paths through the sea, the passageways between Haiti and Miami—and thus the Miami River itself—were imbued with aspects of Haitian maritime life. On the one hand, the routes of these Haitian freighters became associated with Haitian-ness—and by extension, Haitian blackness—from the perspective of those policing them. But they also must be recognized as historically particular geographies of Black, space-making subjects, geographies, as Katherine McKittrick has noted, too often ignored or erased within dominant racial ontologies.[77]

It is important to recognize, then, that it is actual people and groups who co-produced these maritime trade routes (in combination with other non-human entities and forces), connecting distant urban centers across aqueous space and territorial boundaries.[78] The movement across these spaces is in many ways a maritime analogue to the contingent "pedestrian rhetoric" to which de Certeau attributes significant force in the constitution of urban space.[79] Moreover, one can also see these routes as extensions of that urban space. Again, to echo Brenner and Schmid, urban processes are not contained in urban cores. And just as urbanization processes extend across landscapes so do they also extend across seascapes. From this vantage point, one can speak of what Nancy Couling has called, building on Brenner and Schmid, "ocean urbanization," itself a subset of wider processes of extended or planetary urbanization.[80] In Couling's words, the "mesh of activity," the "marine highways," and the "technical sites of extraction" that one finds proliferating at sea all exemplify an expansion of urban processes into ocean realms.[81] Certainly, if one is to accept this definition of extended ocean urbanization, then the freighter routes that connect Haiti's provincial ports to the Miami River must also be considered part of such processes.

And if we can speak of extended urbanization into sea spaces, can we, keeping in mind the earlier discussion of transformations on the Miami River, also speak of an extended, ocean gentrification? To examine the history of Haitian freighter commerce on the Miami River is to witness repeated efforts at displacing certain Haitian populations from the waterway in an attempt to recast the aesthetics and materiality of the river as part of larger projects of capital investment and urban "revitalization."[82] Although these projects do not conform to the traditional notion of gentrification as a middle class endeavor of rehabilitating existing housing stock, the real-estate projects that would eventually take hold along the waterfront in the 2000s are examples of new-build

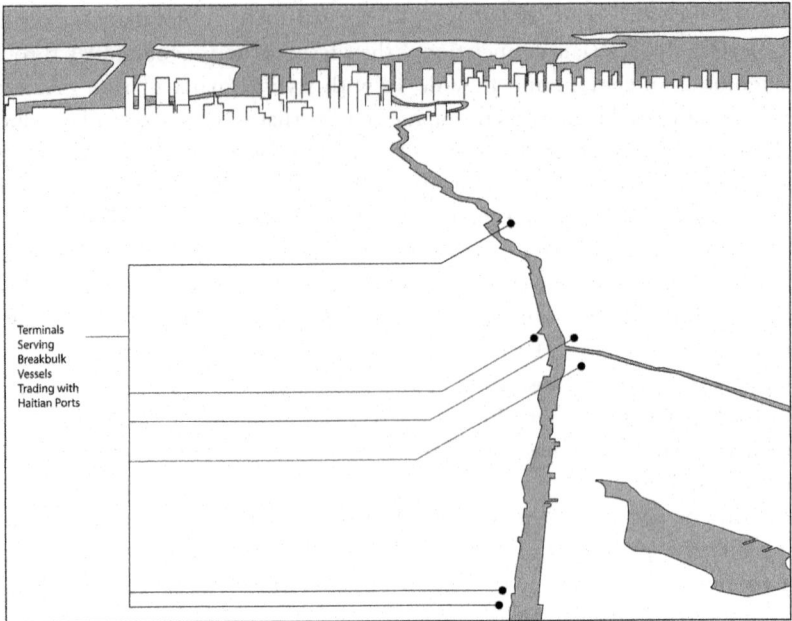

FIGURE 9.3 View of the Miami River from the northwest showing location of terminals serving the Haitian breakbulk trade in 2021. Drawn by author.

and environmental gentrification.[83] The remediation of the toxicity of the waterway—itself often associated during the 1980s and 1990s with the bilge-pumping and waste-dumping of Haitian freighters—and plans to replace, as opposed to restore, the architecture and other aspects of the built environment of its waterfront laid the groundwork not solely for the displacement of working class residents, but also for a continual squeezing of Haitian maritime commerce on the river.[84]

Shipping terminals do remain on the river, and as recently as 2008, Haiti was the second largest export destination for river shipping. Still, many of the smaller terminals serving the Haitian freighters (wood and steel-hulled) have fallen by the wayside, and segments of the Haiti trade were, indeed, blocked from Miami as part of Coast Guard regulatory efforts to "clean up" the waterway, efforts that fit hand in glove with longstanding projects to redevelop the river. While the Miami River Commission has long stated a preference for preserving the river's maritime industry, shipping has increasingly been pushed into a narrow corridor in the upper reaches of the river, outside of downtown, while residential projects proliferate elsewhere along its banks.

Displacements of this sort are not, however, just about remaking the city's *waterfront* properties. They also modify urban *waterways* and the styles of mobility permitted to flourish on them. Moreover, they do the same for the *maritime passages* that exist well beyond the densely populated metropolitan regions with which they are intertwined. This is because for many Haitian shippers, once they are pushed out of the river port, they also end up being pushed off the sea, at least insofar as the routes between Haiti and Miami are concerned. The intersection of Miami port regulation and environmental and new-build gentrification agendas thus constrains, contorts, and remakes the hard fought, oceanic itineraries—and thus, the urbanized maritime geographies—that these Haitian shippers and seafarers created in the northern Caribbean. The exclusions of urban gentrification yield ocean gentrification as well.

While these transformations at sea do not produce the class and ethnoracial homogenization associated with more conventionally imagined urban gentrification, they do give rise to something akin to its displacements. With increasing regulation and redevelopment efforts, a subset of vessel owners and captains have been pushed from the port and thus from the sea. This is possible because the Port of Miami River is the only port that caters to the Haitian breakbulk shipping operations I have been discussing. When Haitian shippers operating on thin margins are forced out of the river, the markets they support either fall to the wayside or are taken over by those who possess greater access to social and money capital. The result is larger and fewer ships than existed in decades past. Moreover, the captains and officers who operate the remaining vessels are in almost all instances no longer Haitian (although the change in staffing demographics has to do with a shifting licensing landscape as well). While three dozen or more Haitian-owned, breakbulk cargo ships engaged in the river's Haiti trade in years past, less than a dozen do so now, and only one is operated by Haitian captains.

The set of logical consequences that attend this history are easy enough to follow: (1) if the circulation of seafarers makes maritime space; (2) if maritime space can be considered part of networks of extended urbanization processes; (3) if gentrification is part of these urbanization processes; and (4) if the displacement of certain Haitian actors from the river also leads to their displacement from the sea routes that connect the river with Haitian ports; then (5) one can also contemplate the exclusion of actors from these very same sea spaces as a form of extended ocean gentrification. Moreover, these exclusions are part of a broader set of migration control efforts aimed at controlling Haitian maritime mobility, even if those interventions, as I have explored at length elsewhere, turn on the balance of sovereign power and judicial constraint in offshore ocean spaces rather than regulatory regimes enforced primarily in the port.[85] My goal is not to collapse land and sea or river and ocean spaces into an undifferentiated whole. I offer a frame in which one can start to examine terra firma, the riverine, and the oceanic as interconnected and mutually influencing in ways related to the complex dynamics of expansive urbanization processes that extend to watery realms.[86]

Conclusion

The Port of Miami River is not a typical port. During the 1980s, it consisted of scattered, private terminals along a narrow 5.5 mile stretch of water. In this sense, the Miami River is a fragmented and mutable "port" that snakes its way through the urban core of the city.[87] Unlike the mega-ports that have been pushed to the suburbs or otherwise hidden out of sight over the past half-century the world over, the small-scale terminals of the Miami River are of the built environment of the city and continue to be embroiled in its tensions.[88]

Various shifting ethno-racializations of river space, including an association of the waterway with Haitian commerce and thus with a particular type of blackness, have been part of Miami's urban tensions for some time, as have efforts at residential and commercial redevelopment along the river—a river increasingly associated with sky-rocketing property values. While the emergence of Haitian trading networks on the river led to its coding as a site of undesirable forms of Haitian commerce in the public imagination, the array of safety regulations, law enforcement actions, and urban redevelopment projects that later emerged would push a dwindling number of Haitian-owned ships into the upper reaches of the river. Condominiums and up-scale restaurants sprang up, in turn, where Haitian freighters once docked. These pressures have reconfigured the river's geography and the contours of its racialization—the linking of its spaces with certain types of racialized subjects.

The reworking of the river is not just a story of urban gentrification. This is because urbanized space is not restricted to the so-called urban core, after all. It extends into the "hinterlands" of land and sea, in this case the maritime routes between Haiti and South Florida, pulling them into the frame of the urban and the urban into the frame of the oceanic. To recognize this is to recognize that transformations of the urban port—transformations often effectuated through legal maneuvers—can lead to transformations of the sea and other distant ports. Drawing on the move to conceptualize these processes in terms of extended and ocean urbanization, I have proposed the possibility of charting the extension of more particular urban processes, such as gentrification, into an extended geography of sea space as well. As certain categories of Haitian actors are pushed from the urban river port, so too are they pushed from the sea routes that connect this port to the provincial harbors of Haiti.

The forms of ocean urbanization that Haitian actors have produced as space-making subjects and the forms of extended ocean gentrification that have played a role in containing and curtailing these geographies are part and parcel of urban struggles. As the urban port is remade, sea lanes and distant ports are also remade. In some ways, this is an obvious claim. But it is, nonetheless, an important one as it allows us to think of an urbanizing geography that extends across these sites of water and firm land and to see not only the extent of the vernacular spatial forms produced in the creation of maritime logistics assemblages, but also how the laws of certain types of ports, in this case the disaggregated river port, remake the sea as a site of dense, consequential human activity.

Acknowledgments

I would like to thank the participants in the Laws of the Sea conference workshops in which an earlier version of this chapter was first presented. I learned a great deal from the interdisciplinary conversations that came out of those meetings. Special thanks are owed as well to Irus Braverman for bringing us all together and for her editorial vision, to Jovan Scott Lewis for introducing me to the Black Geographies literature, to Phil Steinberg for his encouragement with regard to the broader geographic focus of this chapter, and to Liz DeLoughrey for ongoing and always illuminating conversations about critical ocean studies and the Caribbean.

Notes

1 Jeffrey Kahn, *Islands of Sovereignty: Haitian Migration and the Borders of Empire* (Chicago: University of Chicago Press, 2019).
2 Dade County Grand Jury, "Final Report of the Dade County Grand Jury" (11th Judicial Circuit of Florida, May 11, 1992), 1.
3 N.D.B. Connolly, *A World More Concrete: Real Estate and the Remaking of Jim Crow South Florida* (Chicago: University of Chicago Press, 2014); Jessica R. Cattelino, *High Stakes: Florida Seminole Gaming and Sovereignty* (Durham: Duke University Press, 2008).
4 *Payne v. City of Miami*, 53 So. 3d 258, 266 (Fla. 2010).
5 Breakbulk refers to mixed types of individually tallied cargo that longshoremen move and pack into the holds of general cargo vessels. It is distinguishable from containerized cargo or large bulk shipments of single commodities. Marc Levinson, *The Box: How the Shipping Container Made the World Smaller and the World Economy Bigger* (Princeton: Princeton University Press, 2006).
6 Katherine McKittrick, *Demonic Grounds: Black Women and the Cartographies of Struggle* (Minneapolis: University of Minnesota Press, 2006); Katherine McKittrick and Clyde Woods, "'No One Knows the Mysteries at the Bottom of the Ocean,'" in *Black Geographies and the Politics of Place*, eds. Katherine McKittrick and Clyde Woods (Cambridge: South End Press, 2007), 1–13; on Black seafarers, see also Paul Gilroy, *The Black Atlantic: Modernity and Double Consciousness* (New York: Verso, 1993); Michael A. Schoeppner, *Moral Contagion: Black Atlantic Sailors, Citizenship, and Diplomacy in Antebellum America* (New York: Cambridge University Press, 2019); Julius S. Scott, *The Common Wind: Afro-American Currents in the Age of the Haitian Revolution* (New York: Verso, 2018).
7 Neil Brenner and Christian Schmid, "Planetary Urbanization," in *Implosions/Explosions: Toward a Study of Planetary Urbanization*, ed. Neil Brenner (Berlin: Jovis, 2014), 160–163; Nancy Couling, "Formats of Extended Urbanisation in Ocean Space," in *Emerging Urban Spaces: A Planetary Perspective*, eds. Philipp Horn et al. (Cham: Springer, 2018), 149–176; Nancy Couling, "Ocean Space and Urbanisation: The Case of Two Seas," in *The Urbanisation of the Sea: From Concepts and Analysis to Design*, eds. Nancy Couling and Carola Hein (Rotterdam: nai010, 2020), 19–32.
8 Shortly before this chapter went to press, I brought up the idea of ocean gentrification with Naor Ben-Yehoyada, a fellow anthropologist whose work has examined the effects of motorized trawling on region formation in the Mediterranean. In our conversation, he made me aware of his own creative and insightful exploration of the topic of "maritime gentrification" as it has played out in the development of a class of Sicilian "maritime gentrifiers" in Mazara del Vallo. See Naor Ben-Yehoyada, "Maritime Gentrification: Class Formation and the Spatial Aspects of the Labor

Process in the Channel of Sicily, 1955–1990," *Urban Anthropology and Studies of Cultural Systems and World Economic Development* 41, no. 1 (2012): 43–72, 57. While there are some interesting parallels and shared scalar shifts, Ben-Yehoyada's argument focuses on how the expansion of a group into new maritime spaces transforms not only those spaces but also the subjects who now occupy them. In contrast, my own approach in this chapter focuses on the *relation* between specific "urban" *displacements* often associated with particular forms of gentrification and the set of *displacements* in "urbanized" sea space that they entail. These processes are absent from Ben-Yehoyada's account of maritime gentrification, much as a discussion of the gentrification of the gentrifiers, as he puts it, is absent from this chapter. My hope is that the two approaches with their differing foci and scholarly genealogies will be read as complementary and mutually illuminating.

9 Connolly, *A World More Concrete*, 5; Alejandro Portes and Alex Stepick, *City on the Edge: The Transformation of Miami* (Berkeley: University of California Press, 1993), 74.
10 John Lantigua, "Captain Courageous," *Miami New Times*, July 23, 1998, https://www.miaminewtimes.com/miami/Print?oid=6359857.
11 Dade County Grand Jury, "Final Report," 1.
12 Ibid., 2.
13 Barry Klein, "Miami's River of Shame," *St. Petersburg Times* (St. Petersburg, FL), April 30, 1989.
14 John Nordheimer, "Miami's 6-Mile River: Pollution, Aliens, Drugs," *New York Times*, August 1, 1987.
15 Klein, "Miami's River of Shame."
16 Stepick and Portes, *City on the Edge.*
17 Lantigua, "Captain Courageous."
18 John Lantigua, "A New Course: River Improvement Plan Launched," *Miami Herald*, December 9, 1996; Nordheimer, "Miami's 6-Mile River."
19 Lantigua, "A New Course."
20 Nordheimer, "Miami's 6-Mile River."
21 Klein, "Miami's River of Shame."
22 Michel-Rolph Trouillot, *Global Transformations: Anthropology and the Modern World* (New York: Palgrave Macmillan, 2003), 1.
23 See, for example, Jemima Pierre, "Borders, Blackness, and Empire," *Black Agenda Report*, September 22, 2021.
24 I am referring here, of course, to caricatures of the Haitian ritual worlds of Vodou, not its actual practice.
25 Kahn, *Islands of Sovereignty*; Alex Stepick, "Haitian Boat People: A Study in the Conflicting Forces Shaping U.S. Immigration Policy," *Law and Contemporary Problems* 45, no. 2 (1982): 163–196, 187.
26 On stereotypes of Haitian immigrants, see Paul Farmer, *AIDS and Accusation: Haiti and the Geography of Blame* (Berkeley: University of California Press, 1992); Nina Glick Schiller and Georges Fouron, "'Everywhere We Go We Are in Danger': Ti Manno and the Emergence of a Haitian Transnational Identity," *American Ethnologist* 17, no. 2 (1990): 337. With regard to Haitian exceptionalism more generally, see Michel-Rolph Trouillot, "The Odd and the Ordinary: Haiti, the Caribbean, and the World," *Cimarrón: New Perspectives on the Caribbean* 2, no. 3 (1990): 3–12.
27 Kahn, *Islands of Sovereignty*, 13.
28 Alex Stepick and Alejandro Portes, "Flight into Despair: A Profile of Recent Haitian Refugees in South Florida," *International Migration Review* 20, no. 2 (1986): 329–350, 331.
29 Stepick and Portes, *City on the Edge.*
30 A preexisting trade with Haiti existed, but new Haitian players in the shipping business emerged as Haitians established themselves in and around Miami in the 1970s.
31 According to the Coast Guard, the operation was a response to the Senate Report to the 1994 Department of Transportation (DOT) and Related Agencies Appropriations Bill in which Congress had given "firm direction on how to eliminate substandard

ships from U.S. waters." 62 Fed. Reg. 33,445, 33,456 (June 19, 1997). Curiously, there is no reference to substandard vessels in the Senate Report to the 1994 DOT Appropriations Bill cited by the Coast Guard as the primary reason for the inspection changes. Nor have I found any reference to 1994 legislation specifically addressing the issue in the Coast Guard's announcements of changes to its inspection practices and standards that was published in the Federal Register—exactly the place where one would expect to find scrupulous citation to authorizing legal texts. The reasons for this discrepancy, however, are beyond the scope of this chapter.

32 62 Fed. Reg. 33445, 33456 (June 19, 1997).
33 Ibid.
34 Ibid.
35 Lantigua, "Captain Courageous."
36 Fran Bohnsack, "Informal Report on Haitian Embargo for Greater Miami Chamber of Commerce Haiti Committee," May 12, 1994, Folder Title: Summit of the Americas: GMCC, William J. Clinton Presidential Library.
37 Kevin G. Hall, "Vessel Law Threatens Haiti Trade," *Journal of Commerce*, April 21, 1997.
38 Lantigua, "Captain Courageous."
39 62 Fed. Reg. 33455, 33456 (June 19, 1997).
40 Code of Safety for Caribbean Cargo Ships, Feb. 5–9, 1996.
41 "Port State Control in the Caribbean," *FAL Bulletin*, no. 163, March 2000, https://repositorio.cepal.org/bitstream/handle/11362/36257/2/FAL_Bulletin163_en.pdf.
42 William B. Stronge et al., "An Economic Analysis of the Miami River Marine Industry," Center for Urban and Environmental Solutions, Florida Atlantic University, 2008, https://miamirivercommission.org/PDF/EconomicAnalysisoftheMiami%20River42808.pdf.
43 Nordheimer, "Miami's 6-Mile River."
44 Loretta Lees et al., *Gentrification* (New York: Routledge, 2008).
45 Lantigua, "Captain Courageous."
46 Kahn, *Islands of Sovereignty*.
47 For a discussion of the conflation of toxicity and dirt with particular types of bodies in environmental gentrification and neighborhood rebranding efforts, see Leslie Kerns, "From Toxic Wreck to Crunchy Chic: Environmental Gentrification Through the Body," *Environment and Planning D: Society and Space* 33, no. 1 (2015): 67–83. For an account of the racialization of place, see McKittrick, *Demonic Grounds*.
48 Newsweek Staff, "Rolling on the Miami River," *Newsweek*, June 10, 1990.
49 Ibid.
50 Ibid.
51 "Miami River Corridor Urban Infill Plan" (Miami: Miami River Commission, June 2002); Klein, "Miami's River of Shame."
52 "Urban Infill Plan."
53 Martha Brannigan, "Miami River Businesses, Developers Clash," *Associated Press*, June 23, 2008.
54 "Urban Infill Plan."
55 "Urban Infill Plan," 69; "National Drug Control Strategy, HIDTA (High Intensity Drug Trafficking Area) Annual Report," December 2002, 121.
56 Linda Basch et al., *Nations Unbound: Transnational Projects, Postcolonial Predicaments, and Deterritorialized Nation-States* (Langhorne: Gordon and Breach, 1994); Georges Fouron and Nina Glick Schiller, *Georges Woke Up Laughing: Long Distance Nationalism and the Search for Home* (Durham: Duke University Press, 2001).
57 Stefan Helmreich, "Nature/Culture/Seawater," *American Anthropologist* 113, no. 1 (2011): 132–144; Philip Steinberg and Kimberley Peters, "Wet Ontologies, Fluid Spaces: Giving Depth to Volume Through Oceanic Thinking," *Environment and Planning D: Society and Space* 33, no. 2 (2015): 247–264.
58 On the turbulence within ostensibly smooth global supply chains, see Charmaine Chua et al., "Introduction: Turbulent Circulation: Building A Critical Engagement

with Logistics," *Environment and Planning D: Society and Space* 36, no. 4 (2018): 617–629; and Nicole Starosielski, *The Undersea Network* (Durham: Duke University Press, 2015).

59 Aihwa Ong, *Flexible Citizenship: The Cultural Logics of Transnationality* (Durham: Duke University Press, 1999).

60 Jeffrey S. Kahn, "Smugglers, Migrants, and Demons: Cosmographies of Mobility in the Northern Caribbean," *American Ethnologist* 46, no. 4 (2019): 1–12. In this volume, Elspeth Probyn explores a wider array of uneven maritime legal geographies using the concept of the legal "chokepoint."

61 Josiah McC. Heyman, "Ports of Entry as Nodes in the World System," *Identities: Global Studies in Culture and Power* 11, no. 3 (2004): 303–327.

62 Brenner and Schmid, "Planetary Urbanization."

63 Henri Lefebvre, *The Urban Revolution*, trans. Robert Bononno (Minneapolis: University of Minnesota Press, 2003).

64 Brenner and Schmid, "Planetary Urbanization," 163.

65 Neil Brenner, "Debating Planetary Urbanization: For an Engaged Pluralism," *Environment and Planning D: Society and Space* 36, no. 3 (2018): 570–590, 574.

66 Brenner and Schmid, "Planetary Urbanization," 162.

67 Ibid.

68 Marcus Rediker, "Toward a People's History of the Sea," in *Maritime Empires: British Imperial Maritime Trade in the Nineteenth Century*, eds. David Killingray et al. (Rochester: Boydell Press, 2004), 195–222; see also Sharad Chari, "Subaltern Sea?: Indian Ocean Errantry Against Subalternization," in *Subaltern Geographies*, eds. Tariq Jazeel and Stephen Legg (Athens: University of Georgia Press, 2019), 191–209; Elizabeth DeLoughrey, "Toward a Critical Ocean Studies for the Anthropocene," *English Language Notes* 57, no. 1 (2019): 21–36; Jatin Dua, *Captured at Sea: Piracy and Protection in the Indian Ocean* (Oakland: University of California Press, 2019); Epeli Hau'ofa, "Our Sea of Islands," in *A New Oceania: Rediscovering Our Sea of Islands*, eds. Eric Waddell et al. (Suva: Beake House, 1993): 2–16; Renisa Mawani, *Across Oceans of Law: The Komagatu Maru and Jurisdiction in the Time of Empire* (Durham: Duke University Press, 2018); Steinberg and Peters, "Wet Ontologies."

69 Kahn, *Islands of Sovereignty*, 223–224.

70 Michel de Certeau, *The Practice of Everyday Life*, trans. Steven Rendall (Berkeley: University of California Press, 1984), 99.

71 Ibid., 101.

72 Ibid.

73 Kahn, *Islands of Sovereignty*, 215–245.

74 Ibid.; see also Laleh Khalili, *Sinews of War and Trade: Shipping and Capitalism in the Arabian Peninsula* (New York: Verso, 2020), 19.

75 de Certeau, *Practice of Everyday Life*, 99.

76 For parallels with the wooden sail-freighters associated with the transport of Haitian asylum seekers, see Jeffrey S. Kahn, "Racializing Aesthetics: 'Boat People,' Maritime Worlds, and the Metonymy of the Haitian Sloop," *Current Anthropology* (forthcoming).

77 McKittrick, *Demonic Grounds*, xiii, xxv.

78 With regard to the non- or other-than-human elements that play a role in making maritime space, one might look to the interplay of humans and turtles in Annet Pauwelussen and Shannon Switzer Swanson, this volume.

79 de Certeau, *Practice of Everyday Life*, 102.

80 Couling, "Extended Urbanisation in Ocean Space," 151; see also Couling, "Ocean Space and Urbanisation"; Pierre Bélanger and Alexander Scott Arroyo, "Logistics Islands: The Global Supply Archipelago and the Topologies of Defense," *Prism* 3, no. 4 (2012): 55–75.

81 Couling, "Ocean Space and Urbanisation," 24.

82 "Urban Infill Plan."

83 Lees et al., *Gentrification.*

84 Kerns, "From Toxic Wreck to Crunchy Chic."

85 Kahn, *Islands of Sovereignty.*

86 A more detailed look into the particular shifts in these "phatic topoi" at sea is clearly in order, but they are beyond the scope of this chapter.

87 The question of which terminals actually count as "the port" has been disputed. *Payne*, 53 So. 3d at 266.

88 Khalili, *Sinews of War*, 3.

10

MILES AND NORMS IN THE FISHERY OF MARSEILLE

On the Interface between Social Norms and Legal Rules

Florian Grisel

Introduction

In recent decades, lawmakers have paid close attention to tight-knit communities of fishers who use techniques in accordance with ancestral traditions. These small-scale fisheries ("SSFs") are usually defined by the length of the boats (less than 12 or 24 meters) that operate in their waters.[1] Despite their small size, SSFs are deemed to employ about 90 percent of fishers globally and to generate about one-third of the total annual catch of fish.[2] One of the key reasons why these communities have attracted so much attention relates to the traditional techniques that they use and the reduced impact that these techniques have on the environment. For instance, the rate of disposal or waste of fish is about four percent in SSFs, as opposed to 20 to 65 percent for large trawlers in industrial fisheries.[3] For this reason, SSFs are usually deemed to be more selective and protective of fish stocks than large-scale fisheries.

The importance of SSFs and their relatively low impact on the environment have prompted international organizations and national governments to look more closely into their management. In particular, lawmakers have grown increasingly attentive to the specificities of SSFs and the ways in which their communities are often grounded in ancestral systems of local governance that complement and sometimes supersede legal systems.[4]

Global regulators regularly affirm the need to safeguard these local systems of governance. In its Guidelines on Combatting Illegal, Unreported and Unregulated Fishing (2018), for instance, the Organisation for Economic

FIGURE 10.1 General view of Bonne Brise beach, Marseille, France in 2015. Licensed under the Creative Commons Attribution-Share Alike 4.0 International.

DOI: 10.4324/9781003205173-11

Co-operation and Development (OECD) recently noted the importance of SSFs and the need to provide adequate regimes for their management:

> It is estimated that about two-thirds of the world's catches destined for human consumption originate from small-scale fisheries … The size of these estimates suggests the need for adequate MCS (monitoring, control and surveillance) of these activities, so that these catches do not go unreported and CMMs (conservation and management measures) are respected.[5]

The OECD specifically recognizes the "need to tailor the law to allow traditional practices and special exemptions" in these fisheries:

> rules governing small-scale fisheries are often embedded in historical and cultural contexts and it is important to recognise the local specifics of small-scale fisheries. In some cases, countries have found they need to tailor the law to allow traditional practices and special exemptions, in order to assure compliance.[6]

The emphasis placed by global lawmakers on the specific practices and norms of SSFs also appears in the recent FAO's Voluntary Guidelines for Securing Sustainable Small-Scale Fisheries:

> States, in accordance with their legislation, and all other parties should recognize, respect and protect all forms of legitimate tenure rights, taking into account, where appropriate, customary rights to aquatic resources and land and small-scale fishing areas enjoyed by small-scale fishing communities … Local norms and practices, as well as customary or otherwise preferential access to fishery resources and land by small-scale fishing communities including indigenous peoples and ethnic minorities, should be recognized, respected and protected in ways that are consistent with international human rights law.[7]

Similarly, the Ministerial Declaration on a Regional Plan of Action for Small-Scale Fisheries in the Mediterranean and the Black Sea lays out a series of actions to be implemented by 2028.[8] Among these actions, state parties undertake to "reinforce the analysis of legislation and institutional mechanisms which ensure the recognition of relevant small-scale fisher organizations" and to "promote participative management systems, such as co-management bodies, where fisheries management measures and accompanying socio-economic programmes may be established and implemented."[9]

Behind the proliferation of policy recommendations lies a relative consensus concerning the need to recognize and preserve governance systems that are embedded in tight-knit communities with strong cultural traditions. The agenda of global lawmakers therefore seems to be based on the assumption that SSF actors have better knowledge of their own needs and constraints, which makes them better equipped to govern their fisheries than external regulators. This recommendation is largely in line with the prescriptions of most scholars of SSFs. For instance, Berkes et al. argue that "[o]ne of the lessons in the early common property literature was that the legal recognition of communal sea tenure could lead to sustainable resource use."[10] Benkenstein contends that "one of the key developments in fisheries governance in recent decades has been a shift towards a decentralised approach to fisheries management, particularly in the small-scale sector."[11]

It is not entirely clear, however, how the prescriptions of lawmakers can be concretely applied in the local context of SSFs. How can the objective of "tailor[ing] the law to allow traditional practices" or of "recogniz[ing], respect[ing] and protect[ing] local norms and practices" be translated into practice? Regulators seem to view the law as a device that dominates normative frameworks. In this view, norms thrive when the law gives them sufficient space to operate but perish if the legal framework becomes too comprehensive or far-reaching. Regulators therefore approach the law as a device that dominates, protects, and can eventually empower normative frameworks.

The distinction between law and social norms is one that traverses the field of socio-legal studies and, more broadly, of sociology. It appears, for instance, in the writings of Emile Durkheim and Max Weber.[12] The core idea behind this distinction is the observation that legal systems do not exhaust the modes of regulation, and that social norms play an important role that cannot be brushed aside even in modern societies. Despite some important scholarship in the field of legal sociology,[13] the full exploration of this distinction has been carried out by law and economics scholars in the past decades.[14] However, one area that has not been fully explored concerns the linkage between law and social norms. It is generally well-accepted that both types of regulatory systems coexist in society; what is less understood are the ways in which these systems coexist, and whether their coexistence is peaceful or contentious. Most authors consider law and social norms as variables that evolve in reverse order.[15] According to this view, norms thrive when the law gives them sufficient space to operate but perish if the legal framework becomes too comprehensive or far-reaching.[16] This view therefore assumes that law and social norms work best in silos,[17] with the implied understanding that the law dominates the whole normative architecture.

Another area of uncertainty concerns the ways in which global lawmaking can impact local practices in SSFs. In a recent article, Jerneja Penca advocates for an approach which she calls "transnational localism,"[18] where she defines "transnational localism" as the "reinforcement of local-specific approaches (reflecting local ecologies, values, and socio-economic specificities) within a transnational structure that provides support and recognition."[19] Penca notes that "the growing demand for SSF recognition also speaks of the significance of territory in global governance" and that her approach "upset(s) the heavy-rooted assumption of the de-territoriality of transnational law."[20] She finally argues that the management of SSFs is a perfect place for a new approach that requires "matching the demand for the local and transnational at the same time."[21] But is it possible to match the demands for the local and transnational at the same time? Does the transnational have any impact on the local practices of SSFs and, if so, of what kind?

This chapter does not provide a blanket answer to these questions. It seeks instead to highlight different facets of this questioning by examining the interface between local norms and transnational law through a specific case study.[22] This case study focuses on an SSF in the South of France (Marseille), where fishers have elected representatives in an organization called the *Prud'homie de pêche* (herein, "prud'homie") and have entrusted this organization with the task of regulating their fishery since the Middle Ages. In this chapter, I examine the interactions between the prud'homie's social norms and the legal rules enforced by state authorities. For this purpose, I will focus on the example of the territorial delimitation of the sea, a set of rules that gradually applied as a matter of international law before gaining traction under French law, and its impact on the fishery of Marseille.

The empirical evidence used in this chapter is drawn from three sets of data. One set of data is based on archival evidence compiled from a broad range of collections over the past six years. These collections include those of the prud'homie and the national archives, allowing cross-fertilization of data based on each perspective—local/normative, on the one hand, and (trans)national/ legal, on the other hand—that are at the focus of this chapter. Another set of data is based on a series of interviews that I carried out, in person or on the phone, with various actors of the fishery. These actors are situated within and outside the community of fishers and act in different capacities (fishers, state officials, activists, community leaders, et cetera), thus multiplying the vantage points for my analysis. Last but not least, I have gleaned evidence of the regulatory systems at play in the community of fishers in Marseille from ethnographic research that I have conducted over the past few years. As part of this ethnographic research, I have spent time with the local fishers of Marseille, went fishing with some of them, and attended some of their community events (notably religious ceremonies).

Based on this evidence, I argue that the legal rules concerning the delimitation of the sea shaped the community of fishers in ways that constrained social norms and affected their system of communal governance. This analysis suggests that legal rules are not mere containers for social norms, but can deeply shape the identity of close-knit communities. My findings complement those of urban sociologists who highlight the social impact of physical spaces, not as mere containers, but also, and more critically, as shapers of communities.[23] My findings also offer a counterpoint to the dominant view according to which law and social norms interact in opposite directions. My data suggests that law and social norms are part of a whole, rather than separate elements that should be examined apart from each other. To this extent, this chapter contributes to a better understanding of what Probyn and Westholm cast, respectively, as the "jurisdictional twists of ocean legalities"[24] and the "overlap of planning competence" in coastal waters.[25] My argument will proceed in four steps. I will first present my case study and methodology. Then, I will explore the ways in which the legal definition of a three-mile territorial zone catalyzed conflicts between various categories of fishers and framed their local identities. Next, I will show that the extension of this territorial zone to 12 miles in the late 20th century did not affect this frame, which persisted until recently. Finally, I will conclude the chapter by reflecting on the social/legal redux and the need to overcome the terms of this redux.

The SSF of Marseille and its System of Communal Governance

The fishery of Marseille is a prime example of an SSF. In 2015, more than 80 percent of its fishing vessels were less than 12 meters long, and more than 90 percent were less than 25 meters long.[26] The fishers of Marseille operate their boats over an area covering approximately 20 miles of coastline, extending from the calanques of Cassis on the east side to the coastal city of Carry-le-Rouet on the west side. Before the law defined the scope of territorial sea, the outer limits of the fishery were loosely fixed by the practices of local fishers. Like many other SSFs, the fishers of Marseille have developed a system of governance embedded in longstanding traditions. Every year since 1431, they have elected some of their peers to head a special organization called the prud'homie. The four members of the prud'homie, also called the prud'hommes, are elected annually.

The term prud'homie, which comes from the Latin *probi homines*, can be literally translated by "virtuous men." This translation does not fully reflect the fact that, in the medieval cities of Europe, these "virtuous men" offered guarantees of autonomy and fairness for their communities. The prud'homie has played a key role in the regulation of the fishery of Marseille by issuing rules, adjudicating disputes among fishers, and policing their behavior. Its rules are deeply influenced by the social norms of reciprocity and cooperation that the fishers of Marseille cherish. For instance, the prud'homie ensures that fishers do not concentrate their work in the same locations (called posts), that their nets and hooks are limited in size, and that they fish in different posts at various times of the day and year depending on their target species. These rules are also deeply influenced by the fishers' goal of preserving the resources of their fishery. The fishers of Marseille frequently refer to the need to limit the harmful effects of their activities on fish stocks. In addition, the prud'homie has been relatively free from the interference of public authorities in the regulation of its fishery. For instance, even today, the losing parties are not allowed to appeal the prud'homie's judgments before French courts.

The prud'homie therefore provides a case study of a system of private governance deeply embedded in an SSF that has coexisted with a particularly centralized and rigid legal system over the centuries. The prud'homie presents another advantage: its archives are well preserved and supply the material needed for a longitudinal study of an SFF.[27] In particular, this material can be used to examine the relationship between the law and the norms of a small-scale fisher organization. In addition to having rich archival records, the prud'homie still exists today, making it one of the oldest systems of governance in SSFs. In order to extend my historical study of the prud'homie into the present and immediate past, I carried out a series of interviews and ethnographic work among the community of fishers in Marseille, in addition to exploring archival records.[28] Based on this empirical evidence, I retraced the history of this community and the challenges that it faced when regulating the fishery of Marseille.[29] In this chapter, I focus on a historical theme that runs through the last two centuries of the prud'homie's records, namely the impact of maritime delimitations on its regulatory system.

The Three-Mile Limit: A Catalyst for Conflicts between "Grand Art" and "Small Art" Fishers

A brief overview of the legal landscape is necessary to understand the ways in which the prud'homie progressively defined the boundaries of its jurisdiction. French law did not define the scope of its territorial sea until the late 19th century. The main text of French maritime law, the Great Maritime Ordinance of 1681, only defined the "sea shores" as a territory covered by the tide, but this reference was too broad to ground a clear jurisdictional rule.[30] The prud'homie took advantage of these legal uncertainties by defining the spatial scope of its own jurisdiction in broad terms. Its main jurisdictional criterion was, in fact, more personal than spatial, as the prud'homie deemed itself to have authority over all fishers who owned a boat in the port of Marseille, irrespective of how far they operated from the coast. In a codification of its rules in 1725, the prud'homie defined its jurisdiction as extending "*from the Cap de l'Aigle near La Ciotat to the Cap de la Couronne near Martigues*."[31] This definition did not set a water limit, but instead fixed the outer boundaries of the jurisdiction along a stretch of coast centered around Marseille. The same understanding of the prud'homie's territory can be found in a decision of the *Conseil d'Etat* (the French supreme court in the field of administrative law),[32] which refers to the fishers' right to fish from "Cap de l'Aigle to the place named La Couronne."[33] What mattered for the prud'homie was that its fishers were based in Marseille and that they operated within a territory bordered by two lines starting from Cap de l'Aigle and Cap de la Couronne, but without limitations running parallel to the coast.

As shown further below, the gradual definition of the territorial seas along the three-mile limit affected the prud'homie's understanding of its own jurisdiction. The recognition of the three-mile limit under French law was the result of a long and complex process whose roots lay in discussions and conflicts concerning maritime delimitation in the North Sea. In the 17th and 18th centuries,

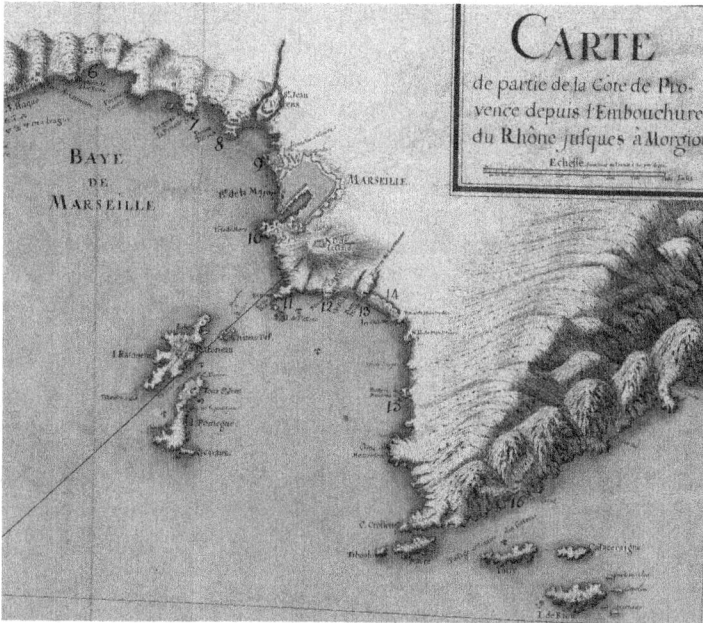

FIGURE 10.2 The Fishery of Marseille (circa 1750). Bibliothèque Nationale de France (BNF Gallica), *Carte de la Côte de Provence depuis l'Embouchure du Rhône jusques à Morgiou* (detail), c. 1750.

states such as Denmark and Norway made claims to jurisdiction over adjacent maritime areas, which resulted in negotiations with other states such as Britain, France, and Holland.[34] During these negotiations, the idea emerged that states, and their fishers, could claim jurisdiction over a three-mile zone around their coast. France recognized this rule for the first time in a treaty signed with Great Britain in 1839, which provided for an "exclusive right of fishery within the distance of three miles of low-water mark."[35] It took several more years for the rule of the three-mile limit to be concretely transposed into French law. In 1862, a decree allowed fishing within three nautical miles of the low-water mark and provided that the state could regulate fishing within this zone.[36] In 1888, a statute once again applied the same limit by prohibiting foreign vessels from fishing within three miles of the French coast.[37] Because France's territorial sea extended up to the three-mile limit, the prud'homie could no longer apply its broad rule of personal jurisdiction, but had to distinguish between fishers operating on either side of the three-mile limit. However, one should not overemphasize the speed with which the new regulatory regime produced its effects. For instance, in a leading textbook on Mediterranean fishing published more than 30 years after the adoption of the decree of 1862, the zoologist Paul Gourret did not refer to the three-mile limit.[38]

The recognition of the three-mile limit coincided with the emergence of a new fishing practice in early 18th-century Marseille. This fishing practice, called "ox-fishing"[39] (*pêche au boeuf*), consisted of a large net dragged by two sail boats.[40] An ancestor of trawling, ox-fishing increased the dragging power of nets many times, and many feared that it would undermine the resources of the fishery of Marseille. For that reason, the prud'homie fined (albeit reluctantly) those fishers who practiced ox-fishing too close to the shores starting in the late 1830s.[41] The prud'homie's goal was to preserve areas that are usually endowed with rich fish stocks and are also traditional spawning grounds for the fish. It is also during this time (the late 1830s) that the prud'homie started distinguishing between two categories of fishers: the "small art" fishers (*petits arts*) who employed traditional techniques (typically, set nets) close to the shore on the one hand, and the "grand art" fishers (*grands arts*) who used dragnets further away from shore (at least in principle).

The distinction between "small art" and "grand art" fishing found an anchor in the three-mile limit. The historical record shows that, as the three-mile limit emerged as a rule of international law and then French law, the prud'homie tried to seclude ox-fishing, keeping it away from the shore. This might have just been a coincidence, but further developments indicate that the definition of "territorial sea" provided a reference for the prud'homie to regulate ox-fishing (and, later, trawling).

The changing legal landscape provides a useful background against which these developments can be tracked. The decree of 1862, which is the first instrument of French domestic law to refer to the three-mile limit, also defined the jurisdiction of the prud'homie in a restrictive manner. Before then, the jurisdiction of the prud'homie had been broadly defined as spreading over the "waters of

FIGURE 10.3 *La Pêche au Boeuf* (Ox-Fishing). Bibliothèque Nationale de France (BNF Gallica). V.F. Garau, *Traité de Pêche Maritime Pratique Illustré et des Industries Secondaires en Algérie* (Algiers: Imp. P. Crescenzo, 1909): 70.

Marseille"[42] or over "territorial waters" (*domaine public maritime*), without further detail.[43] The decree of 1862 filled this void by limiting the jurisdiction of the prud'homie to the seas extending until the three-mile limit.[44]

The Conseil d'Etat later confirmed the limitation of the prud'homie's jurisdiction to territorial waters. In an advisory opinion issued in 1921, the Conseil d'Etat held that the prud'homie could not require payment of membership fees from trawler fishers who operated beyond the three-mile limit, thus implying that the prud'homie lacked jurisdiction beyond this point.[45] In the words of the Conseil d'Etat, "the fishers operating in territorial waters shall participate in the prud'homies, [but] this obligation does not extend to trawler fishers who practice their trade beyond territorial waters."[46] The administrative state interpreted this advisory opinion as a confirmation that the prud'homie lacked jurisdiction over extra-territorial waters (beyond the three-mile limit). The prud'homie took some liberty in applying this jurisdictional rule. For instance, in a dispute that was decided in 1958, one fisher argued that the prud'homie lacked jurisdiction because the disputed events occurred more than three miles from shore.[47] However, the prud'homie brushed aside these jurisdictional objections, which it disregarded entirely, before ruling on the merits of the case.[48] Unsurprisingly, the administrative state was much stricter when it came to construing the prud'homie's jurisdiction. For instance, in 1965, two fishers threatened to bring a case against a trawler fisher before the prud'homie. The trawler fisher complained to the maritime administration, which stated that the prud'homie lacked jurisdiction over the dispute because the fishing incident occurred beyond the three-mile limit.[49]

The division of the community into two groups of fishers, one practicing traditional techniques within the three-mile limit in accordance with the prud'homie's rules (the "small art" fishers) and the other practicing higher-yield techniques with engine trawlers in contravention of the prud'homie's rules (the "grand art" fishers), generated major conflicts within the community and deeply shaped its identity. Even today, most fishers that I interviewed define themselves as "grand art" or "small art" fishers, a strong marker of identity within their community.[50] Most of the conflicts between "grand art" and "small art" fishers arose from the fact that, while they refused to abide by the rules of the prud'homie, "grand art" fishers (typically operating large and powerful trawlers) regularly trespassed on its territory. These increasingly powerful trawlers (200 horsepower on average in the 1960s, 400 horsepower on average in the 1980s, with some trawlers reaching more than 1,000 horsepower) operate within the three-mile limit in order to exploit the rich fish stocks that can be found in coastal areas, and occasionally destroy the smaller set nets used by "small art" fishers.

The fact that Italian immigrants operated most trawlers starting in the 1920s did not help, as ethno-national differences generated additional conflicts with local fishers.[51] In 1927, 200 fishers demonstrated in the streets of Marseille against trawler fishers who operated within the three-mile limit in contravention of the prud'homie's rules.[52] These conflicts persisted for a long time, peaking between the 1960s and the 1980s. In 1980, for instance, the prud'homie sent a letter to the Préfet (the local representative of the central government) denouncing the behavior of trawler fishers who constantly trespassed on the three-mile zone.[53] One of my interviewees, a former member of the prud'homie, who defined himself as a "small art fisher," told me that trawlers did not hesitate to operate as close as one mile to the shore, and that the prud'homie had enormous difficulties in policing their behavior. To illustrate the intensity of the conflicts that arose between the prud'homie and trawler fishers, this former prud'homme told me of the misadventures of a fellow prud'homme who was attacked with an axe while trying to board a trawler operating within the three-mile limit. In a decision from 1973, the Court of Appeal of Montpellier captured the tensions between trawlers and "small art" fishers in a criminal case that was brought by the state against a fisher who trawled in the three-mile zone:

> the goal of the prohibition on trawling within three miles is to shelter from a potentially dangerous technique fish stocks that are located in coastal areas, which shall be essentially exploited with much more limited and restricted means than those of trawlers, that is the small art fishing that is allowed to operate in it … the concurrent presence within this coastal area [the three miles] of trawlers and small art fishers causes harm to the latter fishers, because there are set nets, on the one hand, and dragnets, on the other hand, two incompatible techniques."[54]

The transplantation of the three-mile limit in the fishery of Marseille illustrates how a legal rule cascades down from the global to the local level, and constrains a communal system based on social norms. In particular, the definition of the prud'homie's jurisdiction in terms of the area within the three-mile limit had unintended consequences for the community of fishers. This new territorial boundary redefined their community, significantly weakening its local system of governance and generating important social conflicts within its fishery. One would think that the extension of the territorial sea from three to 12 miles offshore could have solved these conflicts. However, these conflicts, often framed in terms of "grand arts" (trawling) versus "small arts" (traditional techniques), have left deep traces in the community of fishers that persist to this day.

From the Three-Mile Limit to the 12-Mile Limit: Lost in Boundaries

After World War II, several states wished to extend the limits of their territorial sea beyond three miles. This international movement in favor of a wider territorial zone led to the adoption by France of the 12-mile limit in 1971[55] and to the recognition of the same limit in the United Nations Convention on the Law of the Sea in 1982.[56] One could hypothesize that a wider territorial zone should have led to the extension of the prud'homie's jurisdiction and to the resolution of disputes between hostile groups of fishers that operated on either side of the three

miles. In fact, the prud'homie's jurisdiction did expand into the 12-mile territory. However, this expansion did not resolve the conflict between "small arts"
and "grand arts." In addition, it did not affect the importance of the three-mile
limit, which remained a reference point in the heated debates that frequently
arose in the fishery. In 1977, for instance, the prud'homie wrote to the maritime administration to complain about the frequent incursions of trawlers within
the three-mile zone and their frequent conflicts with "small art" fishers.[57] The
prud'homie expressed its fear that "the situation will escalate" and that "young
fishers, when facing ruin and the impossibility of working, will retaliate physically against other fishers and their boats."[58]

While in the 1920s trawlers tried to escape the jurisdiction of the prud'homie
(and the prud'homie tried to assert its jurisdiction over these trawlers), the situation
was reversed in the 1970s. The trawlers decided to increase their political power
by gaining influence within the prud'homie. One of the key political elements at
stake was the possibility for "grand art" fishers to participate in the election of the
prud'homie and be elected as prud'hommes (even though they had not paid the
prud'homie's fees since 1921). The position of "grand art" fishers was vindicated
by the fact that the prud'homie could arguably exercise its powers beyond the
three-mile limit (since the territorial sea extended to 12 miles).[59] This was potentially an opportunity for the prud'homie to regain control over trawler fishers, by
integrating them within its jurisdiction. But the jurisdiction of the prud'homie was
now deeply embedded in the three-mile zone, sometimes called the prud'homie's
waters (les eaux prud'homales).[60] Conflicts between the two groups (the "small arts"
gathered around the prud'homie, on the one hand, and the "grand arts" gathered
around the trawlers, on the other) were so shrill that the prud'homie could not
embrace the integration of trawlers within the community.

The prud'homie made the choice to reject the trawlers from its jurisdiction,
in order to thwart what was seen as a putsch on their part. The community had
excluded the trawlers and there was no turning back. During a meeting held at the
prud'homie in 1980, the conflict was described by a representative of the trawler
fishers as "open warfare" (*guerre ouverte*).[61] Shortly thereafter, the same representative sent a letter to the prud'homie requesting an authorization for trawler fishers
to run in the prud'homie's elections. Unsurprisingly, his request was swiftly turned
down. Although the "grand art" fishers have almost entirely disappeared from the
fishery of Marseille,[62] the conflicts have persisted up until today, framing the identity of the community in terms that cannot be easily overcome.

Conclusion

The empirical study of the prud'homie suggests that the legal apparatus of
French and international law, crystallized in the three-mile limit, had strong
and unintended effects on the life of the fishery of Marseille. The prud'homie
redefined itself along the three-mile limit, casting the groups operating on either
side of this limit under a different name and identity. This limit, however, did
not significantly affect the social practices of the SSF, as suggested for instance

by the constant efforts of trawler fishers to trespass on the three-mile zone. Once the legal definition of "territorial sea" was extended to 12 miles, the conflicts between "grand art" and "small art" fishers persisted in ways that the prud'homie could not easily overcome. The law had introduced a social rift that left a deep imprint on the community.

The social/legal redux on which most of the current policy prescriptions are based fails to capture the huge potential impact of the legal on the social (and vice versa). In other words, policy prescriptions that focus on the distinction between legal rules and social norms might disregard the regulatory challenges raised by the interface between both spheres. The example of the prud'homie and the difficulties raised by the definition of "territorial sea" shows the importance of the link between the legal and the social. Rules that emerge globally can affect local communities in ways that are often invisible, but no less concrete. This link is not unidirectional, however, but can also operate from the local to the global. In the case of the SSF of Marseille, the three-mile limit resulted in longstanding conflicts that redefined in-depth the identity of a community and the powers exercised by the prud'homie over this community. The extension of this limit to 12 miles did not reverse this process. The global rule setting universal limits on territorial seas deeply affected a system of social governance that had prevailed for centuries in the fishery of Marseille, but it could only do so to the extent that it redefined social practices in ways that would be accepted by local communities (as indicated by the contrasting examples of the three-mile and the 12-mile limits). Paradoxically, global policymakers now call for the preservation of a local specificity as shaped and reshaped through legal rules.

The fishers of Marseille have never ceased to be subject to the influence of legal rules when developing their social norms. The case study presented in this chapter therefore encourages one to consider the limits of the social norms/legal rules redux. It also pinpoints the artificial barrier, skillfully maintained by lawmakers, between the legal and the social. The goal of recognizing the autonomy of social norms through legal means might be premised on a fragile, albeit widespread, distinction between the legal and the social. Any efforts to perpetuate this distinction might prove elusive, as illustrated by the case study presented in this chapter. The analytical frame that seems to underlie the work of global policymakers needs to urgently incorporate tools that allow for a better understanding of the interface between social norms and legal rules.

Notes

1 See U. Rashid Sumaila, "Small-Scale Fisheries and Subsidies Disciplines: Definitions, Catches, Revenues, and Subsidies," International Centre for Trade and Sustainable Development, September 2017, https://www.greengrowthknowledge.org/sites/default/files/downloads/resource/Small-scale%20Fisheries%20and%20Subsidies%20Disciplines_Definitions%2C%20Catches%2C%20Revenues%2C%20and%20Subsidies.pdf. Other tests have been developed to define SSFs. See Hillary Smith and Xavier Basurto, "Defining Small-Scale Fisheries and Examining the Role of

Science in Shaping Perceptions of Who and What Counts: A Systematic Review," *Frontiers in Marine Science* 6, no. 236 (2019): 1–19.

2 The World Bank, "Hidden Harvest: The Global Contribution of the Capture Fisheries," Report No. 66469-GLB (May 2012), 22.

3 European Parliament, "Small-Scale Fisheries and the Zero Discard Target" (2015), 15, 23.

4 See, e.g., the example of the fishery of Alanya mentioned by Elinor Ostrom, *Governing the Commons: The Evolution of Institutions for Collective Action* (Cambridge: Cambridge University Press, 1990), 18–21.

5 "Combatting Illegal, Unreported and Unregulated Fishing: Where Countries Stand and Where Efforts Should Concentrate in the Future," OECD, November 2018, https://www.oecd.org/officialdocuments/publicdisplaydocumentpdf/?cote=TAD/FI(2017)16/FINAL&docLanguage=En.

6 Ibid.

7 Food and Agriculture Organization of the United Nations, *Voluntary Guidelines for Securing Sustainable Small-Scale Fisheries in the Context of Food Security and Poverty Eradication* Article 5.4, 2015.

8 Food and Agriculture Organization of the United Nations, *Ministerial Declaration on a Regional Plan of Action for Small-Scale Fisheries in the Mediterranean and the Black Sea*, https://www.fao.org/3/cb7838en/cb7838en.pdf.

9 Ibid., paras. 32–33.

10 Fikret Berkes et al., *Managing Small-Scale Fisheries: Alternative Directions and Methods* (International Development Research Centre, 2001), 177.

11 Alex Benkenstein, "Small-Scale Fisheries in a Modernising Economy: Opportunities and Challenges in Mozambique," *South African Institute of International Affairs* (2013): 1–53, 17. These policy recommendations are also consistent with the conclusions of the broader scholarship concerning the social benefits associated with governance based on social norms. See, e.g., Robert C. Ellickson, *Order without Law: How Neighbors Settle Disputes* (Cambridge: Harvard University Press, 1992); Lisa Bernstein, "Opting Out of the Legal System: Extralegal Contractual Relations in the Diamond Industry," *Journal of Legal Studies* 21 (1992): 115–157; Edward P. Stringham, *Private Governance: Creating Order in Economic and Social Life* (Oxford: Oxford University Press, 2015); Barak D. Richman, *Stateless Commerce: The Diamond Network and the Persistence of Relational Exchange* (Cambridge: Harvard University Press, 2017).

12 Emile Durkheim, *The Division of Labor in Society* (Chicago: The Free Press of Glencoe, 1960), 211; Max Weber, *Economy and Society: An Outline of Interpretive Sociology* (Berkeley: University of California Press, 1978), 311, 312.

13 Stewart Macaulay, "Non-Contractual Relations in Business: A Preliminary Study," *American Sociological Review* 28, no. 1 (1963): 55–67.

14 See, e.g., Robert C. Ellickson, *Order without Law: How Neighbors Settle Disputes* (Cambridge: Harvard University Press, 1992); Lisa Bernstein, "Opting out of the Legal System: Extralegal Contractual Relations in the Diamond Industry," *Journal of Legal Studies* 21 (1992): 115; Edward P. Stringham, *Private Governance: Creating Order in Economic and Social Life* (Oxford: Oxford University Press, 2015); Barak D. Richman, *Stateless Commerce: The Diamond Network and the Persistence of Relational Exchange* (Cambridge: Harvard University Press, 2017); Eric A. Posner, *Law and Social Norms* (Cambridge: Harvard University Press, 2000).

15 See, e.g., Edward P. Stringham, *Private Governance: Creating Order in Economic and Social Life* (Oxford: Oxford University Press, 2015), 194; Bernstein, "Opting Out," 156–157.

16 This view is fully explored in Florian Grisel, *The Limits of Private Governance: Norms and Rules in a Mediterranean Fishery* (Oxford: Hart Publishing, 2021), Chapter 5.

17 Paradoxically, this view still underlies much socio-legal scholarship. See Bruno Latour, *The Making of Law: An Ethnography of the Conseil d'Etat* (Cambridge: Polity

Press, 2010), 262: "One of the main research programmes of the sociology of law, according to which a relationship has to be established between the corpus of rules on the one hand and society on the other, does not withstand examination: law is already of the social, of association; alone it processes more of the social than the notion of society from which it is in no way distinct since it works on it, kneads it, arranges it, designates it, imputes it, makes it responsible, envelops it."

18 Jerneja Penca, "Transnational Localism: Empowerment through Standard Setting in Small-Scale Fisheries," *Transnational Environmental Law* 8, no. 1 (2019): 143–165.

19 Ibid.

20 Ibid., 161.

21 Ibid., 165.

22 Transnational law can be defined as "all law which regulates actions or events that transcend national frontiers." See Philip C. Jessup, *Transnational Law* (New Haven: Yale University Press, 1956), 2.

23 See, e.g., Karin Aguilar-San Juan, "Staying Vietnamese: Community and Place in Orange County and Boston," *City and Community* 4, no. 1 (2005): 37–65.

24 Probyn, this volume.

25 Westholm, this volume.

26 Comité Régional des Pêches Maritimes et des Elevages Marins de Provence-Alpes-Côte d'Azur, "Etat des Lieux et Caractérisation de la Pêche Maritime et des Élevages Marins en PACA," March 2015, 40.

27 The prud'homie donated its archives to the local administration (*Archives départementales des Bouches-du-Rhône*) in 1933. I have supplemented the review of these materials by consulting other sources, such as the national archives of France, the municipal archives of Marseille, the archives of the French Navy, private archives, et cetera.

28 I conducted more than 30 interviews with community stakeholders. Most of these interviews were open-ended. I also spent significant time with the fishers, attending events in their community and occasionally going fishing with some of them.

29 Florian Grisel, "Managing the Fishery Commons at Marseille: How a Medieval Institution Failed to Accommodate Change in an Age of Globalisation," *Fish and Fisheries* 20 (2019): 419–433; Florian Grisel, "How Migrations Affect Private Orders: Norms and Practices in the Fishery of Marseille," *Law & Society Review* 55 (2021): 177; Florian Grisel, *The Limits of Private Governance: Norms and Rules in a Mediterranean Fishery* (Oxford: Hart Publishing, 2021).

30 *Ordonnance de la Marine* (August 1681), Book 4, Title VII, Article 1.

31 *Description des Pesches, Loix et Ordonnances des Pescheurs de la Ville de Marseille*, Archives départementales 250E2, 2.

32 See Bruno Latour, *The Making of Law: An Ethnography of the Conseil d'Etat* (Cambridge: Polity Press, 2009).

33 Decision of the *Conseil d'Etat*, dated November 30, 1622, MA HH370.

34 This process is retraced in H.S.K. Kent, "The Historical Origins of the Three-Mile Limit," *American Journal of International Law* 48, no. 4 (1954): 537–553. See also Claudio Baldoni, "Les Navires de Guerre Dans Les Eaux Territoriales Étrangères," *Recueil des Cours de l'Académie de Droit International de La Haye* 65 (1938): 185–303, 202 et seq.

35 Convention between Her Britannic Majesty and the King of the French, defining and regulating the limits of the Exclusive Right of the Oyster and other Fishery on the Coasts of Great Britain and of France (August 2, 1839), in *Commercial Tariffs and Regulations of the Several States of Europe and America together with the Commercial Treaties between England and Foreign Countries*, Part 4 (London: Charles Whiting, 1842), 51, 54 (Article 9).

36 See *Décret Impérial Sur la Pêche Côtière*, May 10, 1862.

37 See *Loi Ayant Pour Objet d'Interdire La Pêche aux Étrangers dans Les Eaux Territoriales de France et d'Algérie*, March 1, 1888.

38 See Paul Gourret, *Les Pêcheries et les Poissons de la Méditerranée (Provence)* (Paris: JB Baillière et Fils, 1894).

39 Ox-fishing already existed prior to the 1800s. It was strictly regulated throughout the 17th century, but really took off in the 1830s. See, e.g., Decision of the *Conseil du Roi*, September 25, 1725 (prohibiting ox fishing).

40 The term ox-fishing probably owes its name to the fact that two sailboats dragged a net that "plowed" the sea bottom like a pair of oxen in a field.

41 These events are evidenced in the archives of the prud'homie. See *Archives Départementales des Bouches-du-Rhône*, Archives départementales 250E126.

42 Decision of the *Conseil d'Etat*, dated May 16, 1738, Archives Chambre de Commerce et d'Industrie E/159.

43 Decree of November 19, 1859, Article 17.

44 Article 1(2) of the imperial decree implied that the prud'homie could not regulate any type of fishing beyond the three-mile limit.

45 Conseil d'Etat, opinion dated May 11, 1921, No. 178042.

46 Ibid.

47 *Antoine di Mase v. Gabriel Mir*, Decision of the prud'homie dated August 1, 1954, Private Archives.

48 Ibid.

49 Note for the *Administrateur Principal de Martigues*, dated December 1, 1965, Archives départementales 2331W287.

50 I have come across fishers who practiced "grand art" fishing for decades before turning to "small art" fishing when they grew older. This shift in their practice can be explained by age, but also by the fact that "grand art" fishing has become less profitable in recent decades.

51 See Florian Grisel, "How Migrations Affect Private Orders: Norms and Practices in the Fishery of Marseille," *Law & Society Review* 55, no. 1 (2021): 177–202.

52 Letter from the *Commissaire central de Marseille* to the *Préfet des Bouches-du-Rhône*, dated February 17, 1927, Archives départementales 4M2333.

53 Letter from the prud'homie to the *Préfet des Bouches-du-Rhône* (c. 1980), Archives départementales 2331W279.

54 Decision of the Court of Appeal of Montpellier, dated June 20, 1973, *Di Mario*, Archives départementales 2331W268.

55 *Loi no. 71-1060 du 24 Décembre 1971 relative à la délimitation des eaux territoriales Françaises.*

56 United Nations Convention on the Law of the Sea, Article 3.

57 Letter from the prud'homie to the *Directeur Général des Affaires Maritimes de Marseille*, undated (c. 1977), Archives départementales 2331W279.

58 Ibid.

59 See *Note de l'Administrateur en Chef*, dated September 18, 1974, Archives départementales 2331W272; Minutes of the *Comité local des pêches*, dated January 19, 1990), Private archives.

60 Letter from the *Préfet* to the *Directeur des Affaires Maritimes pour la Méditerranée*, dated January 25, 1980, Archives départementales 2331W279.

61 Minutes of the prud'homie, dated January 15, 1980, Private Archives.

62 The reasons why trawler fishers have almost entirely disappeared from Marseille are unclear. The EU's regulating of fishing techniques in the past 20 years is one reason. Another reason is that intensive fishing techniques, such as trawling, are less profitable than in the past due to the exhaustion of fish stocks.

11

DIVIDED ENVIRONMENTS

Scalar Challenges in Sweden's Marine and Coastal Water Planning

Aron Westholm

Planning in the Marine Domain

Since the turn of the millennium, marine spatial planning ("MSP") has become a popular tool for the management of the marine environment. Traditionally, ocean management has been characterized by sectoral division, with little efforts to coordinate the different uses. Some activities, e.g., offshore wind power, have been decided on a permit-by-permit basis, while others, such as fishing and maritime transport, have been controlled through sectoral management regimes.[1] In the early 2000s, the growing pressure placed on the marine environment by human activities prompted a new, more holistic management regime for ocean space. MSP was seen by many as offering an answer to this challenge.[2]

MSP is an integrated and holistic management regime for marine areas, in which all sectors are treated within a single instrument, ensuring that the sum of all activities does not exceed the limits of the marine ecosystem.[3] The definition of MSP is broad and there is no one-size-fits-all design. Any attempt to design a system for MSP needs to be sensitive to the social, cultural, and legal contexts in which it will be implemented.[4] This means that any system for MSP must address differences between various planning regimes and levels of government also within a single country.

My focus in this chapter is on how the division of planning competence between different levels of government within a country affects the possibilities to achieve a functional management of coastal and marine environments. The chapter reveals how each management body has a specific way of imagining and

FIGURE 11.1 The Port of Gothenburg, Sweden, is one of the instances where global and national geographies connect to local geographies and interest. Yet, this is an area of contention as further development of the port can conflict with important marine biodiversity both above and below the surface. Photo by Jan Schmidtbauer Crona, used with permission.

DOI: 10.4324/9781003205173-12

understanding the natural environment. These different imaginaries need to be coordinated and integrated in order to provide a more holistic understanding of the systems they aim to govern.

The Swedish system for MSP serves here to illustrate the challenges in coastal and marine planning. In Sweden, there are two levels of management involved in the planning of marine and coastal areas. First, the central government is responsible for planning the marine areas, from one nautical mile seaward of the baseline, until the end of the exclusive economic zone (EEZ). Second, municipalities are responsible for planning coastal waters out to one nautical mile seaward of the baseline, as well as the entire territorial sea. This division of planning largely follows the design of the EU Directive on MSP,[5] as the Directive covers all marine areas of the EU except for coastal waters covered by a member state's town and country planning.[6]

Interviews carried out in this study reveal that municipal planning entails a specific perspective, *a municipal logic*, that affects management practices and outcomes, but also, consequently, the state of the natural environment. Rather than simply responding to or reflecting factual characteristics of ecosystems, law and policy shapes them and determines their very content. The municipal logic follows both from the regulatory system that governs municipal action, and from

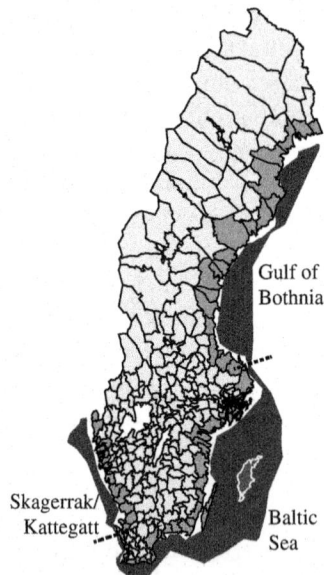

Source: SCB (Map of Sweden), The Swedish
Agency for Marine and Water Management.

FIGURE 11.2 Map of the Swedish proposed MSP areas and coastal municipalities illustrating the difference in planning scale. The dark areas divided by dotted lines are the three proposed national marine plan areas. The light dark areas are the coastal municipalities in charge of planning coastal waters and their parts of the territorial sea. Illustration by Hillevi Duus. Used with permission.

more practical issues such as budgetary and political considerations. My analysis shows that the management level, whether at the municipal or the state level, inevitably leads to different types of management. One is highly detailed, but also explicitly local; the other is more general and with a lower resolution, but is perhaps more apt to deal with challenges on a wider scale. The holistic aspirations of MSP will be hampered if these different types of management are not integrated with each other, as well as with other types of local knowledges.

The chapter starts with a conceptual discussion on the relation between law and coastal landscapes. This is followed by a short introduction to MSP and some of the relevant literature in the field. Empirically, the chapter investigates the Swedish system for MSP to show how different types of planning perspectives and priorities become evident at the different levels of government.

Mapping the Planning Landscape

The relationship between law and nature is a reciprocal one. Through law-making, humans create new, legally defined natural environments (see also Jones, this volume). This delimiting of nature through legal boundaries means that some natural processes become more prominent, while others are ignored. But the process also flows in the other direction. Natural conditions can affect how laws are drafted. In this sense, it is a dynamic process, but one that rests fundamentally on the notion of a natural environment that is defined by legal boundary-drawing. One way in which law shapes nature is through simplification. Nature is complex; natural processes cannot easily fit within the administrative understanding of the world that governs human action.[7] Through simplification, law can make nature understandable, or legible, for administration. Such simplification makes certain things clear so that it becomes easier to systematically order nature. Yet, various aspects of nature are lost in this process.[8] Thus, law entails a specific way of looking at nature.[9]

The aim of this chapter is not to problematize the simplifications of nature as such, nor is it to claim that these simplifications are in fact simple.[10] Instead, I argue that studying the simplistic perception of nature in law can enable a discussion of what is absent from such perception. Such an understanding also allows for a discussion of what could be included to give the perception of nature in law a bit more depth, while still recognizing that some form of simplification is inevitable if nature is to be managed through legal systems. If there is an openness about these simplifications and the reductionist nature of law, it will be easier to discuss what is included and what is omitted, as well as what consequences this might have for the marine environment.

In addition to the simplification of nature by law, the natural world becomes divided and affected by the placing of responsibility for planning on different levels of administration. As Valverde (2009) discusses, while the central state may have one way of seeing, cities, or in my case municipalities, are also characterized by certain ways of seeing and governing.[11] Local governments, such as municipalities, have different approaches to management than regional, national, or international authorities. This may result in a fragmented management, which is not fit to deal with the complex interactions of social-ecological systems[12] on different scales.

To understand how and why the management, in this case planning, is fragmented, it is critical to understand both the legal framework and the role of the administrative bodies performing the planning. Every context produces its own style of management. Law produces different outcomes depending on where it is applied and by whom. This means that while the different laws governing marine and municipal planning may be formally in tune, their implementation can be inconsistent due to differences in management priorities between the levels of government performing the planning. In order to understand the motivation and prioritizations behind municipal or state planning of coastal and marine areas, it is necessary to study the legal system from both a technical, legal perspective and an "extra-legal" perspective.[13] This "extra-legal" study is performed through document analysis of municipal comprehensive plans as well as in-depth interviews with planners. I conducted 23 interviews with planners, ecologists, and politicians at the municipal level as well as at the regional (County Administrative Boards or CAB) and national (the Swedish Agency for Marine and Water Management or SwAM) levels. This data provides knowledge on how the actual planning is performed and hence what law becomes when it meets planning practices, rather than what law "is" or prescribes.

My analysis exposes how municipal and national planning priorities correspond to and interact with each other. The study was carried out in Sweden. However, an examination of different planning levels around the Baltic Sea shows that while every country chooses its own approach to MSP, it is common to have at least two different levels of administration involved in the coastal and marine planning.[14] The challenges arising within the Swedish system are thus translatable to other European planning systems, as the Swedish exclusion of coastal waters in the national MSP closely follows the EU Directive.

Marine Spatial Planning: An Introduction

MSP has been described as a means to provide a vision for the future development of an area, where all affected parties have been able to provide input. This vision can be used to provide information to subsequent permit decisions as well as for stakeholders in their planning of future activities.[15] The outcome of a planning process is, simply put, a map and an accompanying explanatory document, with recommendations on how the marine areas should be used. Some areas will be pointed out as suitable for conservation and nature protection, while others have characteristics that make them viable for offshore wind power, fisheries, et cetera.

The marine plans represent an important aspect of the simplification process of law. The first step is how natural space is divided into planning areas. In the second step, the planned space is zoned and divided into different uses. While this may be necessary from an administrative perspective, the simplification process needs to be attentive to both human and more-than-human lives. When certain areas are excluded and management is fragmented, the image of nature becomes skewed, and perhaps more of an incoherent mosaic rather than a well-composed representation.

In the early years of MSP practice, as well as research on MSP, it was often described as a "rational" technocratic activity.[16] In recent years, however, there has been a critical turn in the MSP literature, questioning the rational ideal and highlighting how MSP tends to uphold existing power structures and further marginalize less economically vested stakeholder groups.[17] While few of these critical interventions have come from legal scholars, there is one intervention that approaches MSP from a legal geography perspective, and critically examines and discusses MSP. However, this intervention provides a more conceptual discussion on law and MSP, and is not based on any particular case study or empirical material.[18] The novelty of the present chapter is that it critically engages with law and MSP on an empirical basis. Furthermore, it highlights that it is not only a matter of the relation between stakeholders and the central government; rather, the different bodies of government, and their ways of understanding, or "seeing," social-ecological systems, also need to be understood in order to provide a more comprehensive understanding of the challenges in MSP.

The process leading up to the adoption of the EU MSP directive accentuated conflicts between the international, national, and local perspectives of management. The same conflicts can be found in the development of the Swedish system for MSP. Both processes identified the same challenges facing the marine environment, and emphasized the need for a holistic perspective, where different sectoral interests are treated as constituent parts of a common pressure. In this holistic perspective, coastal areas were seen as crucial to include as they are important from both a social and an ecological perspective.[19] Still, the final versions of both the EU and Swedish MSP systems exclude coastal waters,[20] begging the question how this holistic mode of planning can be functionally implemented when such important parts of the social-ecological systems are excluded.

In a sense, MSP is a tool for simplification and for creating visions, or imaginaries, of the marine space. These visions are then realized through the plans. The main argument of this chapter is that it matters greatly whose visions they are. While both the EU and the Swedish MSP processes aim to focus on the three-dimensional character of the marine space, municipal planning tends to be preoccupied with the surface. The blue water and archipelago function as symbols for the attractiveness of the municipality, and the traditional marine sectors as providers of an identity. In these visions, the complex ecosystems below the surface are given little attention. To understand the different ways of envisioning coastal and marine areas, and where these visions originate, the following section will provide an overview of the constitutional role of municipalities in Sweden, as well as the regulatory framework governing municipal planning.

Formal Limits of Municipal Planning

To understand the perspectives and priorities of municipal planning in Sweden, it is important to understand the constitutional role of municipalities in the Swedish democratic system. Municipalities in Sweden hold a strong position as autonomous local governments. The municipal autonomy follows from the constitution.[21] This autonomy is not complete, and does not cover all areas of government. Still, in terms of spatial planning, the autonomy is strong and only limited by a few select national objectives.[22]

Municipal spatial planning is performed through the development of comprehensive plans, covering the entire area of the municipality. These are then realized through detailed development plans that cover smaller areas, such as smaller neighborhoods, blocks, or sometimes even larger individual buildings. The municipal comprehensive plans are strategic policy documents, where the long-term land-use development is envisioned and intentions and visions for the municipality are expressed.[23] As such, they provide an insight into the general objectives for municipal development, and give information about how the municipalities understand their coastal areas. Yet, the strategic and general character of the comprehensive plans makes them insufficient as sources for an in-depth analysis of their consequences and impacts. Although they include statements that constitute the municipality's fundamental objectives, they are generally of little guidance when it comes to considerations, weighing of interests, and trade-offs.

There are two levels of management in the Swedish marine planning system: the municipal and the national. The municipal level plans on a local scale, where the geographical boundaries are those of the municipality, and the legal boundaries are set by the Local Government Act and the Planning and Building Act. These two Acts determine the scale and what is (not) relevant for the municipal planning. The national scale is present in the same areas as the local scale, but is defined through the Ordinance on Marine Spatial Planning and the Swedish Environmental Code. The national scale is more dynamic. It is less detailed than the municipal, but it moves from the national perspectives down to the regional and sometimes municipal scale. The municipal scale is detailed and, like the national scale, can move between perspectives. These perspectives, however, span from the regional scale to specific, local, projects within the municipality. The national scale is rarely included. This indicates that both scales are necessary to capture as much of the marine and coastal space as possible. When separated they present fragmented and over-simplified representations of social-ecological systems.

In the following section, interviews with civil servants in the municipalities, municipal politicians and civil servants on the regional (CABs) and national

(SwAM) levels are used to provide an additional layer to the analysis of municipal comprehensive plans, and how these relate to the national planning process.

Themes in Municipal Comprehensive Planning

As seen above, the two levels of planning are regulated in different legal acts. In addition to this, the organizations implementing these acts are different and are driven by different objectives and obligations, which leads to different priorities in planning. If these objectives are not in tune with each other, they will accentuate the problems of fragmentation. As discussed above, the two different types of planning are regulated in different acts and the underlying rationales for planning thus differ between them. This has been described as the Planning and Building Act being built around a *planning paradigm*, and the Swedish Environmental Code building on an *environmental paradigm*.[24] In addition, one of the basic guiding principles for all municipal decision-making stipulates that all decisions need to be taken with the interests of the members of the municipality in mind.[25] While the planning legislation widens this competence to include broader issues and certain types of national legislation, the local perspective is omnipresent in municipal decision-making.

The following paragraphs cover the main interests and perspectives that are highlighted in the municipal comprehensive plans and in interviews with planners. It is key to understand how the municipal planning is formulated, and what the priorities are, not only for the individual municipalities, but also in relation to the national MSP process. As the coastal waters are excluded from the national planning, municipal planning becomes the link between land and sea. In addition, many of the pressures on the marine environment are located in the coastal areas.[26] In terms of applying an ecosystem approach, local planning needs to be connected to national planning to ensure that the entirety of the marine social-ecological ecosystems is accounted for.

In the study of municipal planning, a number of themes recurred both in the planning documents and in the interviews. All of these themes indicate a planning that places a clear focus on strengthening the municipality in relation to surrounding areas, by attracting new inhabitants and creating opportunities for the business sector. Some of the themes related to how the municipalities positioned themselves in relation to neighboring municipalities; national marine planning; the role of the municipality in a larger, holistic picture; and regional aspects. The highly localized focus of municipal planning was perhaps most clearly expressed by one of the respondents, a planner from a small municipality on the west coast of Sweden.

When asked how the national marine plans would affect the municipal planning, the respondent pointed out that the main interests of the municipality were located within the coastal waters. He could see no conflicts or close relations to the municipality in what was taking place in the areas covered by national planning.[27] These areas were thus placed in a clear periphery of municipal planning.[28]

It also indicates that there is little integration between the two levels of planning, even though they both have a responsibility to plan the areas in question.

This type of focus was evident in other interviews as well, although it was often connected to scale too. A number of respondents had concerns regarding the scale of the national plans, namely that the resolution was too low to be able to use in municipal planning.[29] One of the respondents from a regional agency also highlighted that this local perspective was of little surprise, as a basic precondition of the Swedish system is that every municipal decision is driven by economic reasoning. All decisions build on the notion that they contribute to a better economy for the municipality.[30] This shows the problematic aspect of simplified representations of nature, where law has delimited natural systems without giving due respect to ecological boundaries. In these municipal representations, both the more-than-human lives and the ocean were given little to no space. When they were included, it was mainly in terms of how they could strengthen the municipality's economy or position. It was generally not discussed what the effects of planning decisions will be in a wider perspective. This is problematic because, much like the increasing investments in the Miami river had both local and distant consequences for Haiti's trade (in Kahn's contribution to this volume), the local planning decisions in a Swedish municipality can have effects on social-ecological systems far beyond the municipal borders. This is a basic feature of the fluid nature of water.[31]

Another recurring theme, and one that relates to how neighboring areas were treated in the plans, was that of competition. The municipalities all want to grow, and to do so they need to attract businesses and inhabitants. This places them in a competitive position in relation to neighboring municipalities. Sometimes the competition was framed in terms of regional competition, where a few municipalities together formed a region that was placed in a competitive position in relation to other regions.[32] The competitive aspects placed a focus on strengthening the municipalities, or regions, thus placing them in a clear center, while making the interests of other regions or municipalities peripheral issues at best. As succinctly put in the comprehensive plan of the municipality of Pietå: "Cities compete with each other over resources for growth."[33] This competition can be expressed in many ways, but an important consequence is that each municipality looks to their own interests first. One respondent claimed that she could not imagine any local politician who would abstain from planning a new investment in an area just because a neighboring municipality was investing in an adjacent area.[34] Such planning seems contrary to the integrated style of management that is needed to fully capture the complexity of social-ecological systems.

Lastly, ports were another key theme in many of the municipal comprehensive plans. Most of the municipalities included in the study had a port which was important for the municipality's business sector. By framing the port's operations as important from a regional or national perspective, the municipalities could motivate further investments and development. To illustrate this by way of an example, the biggest port in Scandinavia is located in Gothenburg. Nevertheless,

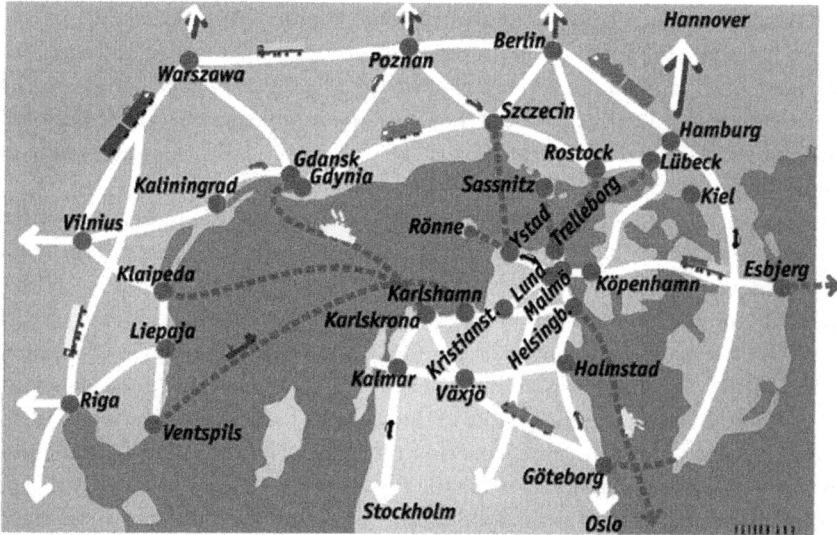

FIGURE 11.3 Map from the comprehensive plan of the municipality of Karlshamn. By flipping the map upside-down and placing Karlshamn in the middle, Karslhamn is promoted as an important node for eastbound transports, while large port cities such as Stockholm and Gothenburg become peripheral. Courtesy of the Municipality of Karlshamn. Used with permission.

the municipality of Malmö promoted their port as the biggest port in Scandinavia for the import of cars.[35] The municipality of Helsingborg promoted their port as the second largest container port in Sweden.[36] In the smaller municipalities, the port could be framed as the largest exporting port in Sweden when it comes to sawn wood products,[37] or that even a small jetty could be one of the most important landing points for shellfish nationally.[38]

All of these themes show how the municipal perspective is highly localized, focusing mainly on issues that will strengthen the economic position of the municipality. This focus pushes the national objectives of a holistic planning to the periphery. Furthermore, it accentuates the simplified idea of the natural environment, where geographical locations outside of the municipal boundaries do not enter into the municipality's field of vision.

In addition to the themes concerning how the municipality related to itself and its neighboring areas, there were also recurring themes of how the municipalities imagined their coastal and marine areas. These themes also seemed to affect the priorities in planning. Mainly they related to questions such as: why is the coast important for a municipality? What values are seen as worthy of protection or development?

"Attractivity" was one of the main themes, recurring in all of the planning documents in relation to the coastal areas. The reasoning behind any decision on how to develop or protect a specific coastal area seemed to be related to the general objective of strengthening the attractivity of the municipality. Attractivity was framed both in relation to new inhabitants or businesses, and existing inhabitants: to make the municipality a place where one wants to live and stay.[39] In the interviews, attractivity was mainly discussed by the respondents from smaller and medium-sized municipalities.[40] For the larger municipalities, attractivity was mentioned, but not as the most prominent interest. Usually, shipping, ports, or other interests had greater economic value to those municipalities.[41]

Another notable recurring theme in the material was the uniqueness of the coastal areas, or the coast as a bearer of the municipality's identity. This identity relates to the idea that the coast contains values that are unique, in both a Swedish and an international context. These unique values and identity bearers differ between the municipalities. As for attractivity, the theme of uniqueness and identity was more pronounced in the comprehensive plans of small/medium-sized municipalities. Most of the plans had statements relating to identity in the sections of the plans covering the coast. The coast represented a historical value, but it could also shape the future of the municipal identity.[42] One such expression of this was the following: "The rural areas and the archipelago are important parts of the identity of the municipality and they create opportunities to conserve and develop the cultural landscape with values for the cultural environment, recreation and biological diversity."[43]

For the larger municipalities, the city life close to the sea was an important factor for the identity.[44] The proximity to water was one of many factors that provided a larger, urban, identity. All mentions of municipal identity had a close relation to the attractivity of the coast. The coastal identity could attract tourists, as well as new inhabitants. In this respect, there were few differences between the variously sized municipalities.[45]

Taken together, all of these themes illustrate the perspectives of municipal planning and how the municipalities imagine the future use of their coastal and marine areas. These areas are visualized as drivers for municipal growth and economic prosperity. The focus is primarily local, with more national objectives being taken into account when they fit into the municipality's vision. Nevertheless, these visions also exhibit a detailed knowledge about the areas covered by planning, something that is far less evident in the proposed national marine plans.

Themes in the National Marine Plans

As discussed above, the national marine planning follows a somewhat different logic from that of the municipal planning. The natural environment has a more prominent position in the legal acts covering the national MSP. This is most clearly exhibited by the fact that MSP is regulated in the Swedish Environmental Code, while municipal planning is regulated in the Planning and Building Act. Another difference between the plans is the scale: the national plans cover far greater areas than a single municipal plan does. The national plans also need to be attentive to two different scales of planning bordering the plan areas. On the landward side of the national plan areas, there are local municipal plans, and on the other side, neighboring countries are also developing their own marine plans.

In relation to neighboring countries, it was stated in the proposed national plans that Sweden was taking an active part in regional cooperation projects. One of these projects was specifically aimed at analyzing and comparing how different Baltic states had applied the ecosystem approach in their MSP. The results were to be used in the continued MSP process in Sweden.[46] The plans of neighboring countries thus informed the Swedish marine planning process. In relation to the local, municipal planning the perspective was different. The connection between the national and municipal plans was acknowledged. It was noted in the national plans that the municipal planning was more detailed closer to land and along the coastline, and that there was potential for a common development of the planning between municipal, regional, and state levels to strengthen the land-sea interaction.[47]

In addition, the marine plans were promoted as a tool to create more clarity relating to sea use. They could support the municipal planning processes and contribute to greater coherence between municipalities and other actors when it comes to considerations between interests.[48] Nevertheless, the perspective was always that information would flow from the national plans down, not the other way around. In this sense, the national plans did not exhibit any strong intentions in terms of integrating the two different levels of planning. Rather, they adopted more of a top-down approach where the national plans would be guiding for municipal planning.

Two recurring themes in municipal planning were the attractivity of the coastline and the coast as an identity-bearer. Neither theme was similarly pronounced in the national plans, although they both occurred. On a national scale, the coast and archipelagos were seen as attractive for Sweden as a whole and contributing to a long-term competitive tourism industry.[49] But attractivity was also discussed more specifically in that the national plans could promote attractivity on a local and regional scale.[50] Granted, these two scales are not mutually exclusive, but the examples illustrate how the national plans switch between scales. This switching is important as it leads to different ways of viewing sectors, or challenges for that matter. On a local level, fisheries were promoted as identity bearers, and as a source of land-based jobs.[51] On an aggregated level, over-fishing was discussed as a problem and one of many stressors for the marine environment.[52] The same was true for tourism, where coastal areas needed to be attractive places in which to live and work, as well as to visit (local scale),[53] while jetties and marinas are negatively affecting the ecosystem in the coastal zones, where many species have their reproduction areas (aggregated scale).[54] This switching between scales shows a promising aspiration to complicate the simplified version of nature that follows the MSP legislation.

Each of the national plans pursued a number of themes, among them recreational and cultural values. However, it was obvious that most places that actually included recreational and cultural values were situated in the coastal waters, outside of the plan areas due to the fragmented nature of the planning system. Thus, the only possible planning measure was to conclude that considerations needed to be assessed in a local perspective. The national plans could not include these values, or plan for them to be ensured. In conclusion, the national planning was not overly concerned with local municipal perspectives. This was because coastal waters were excluded from the planning, but it could also be the result of the scale of the national planning. The holistic nature of the plans does not mesh well with integrating the interests of local scale social-ecological systems.

Conclusion

In this chapter, I have explored how different levels of government within a single country envision the future use of marine and coastal areas in spatial planning. The aim of marine spatial planning is to be a holistic, integrated tool to manage the pressures on the marine environment. But if the different actors performing the planning are not coordinated, it risks becoming a rather fragmented endeavor. Much of the literature taking a critical approach to MSP has focused on the inclusion of stakeholders and how power structures are reproduced rather than torn down in many current MSP regimes.[55] The novelty of this chapter is that it has focused on how different planning authorities also create different imaginaries of coastal and marine environments. When these imaginaries translate into plans and policies, they may either help or hinder ambitions to establish a more holistic planning of coastal and marine areas. This is particularly true in an EU context, where coastal waters are excluded from the EU MSP directive, creating a plethora of different authorities involved in coastal planning around the Union.

As shown by Grisel in his contribution to this volume, legal norms need to be attentive to social norms on a local scale. But this is also true in reverse. As Kahn shows in his chapter, the local is never truly local, but reaches out into a wider geography as well. This needs to be taken into account in the design of legal and management structures. There is no point in trying to distinguish between a local, regional, or national scale, since all of these scales are present simultaneously and constantly interact with each other.[56] Studying the Swedish system for coastal and marine planning shows that when law so clearly distinguishes between scales and strives to draw lines in the fluid marine space, it results in different and not always coherent ways of seeing in management.

Municipal coastal imaginaries are created through the concepts of attractivity and identity. These concepts provide a rationale for further development. This is expressed through the view of neighboring municipalities as competitors, where shopping malls outside of the municipality can be seen as threats to local commerce.[57] With such competition, one municipality is placed at the center, while all others become peripheral. This leads to imaginaries that motivate development and can give the municipality competitive advantages in relation to other, neighboring, municipalities. The attractivity and identity concepts are clear examples of this: they motivate why certain aspects of a municipality should be further developed, as those aspects can give the municipality an advantage in the competition over resources for growth. The national planning process, by contrast, applies a more general conceptualization, which is primarily represented through the application of the ecosystem approach and a general strive for maritime growth. So, there are various reasons why the different levels of management perform in an incoherent manner. These relate to the area being governed, the timescale applied, and who is governing.

This chapter has shown that there are fundamental differences between the municipal and national planning processes. The municipal level promotes local interests, following from both their legal obligations and their local political and economic ambitions. The national planning takes the local perspectives into account. However, local interests are given a relatively peripheral position due to the fact that the national planning excludes the areas where most activities take place. But also because of the geographical scale applied in the national plans, which is not attentive to local variations.

The review clearly shows that while both levels of management have certain advantages, they also have shortcomings. The municipal level is not able to grasp large-scale processes of social-ecological systems. As long as these processes are not in line with the general visions of the municipalities, they are treated as peripheral. The national level, on the other hand, cannot take in the details that are needed to fully understand the intricacies of individual municipalities. It is well equipped, however, to understand the larger processes that affect the environment in the Swedish marine areas.

The case of Swedish planning serves to illustrate the general point that not only is it important to integrate local communities and stakeholders in planning efforts; it is also crucial to understand the different perspectives that follow from choosing different administrative bodies to perform the planning. Law and its implementation is inherently contingent on who is performing the actual implementation. Clearly, any management regime aimed at governing the marine environment needs to be attentive to social-ecological systemic processes taking place at different scales simultaneously, as they are inherently inseparable. This can only be achieved if all levels of management are included in the planning process, and if the marine environment is not represented as an over-simplified image, one that can be functionally divided through the drawing of administrative boundaries.

Acknowledgements

I would like to thank all of the civil servants and politicians who have generously contributed to this research with their time and deep knowledge about Swedish marine and coastal planning.

Notes

1 Frank Maes, "The International Legal Framework for Marine Spatial Planning," *Marine Policy* 32, no. 5 (2008): 797–810.
2 Stephen Jay et al., "Marine Spatial Planning: A New Frontier?" *Journal of Environmental Policy & Planning* 14, no. 1 (2012): 1–5, 1.
3 Charles Ehler and Fanny Douvere, "Visions for a Sea Change: Report of the First International Workshop on Marine Spatial Planning," UNESCO, 2007, https://repository.oceanbestpractices.org/handle/11329/204.
4 See, e.g., Betty Queffelec et al., "Marine Spatial Planning and the Risk of Ocean Grabbing in the Tropical Atlantic" *ICES Journal of Marine Science* 78, no. 4 (2021): fsab006.
5 Directive 2014/89/EU of the European Parliament and of the Council of July 23, 2016 establishing a framework for maritime spatial planning.
6 Ibid., Article 2(1).
7 James C. Scott, *Seeing Like a State: How Certain Schemes to Improve the Human Condition Have Failed* (New Haven: Yale University Press, 1998), 262.
8 Ibid., 11.
9 Nicholas Blomley, "Simplification is Complicated: Property, Nature, and the Rivers of Law," *Environment and Planning A* 40, no. 8 (2008): 1825–1842, 1826.
10 Simplification is a complicated process, a point made by both Scott, *Seeing Like a State*, 81, and Blomley, "Simplification is Complicated."
11 Mariana Valverde, "Jurisdiction and Scale: Legal 'Technicalities' as Resources for Theory," *Social & Legal Studies* 18, no. 2 (2009): 139–157, 139.
12 This concept is used to illustrate how the natural/ecological world is inseparable from the social/human world. Thus, rather than discussing natural systems and social systems, the concept of social-ecological systems is used. See, e.g., Fikret Berkes, Carl Folke, and Johan Colding, *Linking Social and Ecological Systems: Management Practices and Social Mechanisms for Building Resilience* (Cambridge: Cambridge University Press, 1998).
13 Ibid., 154.
14 Aron Westholm, "Appropriate Scale and Level in Marine Spatial Planning—Management Perspectives in the Baltic Sea," *Marine Policy* 98 (2018): 264–270, 264.
15 Fanny Douvere and Charles N. Ehler, "New Perspectives on Sea Use Management: Initial Findings from European Experience with Marine Spatial Planning," *Journal of Environmental Management* 90, no. 1 (2009): 77–88, 79.
16 See, e.g., Fanny Douvere, "The Importance of Marine Spatial Planning in Advancing Ecosystem-based Sea Use Management," *Marine Policy* 32, no. 5 (2008): 761–771, 762.
17 See, e.g., Wesley Flannery et al., "Exploring the Winners and Losers of Marine Environmental Governance/Marine Spatial Planning: Cui bono?/'More than Fishy Business': Epistemology, Integration and Conflict in Marine Spatial Planning/ Marine Spatial Planning: Power and Scaping/Surely Not All Planning is Evil?/ Marine Spatial Planning: A Canadian Perspective/Maritime Spatial Planning – 'ad utilitatem omnium'/Marine Spatial Planning: 'It Is Better to Be on the Train than Being Hit By It'/Reflections from the Perspective of Recreational Anglers and Boats for Hire/Maritime Spatial Planning and Marine Renewable Energy," *Planning Theory & Practice* 17, no. 1 (2016): 121–151, 121; Ralph Tafon, "The 'Dark Side' of Marine Spatial Planning: A Study of Domination, Empowerment and Freedom through Theories of Discourse and Power" (dissertation, Södertörns Högskola, 2019); Wesley Flannery et al., "A Critical Turn in Marine Spatial Planning," *Maritime Studies* 19 (2020): 223–228, 223; Xander Keijser et al., "A 'Learning Paradox' in Maritime Spatial Planning," *Maritime Studies* 19 (2020): 333–346, 333.
18 Mara Ntona and Mika Schröder, "Regulating Oceanic Imaginaries: The Legal Construction of Space, Identities, Relations and Epistemological Hierarchies within Marine Spatial Planning," *Maritime Studies* 19 (2020): 241–254, 241.

19 See, e.g., "COM(2008) 791, Roadmap for Maritime Spatial Planning: Achieving Common Principles in the EU," Commission of the European Communities, 2008, https://eur-lex.europa.eu/LexUriServ/LexUriServ.do?uri=COM:2008:0791:FIN :EN:PDF; Swedish Government Commission Report—SOU 2008:48, En Utvecklad Havsmiljöförvaltning, 2008, 155.

20 The EU MSPD excludes coastal waters "covered by a member state's town and country planning" and is also a framework directive which only stipulates minimum requirements. It is thus possible for member states to include coastal waters in their MSP regimes.

21 The Swedish Instrument of Government (1974:152), Ch. 1 Sec. 1.

22 The Swedish Planning and Building Act (2010:900), Ch. 1 Sec. 2.

23 Ibid., Ch. 3 Sec. 2.

24 Lars Emmelin, *Planera För Friluftsliv: Natur, Samhälle, Upplevelser* (Stockholm: Carlsson, 2010).

25 The Swedish Local Government Act (2017:725), Ch. 2 Sec. 1.

26 See "Marine Spatial Planning—Current Status 2014," Swedish Agency for Marine and Water Management, 2015, Ch. 3.

27 Planner, small municipality in the Skagerrak/Kattegat plan area, in-person interview by author, April 6, 2017.

28 For a discussion on center-periphery relations in law-making, see Boaventura de Sousa Santos, "Law: A Map of Misreading—Toward a Postmodern Conception of Law," *Journal of Law and Society* 14, no. 3 (1987): 279–302.

29 Planner, interview; Ecologist, medium size municipality in the Gulf of Bothnia plan area, in-person interview by author, October 31, 2017; Planning architect, medium municipality in the Baltic Sea plan area, in-person interview by author December 1, 2017.

30 Architect, County Administrative Board, in-person interview by author, April 16, 2019.

31 See also: Philip Steinberg and Kimberley Peters, "Wet Ontologies, Fluid Spaces: Giving Depth to Volume through Oceanic Thinking," *Environment and Planning D: Society and Space* 33 no. 2 (2015): 247–264.

32 Comprehensive Plan of the Municipality of Lysekil—Lysekils Kommun, *Översiktsplan 06 Lysekils Kommun*, 2006, https://www.lysekil.se/download/18.4f9903d01543aee d81352230/1461663384545/%C3%96versiktsplan%202006%20Lysekils%20kommun.pdf, 10; Comprehensive Plan of the Municipality of Malmö—Malmö Kommun, *Översiktsplan för Malmö—Planstrategi*, 2018, https://www.yumpu.com/sv /document/read/19715214/oversiktsplan-for-malmo-op2012-planstrategi-malmo -stad, 23.

33 Comprehensive Plan of the Municipality of Piteå—Piteå Kommun, *Vårt Framtida Piteå—Översiktsplan för Piteå Kommun*, 2016, https://www.pitea.se/contentassets/6f8 58eaa9359447e8f1e83c286bd591e/01-oversiktsplan-planstrategi.pdf, 8.

34 Comprehensive planner, medium size municipality in the Skagerrak/Kattegat plan area, in-person interview by author, October 25, 2017.

35 Municipality of Malmö, *Översiktsplan för Malmö—Planstrategi*, 2018, 9.

36 Comprehensive Plan of the Municipality of Helsingborg—Helsingborgs Kommun, *ÖP 2010—En Strategisk Översiktsplan för Helsingborgs Utveckling*, 2010, https://hels-ingborg.se/trafik-och-stadsplanering/planering-och-utveckling/oversiktsplanering/ gallande-oversiktsplaner/oversiktsplan-2010/planhandlingar/—ÖP 2010.

37 Comprehensive Plan of the Municipality of Varberg—Varbergs Kommun, *Översiktsplan för Varbergs Kommun*, 2010, https://www.varberg.se/download/18.42e 2e0a7143003c9eed68e3/1391705173520/OP_kommunen_antagen_100615.pdf, 21.

38 Planner, interview.

39 See *inter alia*: Comprehensive Plan of the Municipality of Falkenberg—Falkenbergs Kommun, *Översiktsplan 2.0 för Falkenbergs Kommun* (2014), https://vaxer.falkenberg .se/download/18.5660b3216347a1a2d016b72/1526648592412/Antagandehandling

%20%C3%96P%20Falkenberg%20Del%20II.pdf, Part 2, 57. Fifty-two out of 65 municipalities mention population growth as a strategy in their comprehensive plans (studied 2017).

40 Biologist, medium size municipality in the Gulf of Bothnia plan area, in-person interiew by author, April 26, 2017; Comprehensive planner, medium size municipality in the Skagerrak/Kattegat plan area, in-person interview by author, October 25, 2017; Planner, interview; Development secretary, small municipality in the Gulf of Bothnia plan area, in-person interview by author, November 1, 2017; Plan architect, small municipality in the Skagerrak/Kattegat plan area, in-person interview by author, April 24, 2017.

41 Planner, large municipality in the Gulf of Bothnia plan area, in-person interview by author, April 26, 2017; Environmental strategist, large municipality in the Baltic Sea plan area, in-person interview by author, November 29, 2017; Environmental planner and landscape architect, large municipality in Skagerak/Kattegat plan area, in-person interview by author, October 27, 2017; Project coordinator and Water planner, large municipality in the Baltic Sea plan area, in-person interview by author, October 25, 2017; Planner, large municipality in the Gulf of Bothnia plan area, in-person interview by author, November 1, 2017; Planner, large municipality in the Skagerrak/Kattegat plan area, in-person interview by author, November 7, 2017.

42 Municipality of Lysekil, *Översiktsplan 06 Lysekils Kommun*, 2006, 5.

43 Comprehensive Plan of the Municipality of Karlshamn—Karlshamns kommun, *Karlshamn 2030—Översiktsplan för Karlshamns Kommun* (2015) Utvecklingsstrategier, 19, https://www.karlshamn.se/wp-content/uploads/Utvecklingsstrategier.pdf.

44 Comprehensive Plan of the Municipality of Luleå—Luleå Kommun, *Program Kuststaden Luleå* (2013), 18 (new plan adopted 2021 not part of the present analysis), Municipality of Helsingborg (2010), 22.

45 See *inter alia*: Comprehensive Plan of the Municipality of Göteborg—Göteborgs Kommun, *Översiktsplan för Göteborg* (2009), Part 1, 56, https://weblisher.textalk.se/goteborg/op09_del1/; Municipality of Lysekil, *Översiktsplan 06 Lysekils Kommun*, 2006, 5; and Comprehensive Plan of the Municipality of Oskarshamn—Oskarshamns Kommun, *Översiktsplan Oskarshamns Kommun 2050 (Proposal)* (2018), https://www.oskarshamn.se/globalassets/bygga-bo-miljo/dokument/op-fop-o-tillagg/oversikt-splan-oskarshamns-kommun-samradshandling-2021-06-07.pdf, Part 1, 17.

46 The Swedish Agency for Marine and Water Management, *Förslag till Havsplaner för Sverige, Bottniska Viken, Östersjön, Västerhavet—Granskningshandling 2019-03-14* (2019), 27.

47 Ibid., 138.

48 Ibid., 146.

49 Ibid., 159.

50 Ibid., 63.

51 Ibid., 54, 233, 234.

52 Ibid., 163.

53 Ibid., 63.

54 Ibid., 163.

55 See, e.g., Flannery et al., "A Critical Turn"; Tafon, "The 'Dark Side.'"

56 See, e.g., Neil Smith, "Geography, Difference and Politics of Scale," in *Postmodernism and the Social Sciences*, eds. Joe Doherty et al. (Basingstoke: Macmillan, 1992), 57–79.

57 Municipality of Lysekil, *Översiktsplan 06 Lysekils Kommun*, 2006, 8.

12

GOOD HUMAN–TURTLE RELATIONSHIPS IN INDONESIA

Exploring Intersecting Legalities in Sea Turtle Conservation

Annet Pauwelussen and Shannon Switzer Swanson

Introduction

The world of the Bajau in the Asia Pacific is shaped by the entanglement of people and sea turtles, sharing spiritual kinship and companionship through their common migratory and amphibious way of life. The Bajau (or Sama-Bajau[1]) usually identify as an ethno-linguistic assemblage of people dispersed over archipelagic Southeast Asia, including Indonesia, Malaysia, and the Philippines.[2] The region knows a long oral and written history of human–sea turtle interactions, including the consumption of turtle meat and, particularly, eggs as a delicacy and ritual food in ceremonial events, as well as the use of turtle shells for making adornments.[3] Sustaining a society and economy of sea-based mobility, Bajau movements have often followed the migrations of sea turtles and fish, for barter, trade, and livelihood. Places where sea turtles gather in large numbers to feed, mate, and lay eggs have historically attracted Bajau communities to settle and sustain regional networks of trade.[4] Thereby, the turtles' habitats have also shaped the social geographical spaces that constitute Bajau worlds.

Over the last decades, sea turtles have also risen to the center of attention of national and international conservation programs. Three of seven species (including the green sea turtle) are currently classified as "endangered" or "critically endangered" by the International Union for Conservation of Nature (IUCN). Moreover, the animal appeals as a beloved and charismatic megafauna, or "flagship species"—an icon to attract public interest and donor funding, enhancing

FIGURE 12.1 A hawksbill turtle caught at sea is held in a "karamba," or net pen, beneath the house of a fishing family in the Banggai Archipelago, Indonesia. Both green and hawksbill turtles are often held in net pens such as this to grow the turtles to a larger size. In the case of green turtles, this is for eating, while in the case of hawksbill turtles, this is for using the shell to make jewelry and other adornments. Image by Shannon Switzer Swanson, 2017.

DOI: 10.4324/9781003205173-13

its position as a priority species in conservation programs.[5] The abundance of sea turtles in the Indo-Pacific has therefore attracted the interests of marine scientists and conservation agencies such as The Nature Conservancy (TNC), the World Wildlife Fund (WWF), and the Turtle Foundation. The region has become a global conservation priority as one of the world's primary nesting and feeding ground of sea turtles, spurring legal interventions to protect them.[6] Articulated as bans on eating and trading turtles and their eggs and restricting human access to turtle mating and nesting areas, these interventions have intersected—and conflicted—with Bajau engagements with turtles, and with claims to their customary right to do so for their livelihood and social-cultural wellbeing. As we show in this chapter, this situation has given rise to different forms of resistance among the Bajau to contest or circumvent sea turtle protection programs.

While wildlife protection conflicts may inspire analyses of the dialectic and power relations between international laws on the one hand, and customary or Indigenous rights on the other,[7] we are more interested here in unpacking the complex ways in which these coexist and interact in practice. Like in other parts of the world, sea turtle protection programs in the Indo-Pacific encompass a combination of different new and old, international and national, laws and treaties that dictate the protection of these amphibious and migratory creatures in different ways.[8] As noted, these legal interventions intersect with—and are resisted by—legal and normative systems that inform Bajau perspectives on the legality of hunting, collecting, eating, and trading sea turtles and their body parts as a common inherited practice and customary right. However, as we show later, such Bajau customary rights and practices are themselves often a product of historical encounters with colonial systems of governance. Also, acknowledging the entangled nature of Bajau-sea turtle coexistence also brings into the picture the sea turtle herself as a legal subject and object. Working from this complexity, this chapter aims to explore the contours of an inclusive analytical approach that takes in the coexistence of different legal systems, while also bringing in the nonhuman as an agent in the social-legal world.

Legal anthropology has conceptualized the existence and interaction of different legal and normative orders as a situation of "legal pluralism." This concept implies a broad and historically informed understanding of legality, expanding the boundaries of law to embrace "a variety of more or less formalized and institutionalized forms of normative ordering in society."[9] The questions "What is law and where is it?"[10] are then empirically addressed, by studying how—in practice—different normative orders are enacted and coordinated in overlapping or contesting regimes of legitimation.[11] In this sense, the term "legality" refers to the state or quality of being lawful in agreement with a legal system, which includes the meanings, practices, and sources of authority that are not necessarily acknowledged by official law.[12] Particularly in the context of marine governance, studies in legal pluralism have shown that the interaction and power relations between different normative orders and governance regimes, as they are practiced in everyday activities of fishing and marine conservation, generate situated notions of legitimacy and justice.[13] This also applies to the case of turtle legalities in Indonesia in which different formal and informal rules and norms in relation to using, trading, and protecting turtles compete, coexist, or otherwise interact.

While legal pluralism allows for understanding intersecting legalities and enacted power relations *in situ*, including customary and Indigenous legal practices, it has been less equipped to deal with nonhuman agency as part of the social-legal world. Legal pluralist accounts have usually classified animals as a resource or property that people claim rights and access to, which makes them legal objects, not subjects.[14] As pointed out by Zoe Todd in the Arctic Canadian context, such classification may sit uncomfortably with Indigenous legalities based on a notion of society as an entanglement between humans and other beings.[15] To include the more-than-human perspective in legal pluralism therefore requires critical reflection on how conservation and wildlife laws and their contestation presuppose different notions of what the animal is in relation to the human. Whereas both conservation agencies and Bajau may treat sea turtles as highly valuable or even iconic creatures of ocean life, they do not necessarily share the idea that they should be "managed" or "saved," reflecting different normative and ontological notions of what constitutes a good human–turtle relationship.[16]

The case of intersecting legalities around sea turtle conservation in Indonesia gives rise to an alternative theoretical exploration of the social and political dimensions of more-than-human legalities in the governance of marine wildlife. Inspired by Indigenous and feminist critiques by Zoe Todd and María Puig de la Bellacasa, we suggest that an understanding of human–turtle entanglement through historically embedded ethics of care allows for engaging the sea turtle in a way sensitive to power relations.[17] Our argument builds on our own long-term ethnographic engagement with Bajau families and sea turtles in and between the coastal and marine spaces of Kalimantan and Sulawesi in Indonesia. The first author carried out 18 months of fieldwork in 2011–2013 and several shorter visits 2009–2019, staying and traveling with Bajau families in Berau, Makassar, the Masalima Archipelago, and across the Malaysian border. The second author carried out 15 months of fieldwork in Central Sulawesi in 2016–2019, living and traveling with fishing and farming families in the Banggai Archipelago.

Both authors also regularly interviewed staff of governmental and non-governmental organizations (NGOs) in the region, and carried out participatory research in their conservation activities. All names are pseudonyms, and we have openly discussed the objective of our research practices with the people who took part in it. We also acknowledge our privileged status as outsiders working among these communities. While we aim to faithfully report what has been openly shared with us, we do not purport to speak on behalf of the Bajau people (nor sea turtles) as a whole.

In the next section, we turn to the Berau coastal area as a place where different legal approaches to consuming turtle eggs sparked conflict between governmental departments, NGOs, and Bajau communities. This is followed by an elaboration of how Bajau people navigate and resist turtle protection laws. We

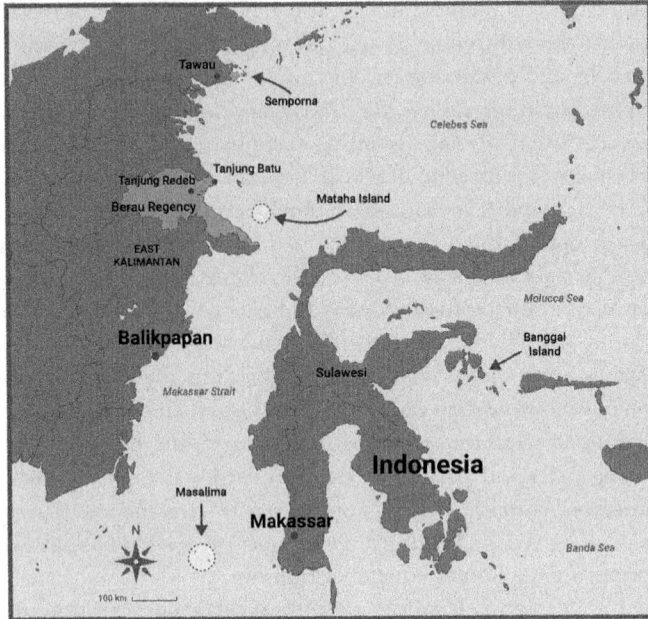

FIGURE 12.2 Map of East Borneo/Kalimantan and Sulawesi indicating fieldwork locations. Map by Ben Swanson. Used with permission.

then dive deeper into the varying turtle–human entanglements, and analyze how they produce different notions of what is a good human–turtle relation. Such notions underpin different legal and normative approaches to caring for them, and sketch the affordances of a more-than-human legal pluralist approach in marine governance.

Legal Complexity in Human–Turtle Relations in Indonesia

"It used to be 'Pulau Telur' ('Egg Island'), but it's gone now, the Germans took it," Arif said, steering his boat to the shore of the island, next to the speedboat of the Germany-based NGO, the Turtle Foundation. In 2013, the first author visited Mataha Island, a small island off the coast of the Berau district in East Kalimantan. She traveled with two long-term friends, Arif and Alisha from the neighboring island Balikukup with a majority Bajau population. On Balikukup, Mataha was and still is known as one of the *pulau telur*; islands where green sea turtle females congregate to lay their eggs. The island used to be a popular place for collecting turtle eggs by Bajau men and women, before the Turtle Foundation extended its turtle protection program in Berau to Mataha. "It is very difficult for us to see how this is a good thing," said Arif, pointing to the monitoring station and hatchery, the only human buildings on the uninhabited island; "NGOs coming from far away, to keep us from doing what we used to do for a living: eating and selling turtle eggs."

The hatchery is a sandy surface of five-by-five meters, surrounded by a wooden fence to protect the relocated eggs from being taken. During the 2013 visit, four men were stationed on Mataha to take care of turtle eggs day and night. One man offered a brief tour and explained his job:

> Every night, we take turns walking around the island, to see if there are new nests. If they are too close to the water, we take the eggs out, and move them to the hatchery. Usually, there are about ten new nests every night. But there is a season to it. In August and September, we see up to 30 sea turtles coming to the beach to lay their eggs.[18]

While the man pointed at the 30-something sticks in the sand, indicating relocated nests, one tiny big-eyed baby turtle crawled around, flapping its fore flippers, making its way to the sea. Alisha remarked it looked a bit clumsy: "why don't we bring it to the sea?" she asked. The NGO man replied: "We let them find their own way, keep it as natural as possible. Those are our instructions. Only when they don't manage and get lost, we sometimes give them a helping hand."

The start of the turtle's life journey is treacherous, as the majority of the hatchlings are eaten by birds, crabs, and fish before they reach deeper waters. Without human intervention, only about one or two out of 1,000 hatchlings survive into adulthood, the NGO man explained: "It's very important we protect these nests. Our data says that the number of nests has been increasing since 2008. That is because we now protect these nests. Once the baby turtles have grown, they will come back in the future to lay their eggs here again." Sea turtles travel across oceans for thousands of miles to return to their birth ground, where they meet, mate, and lay eggs on the sandy beaches where they once started their own life journey.

This increasing focus on protecting and monitoring sea turtles and their eggs in Berau is exemplary for a global trend over the last decades in which sea turtles have risen to the center of attention in marine wildlife conservation. Concerns about their survival as a species have spurred turtle protection policies and regulations at national and international levels.[19] For Indonesia, the first step in the process was Indonesia's ratification of the Convention on International Trade in Endangered Species of Wild Flora and Fauna (CITES) in 1978. In 1981 all marine turtle species were listed within Appendix I of CITES, which made trading species like the green sea turtle parts and products across international borders illegal. In 1990, Indonesia established the more sweeping Act No. 5, prioritizing the "Conservation of living natural resources and their ecosystems."[20] A focus on sea turtle protection followed in 1999 with Government Regulation No. 7, declaring it illegal to catch and trade any species of sea turtles and their eggs. Influenced by the WWF, in 2005, Indonesia also signed the Indian Ocean Southeast Asia Sea Turtle Memorandum of Understanding under the Convention of the Conservation of Migratory Species of Wild Animals.[21]

These complex laws reflect the nature of the sea turtle itself with its amphibious and migratory lifestyle that troubles spatial and legal boundaries. Government agencies, with the help of NGOs, enact the laws by heavily monitoring sea turtle life cycles across oceans. Berau is an exemplary case, with hatcheries and ranger stations scattered over its offshore islands. In Berau, these island stations do more than care for eggs and hatchlings; they also keep out Bajau collectors from neighboring islands to stop—in terms of the Turtle Foundation—the "illegal plundering of nests."[22] The criminalization of egg collection by the Bajau is legally grounded in the aforementioned laws and regulations. Still, NGOs do not have enforcement power to enact laws. The responsibility to enforce national laws on the protection of sea turtles as an "endangered species" lies with the BKSDA (*Balai Konservasi Sumber Daya Alam*—Indonesia's official Nature Conservation Agency), with rangers operating under the responsibility of the national Ministry of Environment and Forestry in Jakarta. Since 1982, the BKSDA has been formally in charge of several of the "turtle egg islands"[23] in Berau, to protect and patrol turtle nests, while NGOs tend to the hatcheries.

Despite the clear mandate for forestry rangers to protect turtles, there are limitations to their authority. With only two or three rangers in Berau, they can only focus on "their" islands. Sea turtles move beyond these islands, onto Berau's seagrass meadows, coral reefs, and around the densely populated Bajau islands Balikukup, Derawan, and Maratua. After the Berau coast appeared on the conservation radar as primary nesting ground of sea turtles in the Indo-Pacific, the entire coastal zone of the Berau district was designated as a marine protected area (MPA) in 2005. With the financial and organizational support of the WWF and TNC, the MPA scaled up turtle protection to an integrated ecosystem-based approach, putting 1.27 million hectares of coastal waters, islands, and turtle habitats formally under decentralized management by the district Department of Fisheries and Marine Affairs.[24]

The declaration of the MPA set the stage for a protracted conflict around the customary rights of the Bajau to collect, eat, and trade turtle eggs. The MPA accompanied an intensified turtle conservation program, as a consortium of agencies operating from Berau's capital Tanjung Redeb took charge of studying, monitoring, and protecting the turtle population in the area. Many Bajau regarded this as a harmful and unjust territorialization of their living spaces, including their cultural and economic traditions relating to sea turtles.[25] Referred to as the "turtle problem" (*masalah penyu*), the intrusive nature of intensified turtle protection came to stand for everything that was wrong with marine conservation and the MPA, as the words of a Bajau captain illustrate:

> Why should we listen to them? Imagine! Suddenly they turn up; people from far away, who have never lived here, whom I have never been introduced to, come all the way here to forbid us to take turtle eggs. They don't know anything about us. They did all these studies here, investigating the turtles, the corals ... but didn't ask us. And they are suddenly telling us

what we should or shouldn't do. They forbid us to do what we used to do since the time of our ancestors.[26]

With the "practice since the time of ancestors" the captain refers to the privileged access that the Bajau have enjoyed in Berau for at least a century. The collection and trade of sea turtle eggs was commercialized in the region circa 1876–1882 under the reign of Sultan Hasanuddin.[27] With a livelihood based on fishing, bartering, and trading, the Bajau families in Berau enjoyed exclusive rights to collect and trade sea turtle eggs under the protection of the sultanate. This arrangement further strengthened the central role the Bajau already played in the trade and exchange of valuables, including turtle eggs, across the sea,[28] attracting Bajau families from Malaysia and the Philippines to East Kalimantan.[29]

Subsequently, the movements of sea turtles helped shape the southward expansion of Bajau worlds, creating overlapping living spaces of Bajau and sea turtle communities.[30] By regency regulation of 1880, the right to manage the "turtle egg islands" was auctioned to entrepreneurs (*punggawa*) to which the Bajau egg collectors paid tribute. This system continued during 1901–1945, when the Dutch put the auctioning of the turtle eggs under their colonial administration. This allocated the lease (*pachterschap* in Dutch) of the collection and trade of turtle eggs on Berau's islands to Bajau families through customary management. It thereby sustained the Bajau long-held de facto monopoly on this livelihood practice in the wider maritime region.[31]

This customary management stipulated that ten percent of the turtle eggs collected from a nest were set aside to hatch. The baby turtles were kept in basins on the islands for three months and then set free.[32] After Indonesian independence, the district government continued the auctioning arrangement that included the care system for hatching part of the eggs.[33] As a former egg collector explained: "Before the NGOs came, we had a system to take care of the sea turtles. We took only part of the eggs. Every tenth nest we found; we saved that one. We took care of it, and released the baby turtles."[34] On Balikukup Island, an elderly Bajau woman remarked: "The men, they fish. The eggs and the clams, that's our work. We used to collect the eggs from *Pulau Telur* (Mataha), but the guards won't let us anymore."[35] When asked what she thought of the idea of taking care of sea turtles, she replied:

> That's just the thing. In the past, we women had a system of taking care of the turtles. We took in part of the baby turtles, and brought them up for three months. I was quite busy with it! ... The fisheries office would pay us 30,000 IDR [Indonesian Rupiah] for every turtle's release. This stopped when the NGOs came in. We used to be part of conservation, now we are not anymore.[36]

The history of customary management by the Bajau has shaped political and kinship alliances in the wider region. In Berau, the sea turtle concession had

FIGURE 12.3 A woman in a Bajau village in Berau offers boiled sea turtle eggs. Image by Annet Pauwelussen, 2009.

been firmly based in Bajau family networks. The last famous turtle *pachter*, who held the lease from 1994 to 2006 was the man known far and wide as "Haji Penyu" (Turtle Haji), who, through his extensive Bajau kinship network, was a widely respected businessman and Bajau patron. He also enjoyed extensive political alliances in the district government of Berau, including its Department of Fisheries and Marine Affairs.[37] This created a strong network of resistance to the enactment of the turtle egg ban along the coast of Berau. This historical and political context matters for the way in which eating turtle eggs is legitimized or criminalized in present-day regulations. Considered from a situated perspective of Bajau living in the coastal zone of Berau, recent interventions to separate the Bajau from turtles and their eggs, is by many considered an illegitimate, ineffective, and harmful intervention. The historical context of formal government regulations which allowed Bajau to legally collect turtle eggs legitimizes their perspective until today. At the same time, it highlights the contradictory nature of Bajau's historical legalities and today's national and international conservation interventions, helping to explain why and how Berau evolved into a site of resistance against international and national species protection laws.

Bajau Acts of Resistance and Persistence

In the years following the designation of the Berau MPA, resistance to the turtle egg ban took different forms. As the previous section showed, some took the form of verbal resistance or even outrage, in which the legitimacy and effectiveness of the turtle laws—and by extension conservation regulations—were questioned. Yet most Bajau resistance has taken the form of what James C. Scott has referred as "everyday resistance": the non-compliance, feigned ignorance, sabotage, and other ways in which those lacking formal positions of power resist territorializing state interventions.[38]

This kind of resistance is illustrated by a 2012 journey made by the first author with a Bajau trader selling valuables between Berau and the Malaysian town Tawau.[39] Twice a month, the trader sent her boat loaded with fish and dried clams from Berau to Tawau. Returning from Malaysia, her boat imported a range of goods on order, including sea turtle eggs, which she only began importing once the ban on collecting eggs in Berau was enforced. While her Bajau boat crew—all kin—moved her load through the coastal trading frontier of Northeastern Borneo, the tradeswoman moved along separately, taking care of business with buyers and suppliers *en route*. The following fieldnotes narrate a stressful moment in Tarakan—along the way from Tawau back to Berau—when the load was being detained:

> Ibu (Mrs.) T is pacing around, mobile phone in hand. Her turtle egg supplier just called from Malaysia: apparently, her boat is held at customs in Tawau. "The turtle eggs are safe," Ibu T says to me, visibly relieved. She explains that luckily the egg supplier was late this time, she (the supplier) just arrived when the border police was busy inspecting Ibu T's boat at Tawau's harbor, detaining boat and crew as permit documentation was not in order. Ibu T continues pacing around, now calling her uncle in Tawau who is a government official, to solve the situation with the police.[40]

At the time of writing these fieldnotes, the first author and Ibu T had already left Tawau the day before, crossing the Indonesian border to move ahead of the load. Meanwhile, they were staying with T's brother in "Kampung Bajau"—a slum-like stilt house quarter in island-city Tarakan, a regular stop-over. Here, Ibu T was waiting for the turtle eggs to arrive.

> "Things have changed," T's brother explains while Ibu T is on the phone again. "In the past, it was easy to get turtle eggs from Berau, it was one of the main turtle egg trading centers of southeast Asia. But nowadays, it's very hard to get turtle eggs from Berau. We have to import them from Semporna now [Sabah, Malaysia]."[41] Ibu T worries over the eggs … After five to six days, the colour of the eggs changes from white to yellowish. "They can expire, the price will plummet."[42]

Some of the eggs Ibu T imported at that time were for a local seller in Tarakan. The other part she planned to bring back to Berau, where her cousin had ordered the eggs for a wedding ceremony.

"They count on me. A wedding without turtle eggs is not a Bajau wedding," she says. She calls the egg supplier from Semporna again and orders her to hire a speedboat—"I'll pay you next time!"—to transport the eggs to us immediately. Ibu T gives instructions: "[U]se cardboard boxes and send them to the Tarakan speed boat terminal. Please make sure to tie the boxes carefully, so they won't slide. Have you paid the police already?"[43]

The main issue for Ibu T was to get the eggs safely across the border between Malaysia and Indonesia and the security forces active there. Ibu T dictated the exact route for the speedboat, over a river flanked by thick mangrove forests, where her son was stationed as police officer at the time. As the first author wrote in her notebook:

> Restrained excitement when the boxes arrive the next day in Tarakan; filled with black plastic bags, each containing 55 turtle eggs, still covered with sand from the beaches where they were dug out. There are over 30 bags, totaling around 1,650 eggs. Ibu T opens several of the bags to check if they are undamaged. She then proceeds to unpack the bags. With great care, she inspects the eggs with eyes and fingers, sometimes smelling the eggs, after which she puts them into two separate boxes. One box goes to the Tarakan trader, another box stays with us, and is moved into my bedroom. Tonight, I will sleep with 500 turtle eggs.[44]

Bajau traders like Ibu T skillfully navigate a dynamic, plural, and spatially dispersed lawscape to continue eating and exchanging turtle eggs throughout Bajau kinship networks. In line with pluralist approaches in legal anthropology, this shows that while nation-states can make formal laws, the extent to which these are enacted in practice is conditioned by how people "on the ground" and "at sea" understand, value, and "work with" them.[45] Among Southeast Asia's peasant communities, the "right to subsistence" and the "norms of reciprocity" often precede formal rules of resource use and access.[46] This is especially the case where enforcement is sparse and intermittent, and mediated by patrons and officials who themselves may prioritize the unwritten rules of being loyal to kinship over formal procedures.[47] Ibu T knew her practice was illegalized, but she also considered it legitimate to trade eggs when they were for a wedding. In Ibu T's view, supplying eggs is about more than profit: it also serves the stability and survival of long-standing Bajau cultural traditions that she felt have been (unjustly) disregarded by terrestrial Indonesian society.

Importing the eggs from other places that are less protected, she also builds on—and sustains—Bajau alliances of kinship and trade that have taken shape over centuries of sea-based movements. By engaging with the law while simultaneously flouting it, Ibu T in essence sustains an alternative legality based on customary laws evolved in the practice of turtle egg trade in Eastern Kalimantan over the past centuries, and embedded in a colonial state legal system. This effectuates a situation of legal pluralism—a parallel legal system as a living legal realm intersecting with formal or written legal systems outlined in (currently enacted or prevailing) official documents.[48] Still, it is important to point out that while the coexistence of different legal systems leads to a situation of legal complexity, this need not necessarily lead to contestation in practice. Different narratives of what is just or legitimate may be spatially distributed. This becomes clear once we shift focus to the ways Bajau engage with sea turtles in other places in Indonesia, where "turtle bans" have not been enforced.

Bajau enclaves in Sulawesi archipelagos, Masalima and Banggai, for example, have continued hunting and eating turtles, as well as using their shells for jewelry and gifts, in relative indifference to formal wildlife laws and policies. In Masalima it is common for Bajau communities to eat the meat of green sea turtles and trade part of it to Makassar and Bali, where it is in demand as a delicacy and ceremonial food. Hunting turtles is an acquired skill for Bajau fishers in Masalima, for which they use special gear to catch and pull the turtles to the boat. Turtle meat barbeques in Masalima are usually lively gatherings during which this protein-rich food is shared with family and friends.

Similarly, in the Banggai Archipelago of Central Sulawesi, the second author often observed fishing families catching green and hawksbill sea turtles and rearing them to a larger size in net pens, known as "*karambas*" beneath their homes to either sell or use for their meat and shells. They would often cook the meat into a spicy curry to be shared with friends and family for special occasions, while they would boil the shell, shape it, and carve it into bracelets, pendants, rings, and other jewelry and adornments. In addition to these uses, live baby turtles would be gifted to the young children of families. For example, the second author was out at sea with a line fisherman who primarily caught snapper for the food fish market, when he happened upon a baby green sea turtle swimming at the surface of the open ocean. He angled the boat toward the turtle and casually scooped it up with his hands. He then filled his "*gabus*" (Styrofoam container) with sea water and placed the turtle inside. Upon returning to his village, he gifted the turtle to his cousin's five-year-old son. The turtle quickly became a focal point of the "*dusun*" (neighborhood), drawing extended family and friends to stop by and feed and play with the turtle, until the boy's mother released the turtle back to the sea a month later. These examples show again the diversity of human–turtle entanglements that extend beyond eating practices to include carving, gift-exchanging, playing and feeding. As such, they sustain social ties amongst the Bajau, as well as between Bajau communities and sea turtles.

When compared to Berau, the way Bajau in Masalima and Banggai sustain their turtle engagement in relative indifference to official wildlife laws is partly explained by the local absence or silent support of government officials or conservation managers. It also shows the situatedness and historical (and colonial) entrenchment of legal disputes around marine conservation.[49] Selective and localized enforcement of "global" turtle laws can add to the perception of unfair intrusion in customary affairs in Berau.[50]

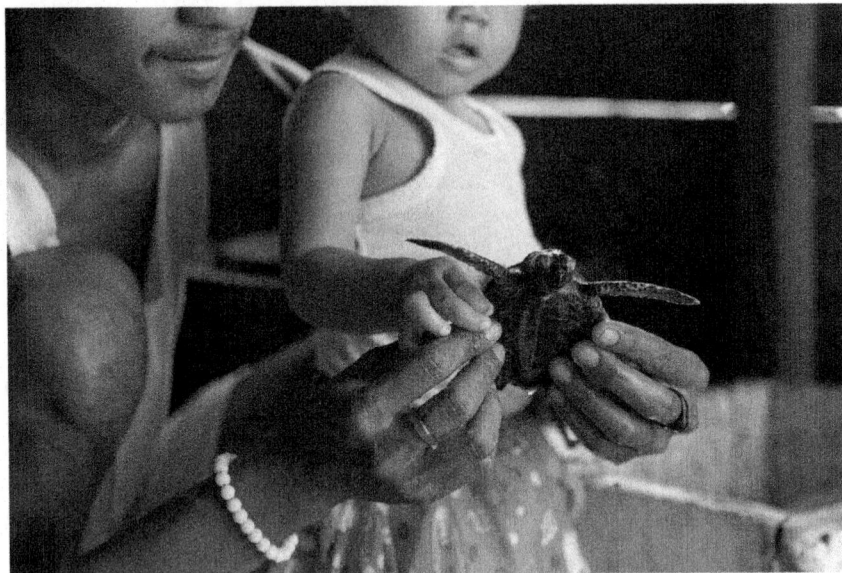

FIGURE 12.4 A fisherman came across this baby green turtle at sea and brought it home as a gift for his cousin's son. The father and daughter holding the turtle in the image are uncle and cousin to the boy to whom the turtle was gifted. The two rings and bracelet the father is wearing are made from the shell of a hawksbill turtle. Image by Shannon Switzer Swanson, 2019.

Still, it becomes clear how such customary or Indigenous legalities are themselves shaped or partly invented in interaction with different (colonial) legal systems with which the Bajau have come to deal.[51] In this, Bajau people navigate different legal orderings while also sustaining their own narratives of what is right and just in how they relate to turtles and each other. Both in Berau and Banggai the Bajau refer to long-standing cultural practices that tie them to sea turtles. Yet while the Bajau in Berau emphasized historical customary rights, the Bajau in Banggai generally believed they were exempt from turtle protection laws due to their self-identification as an Indigenous group.[52] And while the Bajau in Berau became involved in a partly commercialized system of egg trade mediated by entrepreneurs, the Bajau in Banggai mostly engage with sea turtles on a more casual basis, outside market exchanges.

These differing relationships problematize a clear delineation of the Bajau as one group with a primordial normative system that stipulates exactly how the Bajau should engage with sea turtles. At the same time, the sea turtle, in its different forms and qualities, reappears in Bajau practices and narratives as a symbol and agent of reciprocity and gift-giving between and within kinship circles. Through these practices, kinship and community are extended to include sea turtles as an inherent part of social life. As a former egg collector in Berau remarked: "They are part of our identity and livelihood ... They are part of our community here."[53] The sea turtles also shape Bajau bodies by providing a critical source of daily protein and micronutrients.[54] So, the Bajau persist in eating, playing with, and using turtles—not just out of defiance, but because these ways of engaging with turtles are inherent threads in the more-than-human fabric that weaves and holds together their relationship with each other and the sea.[55]

What Is a Good Human–Turtle Relationship?

In the previous sections, Bajau relational and embodied approaches to engaging with turtles appear to conflict with the "turtle protection" approach of overlapping conservation institutions. While this might give the impression of a singular "turtle protection" approach, these institutions themselves may build on different perspectives of what good turtle protection is. In 2012, a three-day conservation workshop brought together conservation partners, government agencies, and community representatives around the issue of protecting species and areas in the Berau coastal zone. Regarding the protection of turtles and their eggs, these different participants were considerably divided in what kind of human–turtle relations they saw as good and legitimate. At the beginning of a session dedicated to species protection, a forestry officer started by outlining the laws on the protection of endangered species, emphasizing the ban on using any part of sea turtles for consumption and commercial purposes. During the presentation, a young fisheries officer stood up and asked:

> I would like to know how Forestry is going to seek connection with the Bajau communities here. It's not logical for them that one species needs to be strictly "protected," let alone one that is so important to them. That feels like discrimination. The problem is, the more rigorous you enforce these bans, the less these people are willing to collaborate. They will do it secretly. It's like closing the door on the local population here.[56]

The forestry officer responded that, of course, conservation has to be community-supported. The fisheries officer interrupted:

> Yes, but how will you engage the communities in turtle protection? They have a right to know this. They were on our side, with the tenure system. But what now? Will there be a new form of collaboration?[57]

The forestry officer responded:

> We are a technical body. Our work is to keep safe the endangered species listed in Law 5 from the year 1990. We think the Bajau can be involved through ecotourism; guiding tourists to watch turtles. If they just stop taking the eggs, they will reap the benefits in the long run. If there are a lot of sea turtles, there will also be a lot of tourists.[58]

Through these excerpts from a workshop discussion, it becomes clear how agents of the departments of forestry and fisheries talk about turtles in different terms, in turn informing different ethics of, or approaches to, taking care of sea turtles. The fisheries department sees turtles as "fish," as a "resource" that needs to be sustainably managed, preferably in collaboration with fishers and local communities. Their staff is local, partly Bajau, and they feel caught between Bajau community interests and conservation interests introduced and lobbied by WWF. They are in charge of the management of marine resources and the MPA for which they need to follow guidelines and decentralized fisheries laws on community-based conservation that offer room for customary and Indigenous legal systems.[59]

In contrast, the forestry department (BKSDA) sees turtles as an endangered species that needs strict protection. They enact a species protection law that is superior to the regional laws on which the fisheries department base their community-based management policies. Another forestry officer has explained:

> You could think that with the new MPA we have to hand over our conservation tasks to the fisheries office. But we don't feel they are up to it. Their interest is with fisheries. But sea turtles aren't fish, they are reptiles. They can make up new laws in fisheries, but with turtles we will work with Law No. 5 from the year 1990. This law is still operational. And as long as it is, and [Fisheries] have not come up with a clear plan for how they are going to protect these poor animals, we will keep to our forestry law. The list is based on CITES. This is not just Indonesian law; it is international law.[60]

The protectionist perspective of the BKSDA classifies sea turtles as vulnerable reptiles and celebrity megafauna in need of care in the form of protection from human interference. Their view aligns with that of the WWF and the Turtle Foundation, but not necessarily with the ecosystem approach of TNC in which turtles are rather one node in a complex and interactive web that includes social and ecological elements. During the first author's fieldwork in 2011–2013, this difference in priorities led to repeated tensions in the collaboration between the WWF and TNC in Berau. Their conflicting approaches show again how sea turtles are particularly "troubling" legal objects because of their mobile and amphibious nature. They transgress administrative boundaries between land and sea and the corresponding governmental departments and law enforcement agencies. Moreover, sea turtles also express a certain "amphibiousness" as plural objects (and subjects), engaging different yet intersecting ontological and normative systems.[61] This also troubles any neat division between "Bajau" and "conservationist" legalities by pluralizing perspectives in and between.

These different perspectives illustrate different approaches in sociolegal theory and methodology regarding the role of nonhuman animals in the constitution of law. The approach of both the fisheries department and TNC gives room for legal pluralism in turtle conservation, allowing human–nature interference as long as resources are sustainably managed. They correspond with ecological approaches to environmental governance that acknowledges human interdependence with ecosystems. Fisheries governance usually classifies and manages animals in terms of property, acknowledging legal pluralism in its management. The approach of the forestry department and of the WWF and the Turtle Foundation is one of species protection prioritizing sea turtles as charismatic animals in need of saving. In this, it aligns more with current legal debates around the rights and well-being of animals, that orient toward common ownership in the form of a "global custodianship" shared across geographies and institutions.[62]

Still, both approaches render the nonhuman passive, disregarding the diverse and embodied ways in which human–nonhuman entanglement takes shape, for example with dogs, trees, and corals.[63] Advocating the interests of the nonhuman also requires critical reflection on the politics involved in the way (certain) animals come to be objects in legal systems and the classification practices involved. The classification of turtles as "endangered" prioritizes this species over others based on values of rarity and vulnerability, and there are politics involved in how certain humans and organizations get to define what animal is worthy of being protected, and what kind of human–animal entanglement this protection enables and reinforces.[64] Through these acts, saving turtle lives becomes a matter of biopolitics, inscribing a particular calculus on which life is (more) worth saving and through which means.[65] Reflecting on such unequal relations of power configured into what and whose definition and valuation of marine turtles is

FIGURE 12.5 A spearfisherman encounters a large green sea turtle while out compressor diving in the Banggai Archipelago, Indonesia. He uses the tip of his spear to nudge the turtle, but then lets it go on its way. Though not necessarily targeting turtles, fishers share the same spaces with them while on and in the ocean every day. Thus, they learn sea turtle behavior, habits, and preferences. Image by Shannon Switzer Swanson, 2019.

prioritized begs the question: is there room for Bajau-turtle entanglement and care practices on Bajau terms? In light of the asymmetric relation between human and animal in which wildlife protection is enacted, the case of the Bajau highlights the underlying politics that determine who gets to define what is a good human–turtle relation in the legally pluralistic seascape of Indonesia.

Lawful Injustice and Ethics of Care

Species and habitat protection laws have intervened in and shaped human–turtle relations, not only in Indonesia, but also in locales across the globe. International conventions have spawned species protection laws that criminalize local and Indigenous people's long-standing practices of eating and living with sea turtles.[66] This chapter shows how, on the basis of historical practice, endangered species bans are resisted as violent intrusions, but also how they are circumvented or accommodated in everyday practice. Political ecologist Lisa Campbell has found that among conservationists the notion of sea turtles as a global resource in which "everybody on the planet has a stake" effectively pushes aside the customary claims of the human communities whose daily lives and livelihoods are most affected.[67] While privileging turtle habitats, mobility, and wellbeing over that of (certain) human ones, this turtle-as-a-global-commons perspective fails to acknowledge the structural inequities between different kinds of human communities. It also ignores the biopolitics involved in the classification of certain animals as worth saving, making it still necessary to invoke Donna Harraway's entreaty to ask "what counts as nature, for whom, and at what cost?"[68]

With these imbalanced geo-political power dynamics, it has fallen in the hands of local communities to "exempt" themselves from these laws either through proper legal channels or by practicing legal pluralism in creative ways. For example, Aboriginal communities in northern Australia successfully fought for their traditional rights to use and eat sea turtles.[69] Similarly, in Ostional Costa Rica, local communities were able to keep access to Olive Ridley turtle eggs.[70] By contrast, the Bajau have not enjoyed the exceptional status of Indigenous or traditional (adat)[71] communities in Indonesia.[72] Instead, the sea-oriented Bajau have been, and often still are, perceived by the Indonesian state as outlaws whose lack of attachment to a land-based territory forecloses their ability to claim indigeneity.[73] This makes the Bajau doubly excluded. Ironically, while the wide-ranging habits of sea turtles have afforded them international protection, the mobile and sea-based lifestyle of the Bajau seemingly has the opposite effect, historically preventing them from organizing to be formally acknowledged as a cohesive Indigenous group with customary rights.[74]

Acknowledging the plight of the Bajau and their claims to turtles should not necessarily lead to the romantic notion of them being "ecologically noble"[75] and morally superior. In Berau, turtle egg exploitation arguably evolved into a commercial business operated by a regional (Bajau) elite to the detriment of the sea turtle population. Severing the Bajau from their involvement with sea turtles by conservation intervention does, however, ask for a critical reflection on the harmful effects (for humans and turtles) of the alienation process that wildlife bans can generate, particularly when based on a modernist separation of humans from nature.[76] In Banggai and Masalima, where bans are evidently not enforced, Bajau families eat and use turtles as part of a wider notion of reciprocity, acknowledging their spiritually and materially entangled worlds—perhaps a case of what Susan Reid refers to as being "a more ethical predator."[77] By contrast, in Berau angered Bajau men are now frequently reported to catch egg-bearing turtles before they come on land and cut open their bellies to take their eggs. The system of caring for ten percent of the eggs and baby turtles is no longer in place, which Bajau in the area claim has resulted in a more rapid deterioration of turtle populations in Berau since NGO intervention. Although the effects of different conservation and care systems on Berau's turtle population are contested, many Bajau families feel that—since the Bajau no longer care for the turtles—the animals may no longer come back. While the movements of turtles have shaped Bajau living spaces, the Bajau in turn have also shaped the living spaces of sea turtles through life-supporting care practices.

How can we engage productively and politically with such more-than-human entanglement and "conviviality"[78] through a legal lens? Legal pluralism scholarship has shown how legal and normative systems are diverse, overlapping, and contesting, and how people like the Bajau skillfully navigate their in-between spaces. But this scholarship has yet to include the nonhuman as taking an active part in the social-legal world. In response to the anthropocentric and humanist basis of legal systems, several approaches have enriched sociolegal scholarship over the last decades by explicitly considering nonhuman animals as legal persons and subjects into the constitution of law.[79] Notable examples are the attribution of rights or legal personhood to nonhuman animals, and the rights of nature approach, delivering "judicial protection of nature for the sake of nature itself."[80] Still, in the practice of marine wildlife management, the "sake of nature" is selectively represented by only certain human persons or organizations,[81] and is usually embedded in asymmetric power relations between them.

So, while these approaches allow animals to enter the legal scene as "subjects," there are still vital biopolitical questions of what nonhuman animals are deemed worthy of being given a right or personhood, and on whose terms? Or in our case, more specifically, whose version of what counts as good—or harmful—human–turtle relations and care practices. Where conservation perspectives may consider eating turtles as a harmful interference in species wellbeing and survival, for the Bajau, harm is in the alienation between humans and turtles. Thus, their embodied and social practice of eating and sharing turtles and their eggs constitutes a social and moral world in which both humans and turtles participate as beings or subjects, and sea turtles co-constitute Bajau life and wellbeing.

Human–turtle relations in Indonesia could then—after Zoe Todd—be seen as troubled "sites of engagement,"[82] narratives and embodied practices in which different enacted versions and ethics of human–turtle relations intersect. This creates or sustains a situation of legal pluralism wherein we find not only the coexistence of different legalities, but also the intersection of different ethics of care.[83] Following human–turtle relations as historical and contemporary sites of engagement involving Bajau, NGO and governmental agents, as well as Dutch colonial administrators, the case of turtle conservation in Indonesia shows how notions of justice pertaining to turtles bring the turtle itself as both an object and a subject into the picture. Acknowledging the entanglement of human and sea turtle agency can enrich the discussion of marine legalities to engage more reflexively with the question of what, when, and where is a good and just human–turtle relation in situated practices? Taking seriously such ontological and ethical pluralism undergirding legal contestations in wildlife protection is vital for understanding the complexity of disputes around marine wildlife, and a first step toward fostering coexistence or dialogue across different legalities with a potential for mutual support.

Envisioning such more-than-human or "lively"[84] legalities may require a radical re-examination of identifying with self and other, as Gumbs does, for example, in her anthology that explores ways of relating to marine mammals as kin.[85] Ethnographic, feminist, and Indigenous perspectives and critiques can provide inspiration to do such radical re-examination in legal scholarship, while we acknowledge the need to do so carefully so intellectual inspiration does not lead to appropriation. As a way forward we suggest that a careful and respectful exchange between feminist, Indigenous, and legal pluralist perspectives that take in situated legalities can facilitate an "enriched understanding of interspecies relations, the lives of animals and humans, as well as broader societal relations of power" in legal scholarship and wildlife protection.[86] Doing so can help the various groups engaging with sea turtles to "stay with the trouble,"[87] with the murkiness and entanglements of human–turtle relations, to gain a more nuanced and enriched understanding of their historically embedded ethics of care and justice.

Acknowledgments

Both authors would like to thank the many people who took the time to share their invaluable thoughts and opinions with them both during fieldwork and in writing this chapter. They are grateful to their families for their faithful support through long absences and late nights of work, and to Irus Braverman, Leontine Visser, and Veerle Boekestijn for their thoughtful feedback on earlier versions of the chapter. The first author would also like to thank the Wageningen School of Social Sciences and the Nippon Foundation Ocean Nexus Center at the University of Washington for funding and supporting her research. The second author thanks the National Geographic Society, the Stanford School of Earth, and the Emmett Interdisciplinary Program in Environment and Resources for funding her research, as well as the Indonesian Nature Foundation (LINI) for logistical and moral support in the field.

Notes

1 Autonyms vary by region and context. In the Banggai Archipelago, when speaking with non-Bajau, they refer to themselves as Bajo; however, when speaking amongst themselves, they refer to themselves as Sama—like in the southern Philippines, the Sulu Archipelago, and elsewhere in North Sulawesi; Celia Lowe, *Wild Profusion: Biodiversity Conservation in an Indonesian Archipelago* (Princeton: Princeton University Press, 2006). In Berau they refer to themselves in both cases as Bajau. For simplification, and because most of the ethnography is based on Berau, we use Bajau in this chapter.

2 Annet Pauwelussen, "The Moves of a Bajau Middlewoman: Exploring the Disparity Between Trade Networks and Marine Conservation," *Anthropological Forum* 25, no. 4 (2015): 329–349. Clifford Sather, *The Bajau Laut: Adaptation, History, and Fate in a Maritime Fishing Society* (Kuala Lumpur: Oxford University Press, 1997).

3 Rini Kusumawati, "Networks and Knowledge at the Interface: Governing the Coast of East Kalimantan, Indonesia" (PhD dissertation, Wageningen University, 2014); Arlo Nimmo, *The Sea People of Sulu: A Study of Social Change in the Philippines* (San Francisco: Chandler Publishing, 1972); Regina Woodrom Rudrud, "Forbidden Sea Turtles: Traditional Laws Pertaining to Sea Turtle Consumption in Polynesia (Including the Polynesian Outliers)," *Conservation and Society* 8 (2010): 84–97; Clifford Sather, "Commodity Trade, Gift Exchange, and the History of Maritime Nomadism in Southeastern Sabah," *Nomadic Peoples* 6, no. 1 (2002): 20–44.

4 Sather, "Commodity Trade"; Eric Tagliacozzo, "Navigating Communities: Race, Place, and Travel in the History of Maritime Southeast Asia," *Asian Ethnicity* 10, no. 2 (2009): 97–120; James Warren, *The Sulu Zone 1768–1898: The Dynamics of External Trade, Slavery, and Ethnicity in the Transformation of a Southeast Asian Maritime State* (Singapore: NUS Press, 2007).

5 Jack Frazier, "Marine Turtles: The Role of Flagship Species in Interactions Between People and the Sea," *Mast* 3, no 2. (2005): 5–38.

6 Imam Musthofa Zainudin et al., "Interaction of Sea Turtles with Indonesian Fisheries – Preliminary Findings," *Indian Ocean Turtle Newsletter* (2007): 1–10. Irawan Asaad et al., "Delineating Priority Areas for Marine Biodiversity Conservation in the Coral Triangle," *Biological Conservation* 222 (2018): 198–211.

7 See, e.g., discussions in the Introduction as well as Braverman and Reid, this volume.

8 Hector Barrios-Garrido et al., "Conflicts and Solutions Related to Marine Turtle Conservation Initiatives in the Caribbean Basis: Identifying New Challenges," *Ocean and Coastal Management* 171 (2019): 19–27; Jarina Mohd Jani et al., "To Ban or Not to Ban? Reviewing an Ongoing Dilemma on Sea Turtle Egg Trade in Terengganu, Malaysia," *Frontiers in Marine Science* 6, no. 762 (2020): 1–18.

9 Rutgerd Boelens et al., "Legal Complexity in the Analysis of Water Rights and Water Resources Management," in *Liquid Relations: Contested Water Rights and Legal Complexity*, eds. Dik Roth et al. (New Brunswick: Rutgers University Press, 2005), 6. Useful reviews of debates in legal pluralism are given by Paul Schiff Berman, "The New Legal Pluralism," *Annual Review of Law and Social Science* 5 (2009): 225–242 and (particularly for customary law) Keebet von Benda-Beckmann and Bertram Turner, "Legal Pluralism, Social Theory, and the State," *Journal of Legal Pluralism and Unofficial Law* 50, no. 3 (2018): 255–274.

10 Boaventura de Sousa Santos, "Law: A Map of Misreading. Toward a Postmodern Conception of Law," *Journal of Law and Society* 14 (1987): 279–302, 281.

11 Laurent Thévenot, "Which Road to Follow? The Moral Complexity of an 'Equipped' Humanity," in *Complexities: Social Studies of Knowledge Practices*, eds. John Law and Annemarie Mol (Durham: Duke University Press, 2002), 53–87.

12 Paul Schiff Berman, *Global Legal Pluralism: A Jurisprudence of Law Beyond Borders* (Cambridge: Cambridge University Press, 2012).

13 Janne R. Rohe et al., "A Legal Pluralism Perspective on Coastal Fisheries Governance in Two Pacific Island Countries," *Marine Policy* 100 (2019): 90–97; Joeri Scholten

and Maarten Bavinck, "Lessons for Legal Pluralism: Investigating the Challenges of Transboundary Fisheries Governance," *Current Opinion in Environmental Sustainability* (2014): 10–18.

14 Wendy A. Adams, "Human Subjects and Animal Objects: Animals as 'Other' in Law," *Journal of Animal Law and Ethics* 3 (2009): 29–51.

15 Ibid, 50; Zoe Todd, "Fish Pluralities: Human-Animal Relations and Sites of Engagement in Paulatuuq, Arctic Canada," *Inuit Studies* 38, nos. 1–2 (2014): 217–238.

16 For this argument's application to corals see Annet Pauwelussen and Gerard Verschoor, "Amphibious Encounters: Corals and People in Conservation Outreach in Indonesia," *Engaging Science, Technology and Society* 3 (2017): 292–314, and to elephant conservation see Charis Thompson, "When Elephants Stand for Competing Philosophies of Nature: Amboseli National Park, Kenya," in *Complexities: Social Studies of Knowledge Practices*, eds. John Law and Annemarie Mol (Durham: Duke University Press, 2002), 166–190.

17 María Puig de la Bellacasa, *Matters of Care: Speculative Ethics in More Than Human Worlds* (Minneapolis: University of Minnesota Press, 2017); Todd, "Fish Pluralities."

18 Local staffmember, Turtle Foundation, personal communication with first author, Mataha, April 9, 2013.

19 Chris Wold, "The Status of Sea Turtles Under International Environmental Law and International Environmental Agreements," *Journal of International Wildlife Law and Policy* 5, nos. 1–2 (2002): 11–48; Barrios-Garrido, "Conflicts and Resolutions."

20 Zainudin, "Interaction of Sea Turtles"; Lalita Gomez and Kanitha Krishnasamy, "A Rapid Assessment on the Trade in Marine Turtles in Indonesia, Malaysia and Viet Nam," TRAFFIC, 2019, https://www.traffic.org/site/assets/files/12524/se-asia -marine-turtle-trade.pdf.

21 Ibid.

22 According to the Turtle Foundation, this plundering by the local population of Berau has decimated the sea turtle population to ten percent of what it was 70 years ago. "Program Indonesia: General Information and Background," Turtle Foundation, accessed October 14, 2021, https://www.turtle-foundation.org/en/program -indonesia/.

23 Two of these islands (Sanggalaki and Semama) have national park status under Forestry law. Budi Wiryawan et al., *Menuju Kawasan Konservasi Laut Berau, Kalimantan-Timur: Status Sumberdaya Pesisir dan Proses Pengembangan* (Jakarta: TNC-WWF-Mitra Pesisis/ CRMP II USAID, 2005).

24 Bambang Gunawan, "Shrimp Fisheries and Aquaculture: Making a Living in the Coastal Frontier of Berau, Indonesia" (PhD dissertation, Wageningen University, 2012); Rini Kusumawati and Leontine Visser, "Collaboration or Contention? Decentralised Marine Governance in Berau," *Anthropological Forum* 24, no. 1 (2014): 21–46.

25 Kusumawati, "Networks and Knowledge"; Rini Kusumawati and Leontine Visser, "Capturing the Elite in Marine Conservation in Northeast Kalimantan," *Human Ecology* 44 (2016): 305–307.

26 Bajau captain, personal conversation with first author, Balikukup, February 21, 2012.

27 Kusumawati, "Networks and Knowledge," 59, 67.

28 Tagliacozzo, "Navigating Communities"; Warren, "The Sulu Zone," and specifically for Berau: Annet Pauwelussen, "Community as Network: Exploring a Relational Approach to Social Resilience in Indonesia," *Maritime Studies* 15, no. 1 (2016): 1–19.

29 C. Ensing, "Memorie van Overgave van de onderafdeling Beraoe," *Koninklijk Instituut voor de Tropen: Memories van Overgave 1852–1962* 70 (1937): 1078.

30 Nimmo, *The Sea People of Sulu*; Sather, "Commodity Trade"; as also shown for another region in Indonesia by Lance Nolde, "'Great is Our Relationship with the Sea': Chartering the Maritime Realm of the Sama of Southeast Sulawesi, Indonesia," *Explorations* 9 (2009): 15–33.

31 Ensing, "Memorie van Overgave"; Kusumawati, "Networks and Knowledge."
32 This care overlaps with Susan Reid's notion of being "a more ethical predator" as the Bajau seem to acknowledge that "eating the other entails taking in, at least partially, the worlds that constitute them" this volume, 81.
33 Ibid.
34 Former egg collector, conversation with first author, Tanjung Batu, June 10, 2012.
35 Bajau fisherwoman, conversation with first author, Balikukup, February 19, 2012.
36 Ibid.
37 Based on multiple conversations during fieldwork. Kusumawati, "Networks and Knowledge" and Kusumawati and Visser, "Capturing the Elite" provide detailed accounts of the role of Haji Penyu and the regional elite in marine governance in Berau. Until turtle egg collection was formally banned in 2006, the turtle egg business was the primary revenue resource (80 percent) for the Berau district government, which explains the reluctance even within Berau government to enforce the ban.
38 James C. Scott, *Weapons of the Weak: Everyday Forms of Peasant Resistance* (New Haven: Yale University Press, 1985).
39 Pauwelussen, "Moves of a Bajau Middlewoman."
40 Annet Pauwelussen, fieldnotes, Tarakan, June 30, 2012.
41 See Figure 12.2.
42 Pauwelussen, fieldnotes.
43 Ibid.
44 Ibid.
45 Benda-Beckmann and Turner, "Legal Pluralism."
46 James C. Scott, *The Moral Economy of the Peasant. Rebellion and Subsistence in Southeast Asia* (New Haven: Yale University Press, 1976).
47 Pauwelussen, "Community as Network" and "Moves of a Bajau Middlewoman."
48 Sociolegal Scholar Eugene Ehrlich conceptualized this as "living law." For a discussion of this concept and how it has been developed in later writing, see David Nelken, "Eugen Ehrlich, Living Law, and Plural Legalities," *Theoretical Inquiries in Law* 9, no. 2 (2008): 443–471.
49 Christine Walley, *Rough Waters: Nature and Development in an East African Marine Park* (Princeton: Princeton University Press, 2004) shows this colonial embeddedness of MPAs in an East African context. See also Paige West, *Conservation is our Government Now* (Durham: Duke University Press, 2006) and Dan Brockington et al., *Nature Unbound: Conservation, Capitalism and the Future of Protected Areas* (London: Routledge, 2008) for relevant critiques of conservation.
50 Due to the sea-based mobility and connectivity of Bajau worlds, people usually have detailed knowledge of such local differences in law enforcement.
51 This interwoven nature of customary law with colonial state law is central to discussions in legal pluralism: Benda-Beckmann and Turner, "Legal Pluralism."
52 Noted from second author's in-person conversations with multiple members of fishing families in the Banggai Archipelago during fieldwork.
53 Former egg collector, personal conversation.
54 Martin Quaas et al., "Fishing for Proteins" WWF Germany, 2017, https://www.researchgate.net/publication/328428806_Fishing_for_Proteins_How_marine_fisheries_impact_on_global_food_security_up_to_2050_A_global_prognosis.
55 See Pauwelussen and Verschoor, "Amphibious Encounters" for this argument related to coral reefs. For the way sea turtles figure as expression of life and consciousness in situated and Indigenous oral histories see Frazier, "Marine Turtles."
56 Annet Pauwelussen, fieldnotes, Tanjung Redeb, February 7, 2012.
57 Ibid.
58 Ibid.
59 Pauwelussen, "Community as Network"; Gunawan, "Shrimp Fisheries"; Kusumawati, "Networks and Knowledge."
60 Forestry officer, conversation with first author, Tanjung Redeb, January 31, 2012.

61 Annet Pauwelussen, "Amphibious Anthropology: Engaging with Maritime Worlds in Indonesia" (PhD dissertation, Wageningen University, 2017); Pauwelussen and Verschoor, "Amphibious Encounters." This also resonates with Todd, "Fish Pluralities," 218, which shows the "slipperiness" of fish as beings, existing as "simultaneously different entities" that "challenge existing articulations of human-environment relationships."

62 Irus Braverman, "Law's Underdog: A Call for More-than-Human Legalities," *Annual Review of Law and Social Science* 14 (2018): 127–144, 133.

63 Eva Hayward, "FINGERYEYES: Impressions of Cup Corals," *Cultural Anthropology* 25, no. 4 (2010): 577–599; Eduardo Kohn, *How Forests Think. Towards an Anthropology Beyond the Human* (Berkeley: University of California Press, 2013).

64 For a discussion of how such classifications order human–nonhuman relations see Irus Braverman, "The Regulatory Life of Threatened Species Lists," in *Animals, Biopolitics, Law: Lively Legalities*, ed. Irus Braverman (London: Routledge, 2016), 19–36.

65 Braverman, "Law's Underdog," 139.

66 Lisa Campbell, "Use Them or Lose Them? Conservation and the Consumptive Use of Marine Turtle Eggs at Ostional, Costa Rica," *Environmental Conservation* 25 (1998): 305–319; Rudrud, "Forbidden Sea Turtles."

67 Lisa Campbell, "Local Conservation Practice and Global Discourse: A Political Ecology of Sea Turtle Conservation," *Annals of the Association of American Geographers* (2007): 313–334.

68 Donna Haraway, *Modest_Witness@Second_Millennium.FemaleMan_Meets_OncoMouse: Feminism and Technoscience* (New York: Routledge, 1997). See also Lowe, *Wild Profusion*.

69 Rod Kennett et al., "Indigenous Initiatives for Co-Management of Miyapunu/Sea Turtle," *Ecological Management and Restoration* 5 (2004): 159–166.

70 Campbell, "Use Them."

71 *Adat* is often described as customary law of Indigenous people in Indonesia; unwritten traditional code governing personal conduct as part of the community. The Bajau have been structurally excluded from claims to indigeneity and customary tenure due to their mobility at sea and lack of land ownership. See Suraya Abdulwahab and Celia Lowe, "Claiming Indigenous Community: Political Discourse and Natural Resource Rights in Indonesia," *Alternatives* 32, no. 1 (2007): 73–97.

72 We have not found any formal legal exceptions that allow turtle hunting or egg consumption in Indonesia. However, Michael de Alessi, "Archipelago of Gear: The Political Economy of Fisheries Management and Private Sustainable Fisheries Initiatives in Indonesia," *Asia and the Pacific Policy Studies* (2014): 576–589, discusses overlapping fisheries regulation in Indonesia and the recognition of local tenure.

73 Julian Clifton et al., "Statelessness and Conservation: Exploring the Implications of an International Governance Agenda," *Tilburg Law Review* 19, nos. 1–2 (2014): 81–89.

74 One possible exception is in the Torres Strait, detailed in Natasha Stacey, *Boats to Burn: Bajo Fishing Activity in the Australian Fishing Zone* (Canberra: ANU Press, 2007).

75 Hames Raymond, "The Ecologically Noble Savage Debate," *Annual Review of Anthropology* 36 (2007): 177–190.

76 See also Bram Büsscher and Robert Fletcher, *The Conservation Revolution: Radical Ideas for Saving Nature Beyond the Anthropocene* (London: Verso Books, 2020).

77 Reid, this volume, 81.

78 Ibid.

79 Braverman, "Law's Underdog."

80 Erin Daly, "The Ecuadorian Exemplar: The First Ever Vindications of Constitutional Rights of Nature," *Review of European, Comparative & International Environmental Law* 21, no. 11 (2012): 63–66, 63.

81 Erin L. O'Donnell and Julia Talbot-Jones, "Creating Legal Rights for Rivers: Lessons from Australia, New Zealand, and India," *Ecology and Society* 23, no. 1 (2018): 7.

82 Todd, "Fish Pluralities."

83 "Care is everything that is done (rather than everything that 'we' do) to maintain, continue, and re-pair 'the world' so that all (rather than 'we') can live in it as well as possible." Puig de la Bellacasa, *Matters of Care*; 103.

84 Irus Braverman, ed., *Animals, Biopolitics, Law: Lively Legalities* (London: Routledge, 2016).

85 Alexi Pauline Gumbs, *Undrowned: Black Feminist Lessons from Marine Mammals* (Chico: AK Press, 2020).

86 Alice J. Hovorka, "Feminism and Animals: Exploring Interspecies Relations Through Intersectionality, Performativity and Standpoint," *Gender, Place and Culture* 22, no. 1 (2015): 1–19.

87 Donna Haraway, *Staying with the Trouble: Making Kin in the Chthulucene* (Durham: Duke University Press, 2016).

Andreas Philippopoulos-Mihalopoulos

AFTERWORD

We Are All Complicit: Performing Law through Wavewriting

Andreas Philippopoulos-Mihalopoulos

Performing law is unexpectedly similar to performing water. We do not even need to *become* law or water. We *are* it. Our bodies are water,[1] our bodies are law.[2] They move together in conflict and confluence, a Spinozan parallelism, now they meet, now they don't, but they are always here, evolving in tandem. Laws and waters are co-extensive in "us"—us the humans, us the posthumans, us the never-just-humans, us the enlightened, us the slaves, us the animal, us the skin that does not separate, us the skin that discriminates, us the fear, us the dead, us the minor juris-prudence, the end of a planetary turn around an anthropocentric pivot.

In this short Afterword I carry on my explorations of the connection between law and water. I do this partly through a poetic type of writing I have called *wavewriting* (more on this below); and partly by reflecting on, or at least drawing from, a specific performance atmosphere. I gave that performance online, the finale of our extended meetings as a group that eventually led to this collection.[3] It was perhaps an eccentric addendum to the scholarly encounters, aiming at facilitating new synapses and confounding existing ones. I tried to put forth the agency of water and create an atmosphere of confluence yet guilty complicity, an action along the lines of Susan Reid's *material predation* (this volume) and Renisa Mawani's ocean as method, seeking to bring forth the blackness and brownness of the colonially striated waters.[4] But why would I not present an academic paper on ocean law? Because, since I also identify as an artist whose performance prac-tice often channels legal and other theory into embodied experience, I consider it my ethical responsibility to enable a confluence between the academic and the artistic. To show how the artistic can also be academic, and the academic can find other ways of reaching out. On a more personal note, art is my escape from law.

FIGURE A.1 *Slashing Waters*, an online performance for the Laws of the Sea workshop by Andreas Philippopoulos-Mihalopoulos, May 27, 2021. Screenshot 1 by author/performer.

DOI: 10.4324/9781003205173-14

Just like water used to be my escape from law. Because I genuinely thought that water cleanses the law, leaves law behind, remains untouched by the law. How deluded I was. And what a banal delusion at that.

My delusion was quite simply that law deals with land, territories, countries, and continents. But not with water. Not with smooth ocean spaces. I was not alone in thinking that of course. We know from Carl Schmitt that lines are carved on the earth, and that these lines signal the beginning of the law.[5] Like Schmitt, I thought that water, and especially oceanic water, escaped that. It couldn't be striated in the same way as land. And even when law deals with water, it does it in a land-based rationale. Cut, split, occupy. But water never stays still. Lines move, boundaries bleed, surfaces get swallowed up. Waves everywhere, lines nowhere.

Unlike Schmitt, I cherished this quality of the water. It was my way of keeping water away from my legal academic career and thought. I was happy I could shield it away from the law. I was painting it or sculpting it, performing it or conjuring it, installing it, and ingesting it. But none of that felt like law.

The first time I decided to deal with water academically was only recently, in a text on water from a law and literature perspective.[6] My initial intention was to show how different, indeed incongruous, the two were (a bit like the dam I used to keep between my legal theory research and my art practice). However, while researching on the topic and especially in terms of literary fiction, I encountered something very different. I saw that law is fundamentally aquatic, literally and metaphorically.[7] It can drown you as well as help you float. It has no hard edges. It wraps around and slides into bodies, making them move in certain ways. It is a shapeshifter.[8] Law dwells in water, using its surface tension to magnify itself yet also to diffuse itself to an imperceptible molecular consistency. Their frequent literary encounters reveal how water for law has always been the great unknown, some sort of legal unconscious, always within law yet inaccessible, laughably striated by tellurian lines of sovereignty, colonialism, and imperialism. Literature allows that access to the legal unconscious, the flooded Jungian basement swarming with all the affects that our legal institutions cannot or will not deal with. Water haunts the law. It deluges it uncontainably, flooding statutes and treaties with washed out bodies of refugees, melted icecaps, itinerant plastic islands, upturned bodies of fish, cruise ships of the damned that cannot dock anywhere for fear of spreading the virus,[9] stranded sailors onboard legally forsaken cargo ships gazing wistfully toward the nearby shore,[10] unborn babies of a black Atlantic moving beyond breath in a deep-sea redress.[11]

The performance I gave as part of the preparatory workshop for this volume was an emergence of this thought process: a slow, perhaps unwilling realization that water and law are confluent, if not co-emergent and indeed ontologically co-extensive. I was pulling down my shield and was finally allowing law and water to flow into each other; or, to put it differently, I was finally acquiescing to these risky nuptials between legal rationality as a conscious career choice, on the one hand, and art as an unconscious constant companion, on the other.

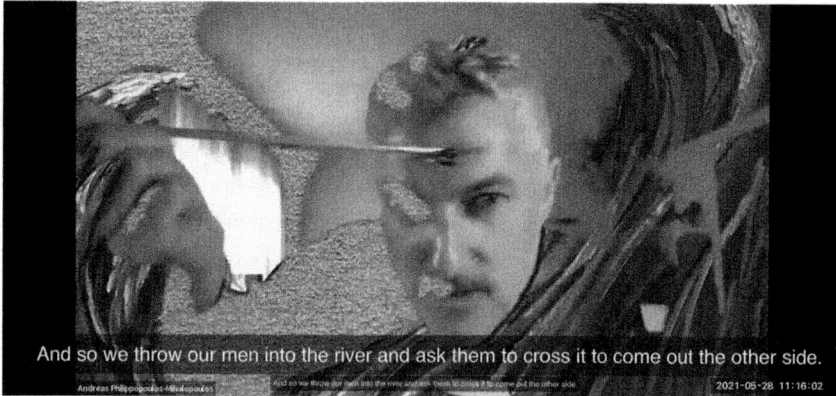

And so we throw our men into the river and ask them to cross it to come out the other side.

FIGURE A.2 *Slashing Waters*, an online performance for the Laws of the Sea workshop by Andreas Philippopoulos-Mihalopoulos, May 27, 2021. Screenshot 2 by author/performer.

Performing water has its own laws. Especially when done online: don't splash too much, avoid spilling onto the keyboard, keep the water visible. But also: do not explain, keep it unconscious, keep it sensorial and tactile. So even in water, one cannot get away from law. *Especially* in water one cannot get away from law. Let me therefore reverse my previous deluded assertion: on land, laws are everywhere, their omnipresence banalized, becoming white noise. In water, law becomes water itself: elusive, dissimulated, spread, layered, with a depth that belies its surface. I think I knew this when I started this journey of aquatic law, yet perennially postponed the time I would eventually have to bring together water and law—for fear of drowning no doubt. I thought I wanted to escape law via a waterway. But I also knew that one must bring the two together in a way that goes beyond striation, measurement, calculation. Rather than trying to fit one into the other, I needed to allow them to bleed into each other, Venetian colors that leak out of their Florentine outlines and become rapids, canals, lagunas.

Law does not know how to leak. Well, it does. But it doesn't admit to it.

We must listen to the water. There is a law, the standard law of the standard person, that imposes itself on water with abusive force. This law measures and traps water, drills through it, pollutes it, drains it, changes its route, fills it with plastic. We must listen to the water. Not measure it or trap it, not drill through it or analyze it. We have done enough of that, and it has led us here. We must listen to the water. Not romanticize and exoticize it. Not essentialize and beatify it. There is a law, the standard law of the standard person, that imposes its delusions on water with abusive force. This law separates us from the water, allows us to forget that we are all bodies of water, armors our skin with imaginary impermeability. We have done enough of that, and it has led us here. We must listen to the circularity of the abyssal and the heavenly,[12] the continuum between rivers and oceans, earth core water and interplanetary seas, our organs and the rains, the never breaking cycle of aquatic molecules.

We must listen to the water:

It is such a bore when you try to impose your sense of rhythm on me: ebb and flow, streams and currents, waves and tsunamis. You study my comings and goings and jot them down in order to catch them next time, or to not be caught unaware by me. You use me to relax or stimulate yourselves, you count waves to sleep, you count oceans as if they were human syncopated breaths. What annoys me most is your obsession with enclosing me, channelling me, controlling me, putting me in narrow waterways or large deposits, trapping me in reservoirs, framing me behind long dams or in tiny tubes. No, you don't understand—you don't hurt me, you cannot hurt me when you do these things. It is your attitude that irritates me, that bossy pitiable macho egomania with which you think you can control me, and your automatic tendency to measure me, weight, speed, density, frequency.

But human rhythms are so shallow I can barely hear them. They ride a different temporality than I do, a temporality of day and night, forgettably short seasons, unregistrable geological epochs. You even talk about my history and my composition in ways that make sense to you, and yeah fine, carry on, see if I care. You want me to obey the rhythm that you impose on me—see how you can never see beyond your little selves? —a breath of rising and falling, a living regularity because life is regular and water is life, right? Ok. But my breath is not of your life, or of any life. My breath is time. My breath is of a universe that hosts me in globular suspension between planets, in vast clouds of a rain that will never fall, in fathomless oceans suspended in space and floating about unsupported: my breath is there, rounded up in a water that you will never drink. My breath is polarized, spread across aeons, breathing in when nothing was impossible and breathing out when the possibilities will have shut forever.

Okay, you cannot conceive this. Let's focus on your planet, that hydrospheric apparition on stilts in the great hall of cosmic gossip. Even then, you think of me as percentages, oh wow so much of the surface of the earth, no really, so much of a human body, eye-rolling stuff. But just shut up and listen for a moment and you might just about understand that my breath is deeper and more cavernous than your deepest history. My breath is caught in the rattle of a dying sun, hidden under strata of a geology that ignores you, deep in the center of what you call your planet and with whose body you will never manage to sleep. My breath is liquidity in waiting, tangled with chunks of eternity.[13]

I now know that my real desire has always been to depart from the law and yet to carry on floating on the law. Is there anything more despicably, hilariously legal than wanting to do both? To open up the floodgate but just enough so that the law can control the flow? Isn't this what I do with my art and my legal scholarly writing too? Keep my fingers plunged in different pies?

Perhaps there is a somewhat nobler idea behind all this, a real ethical cry and not just a desire to escape (although nothing nobler than realizing the need to withdraw). Let me call this idea *wavewriting*. A writing of waves, on waves, with waves. A writing in waves. A writing that is erased by its own ebb. A writing that swirls, never connects, never concludes, does not offer a guilty or non-guilty end-point but only the unnegotiable certainty of an eternity of deferral. Wavewriting is a material metaphor to start with.[14] A tidalectics of diffraction.[15] But also, away from any figure of speech, an ocean methodology,[16] a seascape epistemology,[17] a wet ontology,[18] an amphibious legal geography,[19] an enquiry of submerged perspectives.[20] It is persistent, fathoming, layering, uncovering. Like Michel Serres's *la belle noiseuse*,[21] the repetitive ruckus of waves, the noise that reveals existence. It is also about a readiness to ride a certain wave. It is collective, and at ease with losing control and becoming one with the elemental. Do you see? How legal is indeed wavewriting! And how truly constant. For we never truly depart from the law, however much we might recede. All legal writing is wavewriting. But more than that. Every text is turning blue.[22] Every law is turning blue.[23]

This is not just a fad. "Despite international efforts and tireless research, there is no permanent solution—no barriers to erect or walls to build—that will protect us in the end from the drowning of the world as we know it."[24] This wave might well be the last. From water scarcity, droughts, and gigantic global fires to flooded cities, melting icecaps, submerging islands, and drowned states, water is becoming the determining element of our century, asphyxiatingly present and scorchingly absent. Rivers, underground water reserves, oceans: they are all claiming their textuality.[25] Two provisos. First, the disengagement of the legal from the purely tellurian, especially when it comes to questions of sovereignty and jurisdiction.[26] And second, it is not about fully knowing the water. It is not about surface and depth control.[27] It is about maintaining the aquatic unknowability while acknowledging the affinities with our own, more proximate bodies of water.[28] Acknowledging the continuum between our body and the hydrosphere. Wavewriting, and legal wavewriting specifically, requires a radical immersion of the kind that Susan Reid invites us in: "In these early days for ocean justice, I propose thinking with the ocean's midnight aphotic depths, invoking it to bubble up through a juridical imaginary that would not deny its lively worlds and our relations with them."[29]

For the performance, I wavewrote on two screens: on the first one, I was writing on my face and with my face, erasing the skin and dissolving it into the digital, trying to reach my interlocutors like an electrocuting eel, slithering past firewalls and conference expectations. On the other, I was waterwriting a wave of literary texts that swirled on a simple kitchen bowl filled with water—a technique Jan Hogan and I used in our first joint show on *Tracing Submergence* in 2020.[30] The two screens performed the paradox of wavewriting itself, the double presence, the *this* and the *that* of a law that must perforce keep its aquatic subconscious distant, a different continent all together, a thing to draw resources from, a *terra nullius* teaming with bodies violently woken up to the reality of another law. I pour water out of law, and law into water. I try to listen to the water.

FIGURE A.3 *Slashing Waters*, an online performance for the Laws of the Sea workshop by Andreas Philippopoulos-Mihalopoulos, May 27, 2021. Screenshot 3 by author/performer.

"How does it start the sea has endless beginnings,"[31] Alice Oswald wonders, and I wonder too. What is this legal obsession with the start, the origin, the one who came first? Wavewriting abandons itself into the endlessness of beginnings. Real or false, it does not matter since they lead nowhere anyway. Nowhere, that is, that the law can recognize as the certain outcome of a causal link. What is the rage against the law? That it cannot allow itself to be wrong. Legal wrongs devour the only currency we have: time. The law can never be guilty. It is only another law, a later or parallel law, that can find this law guilty. The whole edifice of legal legitimacy rests on rotting palisades deep in the legal seabed. We need to make waves that smudge the black and white of the law. Well, black and white were never there to start with. The lines between bodies have always run into each other, churned by centuries of grief. I spit on your categories and smudge them with my wet hands. Your law is no longer dry. It drips bleeding.

We the water, we the eternal, we the jellyfish. We the ones that float, we the ones that drown.

We must listen to the water.

My white is blue. So is my slow green, my oily black, my spirited azure or my dirty grey. All blue really. Whenever there is light, I catch it, play with it, absorb it as if I needed it, why not make it happy, light has always been a good friend really. Even so, I do not welcome it all. I choose only the parts I want, picky cobalt peacock me, and then I scatter them around like phenomenological fireworks, dot them like big bangers on the world's retina—see how cool I am, seamlessly moving between philosophical parlance and street. But that's another story, another great quality of mine. We are now talking about my color. So, if there is light, I reflect it all blue and cocky. If there is no light around, I wait. Aeons of waiting, knitwork of a universe that forgets its own self. But light always comes.

So it is blue even when my white mountains, ice peaks of my consistency and scraping skies of my polar glory, glisten, slide, and melt. It is blue when my powdery white chasms up crevasses of raw thaw, bubbling up with my seas underneath. It is still blue when it devours your cities and your minds, still blue when it creeps in your mouths yellow with acid and death, gleaming like radioactive enamel spread over your graves. It is still blue when you scatter colorful flowers around your floating dead.

It is blue when red with charcoal frenzy in the deepest core of your planet, and it is blue between your tall buildings on those hot summer evenings when even the breath of your lover is a skin too many. It is blue when you let yourself fly in me, cutting my globules in thick slices, spreading your dream bodies light and wavy across time. It is blue when caught on the wings of a bird, and it is blue when mixed with the green iguanas of the deep. It is blue when you shut your eyes and it is blue when you open them. It is blue when I rush down, shards of transparency drumming the top of your heads like night thoughts. And it is blue when you piss me, yellow reminders of dehydration.

It is blue when my impasto blends the above and the below, sky and sea with their edges always deferred, steam and myopia, the curve of every star, the horizon that opens with every new wave. It is blue, that round thing that moves slowly with you balancing on its crust, a shawl of suspended lakes as deep as the weather trailing around it.

We are nearing the end of the performance. Even wavewriting demands a sort of conclusion. Diffracted and repetitive, rhythmic but without direction, openings and closing of a fist in the middle of a lake. Here it goes—to be read with a staccato voice:

1953 Love Canal, Niagara Falls
21,000 tons of toxic industrial waste released in Niagara Falls, New York, US

(Maybe one could tell by the acceleration. Something in the rhythm, the breaths more bated, the writing on the water more frantic.)

1956 Minamata
Tons of methyl mercury wastewater released into Minamata Bay and the Shiranui Sea, Japan
(By now the water has turned all black, sumi ink made of burnt pine ink, hands and sentences swirl in the bowl.)

1958 Niger Delta Oil Pollution – ongoing

estimated nine to thirteen million barrels of oil released so far into the Niger, Nigeria

(Please pick up a glass of water and place it on the desk where you are sitting.)

1978 Amoco Cadiz
219,797 tons of light crude oil and 4,000 tons of fuel oil released in Porstsall Rocks, France

(Now please gently dip your fingers in it. I am doing it too. Let's all do it.)

1964–1990 Amazon
400 million barrels of toxic oil waste released in the Ecuadorian part of the Amazon

(There is nothing to it, just a bunch of people with their fingers in a glass of water, the same but different water across continents and time zones. We all look at the same screens, one hand in water the other in the air, all listening to the same incantation.)

1989 Exxon Valdez Oil Spill
257,000 barrels of crude oil released in Prince William Sound, Alaska

(We are connected in some way: "he stepped out into the lake whose waters now seemed an extension of his own bloodstream. As the dull pounding rose, he felt the barriers which divided his own cells from the surrounding medium dissolving."[32])

1998 Guadiamar River
4.5 million cubic meters of acidic water with heavy metals and other toxic elements released into the Guadiamar River, Spain

(There is another voice, it is coming from the other screen. It keeps on about the same thing.)

2005 Jilin Chemical Plant Explosions
100 tons of pollutants containing nitrobenzene and benzene released into Songhua River, China

(A wavewriting that keeps on relentlessly eternally returning. The phrase is *we are all complicit*.)

2010 Deepwater Horizon
3.19 million barrels of oil released in the Gulf of Mexico

we are all complicit

2020 Baia Mare Cyanide Dam Spill
100 tons of cyanide released into the Somes, Tisza and Danube rivers, Romania

(Keep your fingers in the water. *we are all complicit*. It is becoming harder. *we are all complicit*. The water is churning out laws, the law is pushing the water out. *we are all complicit*.)

Yamuna River—ongoing
Tons of heavy metals being released into the Yamuna river, India

(The writing cannot take it anymore, the water spills and splashes everywhere, the hand is spread anemone-like. *we are all complicit* on the eve of the planet's death.)

The Great Pacific garbage patch—ongoing

we are all complicit

Coral Bleaching—ongoing

we are all complicit

Desertification—ongoing

we are all complicit

Notes

1 Astrida Neimanis, *Bodies of Water: Posthuman Feminist Phenomenology* (London: Bloomsbury, 2017).
2 Andreas Philippopoulos-Mihalopoulos, *Spatial Justice: Body Lawscape Atmosphere* (London: Routledge, 2014).
3 The video of the performance can be found online at andreaspm.com/performance.
4 Renisa Mawani, *Across Oceans of Law* (Durham: Duke University Press, 2018).
5 Carl Schmitt, *Land and Sea: A World-Historical Meditation*, trans. Samuel Garrett Zeitlin (Candor: Telos Press Publishing, 2015). See also De Lucia, this volume as well as Braverman's Introduction, this volume.
6 Andreas Philippopoulos-Mihalopoulos, "Slashing Waters," in *The Routledge Handbook of Law and Literature*, eds. Peter Goodrich and Daniela Gandorfer (London: Routledge, 2022), 419–437.
7 See DeLoughrey, this volume, on how this is also regularly co-opted.
8 From mist to ice and everything in between: Philip Steinberg et al., this volume.
9 Probyn, this volume.

10 Olivia Lace-Evans, "The Sailor Living Alone on an Abandoned Cargo Ship," *BBC News*, April 6, 2021, https://www.bbc.co.uk/news/av/world-middle-east-56615163.

11 Kodwo Eshun, *More Brilliant than the Sun: Adventures in Sonic Fiction* (London: Quartet Books Limited, 1998); Paul Gilroy, *The Black Atlantic, Modernity and Double Consciousness* (Cambridge: Harvard University Press, 1993).

12 "For terrestrial-bound humans and others, what happens in the abyssal zone influences whether rains come, plants thrive, temperatures are liveable; or whether the ocean provides sufficient food or oxygen to breathe." Reid, this volume.

13 Italicized extracts are quoted from Andreas Philippopoulos-Mihalopoulos, *Our Distance Became Water* (unpublished).

14 See my work on material metaphors in "Flesh of the Law: Material Metaphors," *Journal of Law and Society* 43, no. 1 (2016): 45–65, or indeed a *matterphor* in Daniela Gandorfer and Zulaikha Ayub, "Matterphorical," *Theory & Event* 24, no. 1 (2021): 2–13.

15 Elizabeth DeLoughrey and Tatiana Flores, "Submerged Bodies: The Tidalectics of Representability and the Sea in Caribbean Art," *Environmental Humanities* 12, no. 1 (2020): 132–166.

16 Mawani, *Across Oceans of Law*.

17 Karin Amimoto Ingersoll, *Waves of Knowing: A Seascape Epistemology* (Durham: Duke University Press, 2016).

18 Philip Steinberg and Kimberley Peters, "Wet Ontologies, Fluid Spaces: Giving Depth to Volume through Oceanic Thinking," *Environment and Planning D: Society and Space* 33, no. 1 (2015): 247–264.

19 See Introduction, this volume.

20 Macarena Gómez-Barris, *The Extractive Zone* (Durham: Duke University Press, 2017).

21 Michel Serres, *Genesis*, trans. G. James and J. Nielson (Ann Arbor: University of Michigan Press, 1995).

22 For the blue turn in humanities and social sciences, see, e.g., Steinberg and Peters, "Wet Ontologies, Fluid Spaces"; Stefan Helmreich, *Alien Ocean: Anthropological Voyages in Microbial Seas* (Oakland: University of California Press, 2009). For a different kind of blue, see Gaston Bachelard, *Water and Dreams: An Essay on the Imagination of Matter* (Dallas: Dallas Institute of Humanities and Culture, 1994).

23 Irus Braverman and Elizabeth R. Johnson, eds., *Blue Legalities: The Life and Law of the Sea* (Durham: Duke University Press, 2019).

24 Jeff Goodell, *The Water Will Come: Rising Seas, Sinking Cities, and the Remaking of the Civilized World* (New York: Little, Brown and Company, 2017), back cover.

25 "The ocean is involved in the writing and reading process, affecting how we create and shape both ourselves and our nations." Karin Amimoto Ingersoll, *Waves of Knowing: A Seascape Epistemology* (Durham: Duke University Press, 2016), 93.

26 The classic example is UNCLOS Article 76(1) which defines a state's continental shelf as "the natural prolongation of its land territory" with the result that "Complex, intra-active ocean elements and relations are effectively redacted in a governance framework based on law of the land." Susan Reid, "Solwara 1 and the Sessile Ones," in *Blue Legalities: The Life and Law of the Sea*, eds. Irus Braverman and Elizabeth R. Johnson (Durham: Duke University Press, 2019), 57.

27 We can nowadays control even the most evanescent of water forms: "it *is* possible, to some extent, to control and command ocean waves: to build infrastructures that guard shorelines, to mold beaches that generate waves of stipulated measure and shape, and to engineer devices that 'harness' wave energy." Stefan Helmreich, "Wave Law," in *Blue Legalities: The Life and Law of the Sea*, eds. Irus Braverman and Elizabeth R. Johnson (Durham: Duke University Press, 2019), 168.

28 Mawani, *Across Oceans of Law*.

29 Reid, "Solwara 1 and the Sessile Ones," 47–48. See also Stefan Helmreich, "Seagoing Nightmares," *Dialogues in Human Geography* 9, no. 3 (2019): 308–311.
30 "Tracing Submergence," Kunstmatrix, accessed April 3, 2022, https://artspaces .kunstmatrix.com/en/exhibition/1574403/tracing-submergence-jan-hogan-andreas -philippopoulos-mihalopoulos-please-turn.
31 Alice Oswald, *Nobody* (London: Jonathan Cape, 2019), 13.
32 J.G. Ballard, *The Drowned World* (London: Fourth Estate, 2012), 71.

INDEX